AF527851

Agricultural Extension :
Innovations and Dimensions

Agricultural Extension : Innovations and Dimensions

By

DILEEP KUMAR
KHAJAN SINGH

SATISH SERIAL PUBLISHING HOUSE
403, Express Tower, Commercial Complex, Azadpur, Delhi-110033 (India)
Phone : 011-27672852, Fax : 91-11-27672046
E-mail :info@satishserial.com, hkjain1975@yahoo.com
Website : www.satishserial.com

PREFACE

Last 50 years witnessed a vast expansion of both interest and knowledge in the area of agricultural extension. There has been a widening academic awareness of the importance of issues related to extension such as emerging extension perspectives and changes on the focus of the extension approaches. The out dated information and difficulty in understanding, generally led to problem for the students as leading text books has grown in size and complexity to accommodate new concepts, theories and approaches. This book is an attempt to distill the key areas of extension in a way which will help full time students of the subject. It is designed to give students rapid and easy access particularly to those preparing for competitive exams such as ARS/ NET etc. extension material in a format which facilitate learning and revision. The book is divided into eight chapters which cover material on all most all issues of the agricultural extension. To facilitate easy understanding of the complex concepts these are simplified and special focus is given on the focused point and concrete presentation of the material of the book. Long paragraphs and unnecessary explanation of various areas are being avoided to keep the size of the book compact. Different concepts and theories are cross checked to maintain high standards of accuracy and correctness.

CONTENTS

Chapter-8 : Recent Developments in Agriculture and Extension 477

Chapter 1

Introduction

Extension Terminology

- The term extension was first used to describe adult education programmes in England in the second half of the 19th century; these programmes helped to expand - or extend - the work of universities beyond the campus and into the neighbouring community. The term was later adopted in the United States of America, while in Britain it was replaced with "advisory service" in the 20th century. A number of other terms are used in different parts of the world to describe the same or a similar concept:

  - *Arabic*: Al-Ershad ("guidance")
  - *Dutch*: Voorlichting ("lighting the path")
  - *German*: Beratung ("advisory work")
  - *French*: Vulgarisation ("simplification")
  - *Spanish*: Capacitacion ("training", "capacity building")
  - *Thai, Lao*: Song-Suem ("to promote")
  - *Persian*: Tarvij & Gostaresh ("to promote and to extend")

In the US, an extension agent is a university employee who develops and delivers educational programs to assist people in economic and community development, leadership, family issues, agriculture and environment. Another program area extension agents provide is 4-H and Youth. Many extension agents work for cooperative extension service programs at land-grant universities. They are sometimes referred to as county agents or educators.

Definitions of Extension

There is no widely accepted definition of agricultural extension. The ten examples given below are taken from a number of books on extension published over a period of more than 50 years:

1949 : The central task of extension is to help rural families help themselves by applying science, whether physical or social, to the daily routines of farming, homemaking, and family and community living.

1965 : Agricultural extension has been described as a system of out-of-school education for rural people.

1966 : Extension personnel have the task of bringing scientific knowledge to farm families in the farms and homes. The object of the task is to improve the efficiency of agriculture.

1973 : Extension is a service or system which assists farm people, through educational procedures, in improving farming methods and techniques, increasing production efficiency and income, bettering their levels of living and lifting social and educational standards.

1974 : Extension involves the conscious use of communication of information to help people form sound opinions and make good decisions.

1982 : Agricultural extension is assistance to farmers to help them identify and analyse their production problems and become aware of the opportunities for improvement.

1988 : Extension is a professional communication intervention deployed by an institution to induce change in voluntary behaviours with a presumed public or collective utility.

1997 : Extension is the organized exchange of information and the purposive transfer of skills.

1999 : The essence of agricultural extension is to facilitate interplay and nurture synergies within a total information system involving agricultural research, agricultural education and a vast complex of information-providing businesses.

2004 : Extension is a series of embedded communicative interventions that are meant, among others, to develop and/ or induce innovations which supposedly help to resolve (usually multi-actor) problematic situations.

Origins of Agricultural Extension

Men and women have been growing crops and raising livestock for approximately 10,000 years. Throughout this period, farmers have continually adapted their technologies, assessed the results, and shared what they have learned with other members of the community. Most of this communication has taken the form of verbal explanations and practical demonstrations, but some information took a more durable form as soon as systems of writing were developed. Details of agricultural practices have been found in records from ancient Egypt, Mesopotamia and China going back more than 3,000 years.

It is not known where or when the first extension activities took place. It is known, however, that Chinese officials were creating agricultural policies, documenting practical knowledge, and disseminating advice to farmers at least 2,000 years ago. For example, in approximately 800 BC, the minister responsible for agriculture under one of the Zhou dynasty emperors organized the teaching of crop rotation and drainage to farmers. The minister also leased equipment to farmers, built grain stores and supplied free food during times of famine.[11]

The birth of the modern extension service has been attributed to events that took place in Ireland in the middle of the 19th century. Between 1845–51 the Irish potato crop was destroyed by fungal diseases and a severe famine occurred (see Great Irish Famine). The British Government arranged for "practical instructors" to travel to rural areas and teach small farmer how to cultivate alternative crops. This scheme attracted the attention of government officials in Germany, who organized their own system of traveling instructors. By the end of the 19th century, the idea had spread to Denmark, Netherlands, Italy, and France.

The term "university extension" was first used by the Universities of Cambridge and Oxford in 1867 to describe teaching activities that extended the work of the institution beyond the campus. Most of these early activities were not, however, related to agriculture. It was not until the beginning of the 20th century, when colleges in the United States started conducting demonstrations at agricultural shows and giving lectures to farmer's clubs, that the term "extension service" was applied to the type of work that we now recognize by that name.

In the United States, the Hatch Act of 1887 established a system of agricultural experiment stations in conjunction with each state's land-grant university, and the Smith-Lever Act of 1914 created a system of cooperative extension to be operated by those universities in order to inform people about current developments in agriculture, home economics, and related subjects.

The birth of modern agricultural extension services

The first agricultural extension service of a moden kind came into existence as the result of a crisis and the initiative of the occupant of a high office of authority. The crisis was the outbreak of potato blight in Europe in 1845. In Ireland its effects were particularly severe because the predominantly peasant population relied on potatoes in their diet, and "the potato famine" persisted until 1851. The new British Viceroy appointed to Ireland in 1847, the Earl of Clarendon, soon after his arrival in Dublin wrote a letter (Jones, 1982) to the president of the Royal Agricultural Improvement Society of Ireland (founded in 1841), which acted as the central society for numerous local agricultural societies. This letter, no less than an official directive, urged the society to appoint itinerant lecturers to travel around the most distressed districts to inform and show small farmers, in simple terms, how to improve their cultivation and how to grow nutritious root crops other than potatoes. "Lord Clarendon's practical instructors in husbandry," as they became known, were centrally appointed, deployed, and paid and reported weekly to the society in Dublin, with some local control of their activities being exercised by the major landowners in their areas. Over the four years of its existence, the scheme was funded to about half its total cost by landowners and charitable donations, with the remainder coming from government-controlled funds (Jones, 1979, 1981).

The potato famine also led to consideration being given in Württemberg to employing itinerant farm advisers, but the proposal failed to gain approval (R. Bühler, personal communication). However, from the mid-1850s, first in Württemberg, Hessen, and western Prussia, itinerant agricultural teachers (*Wanderlehrer*) began to be appointed under the auspices of central agricultural societies. After some ten years, the system grew rapidly, influenced in part by the crisis among vine growers resulting from the devastation caused by *phylloxera* aphid infestations, and became formalized (Jones, 1981). Normally, the *Wanderlehrer* spent the summer half of the year travelling around their districts giving talks, demonstrations, and advice to farmers; during the remainder of the year they taught farmers' sons at winter agricultural schools. Although officially they were part of the activities of the agricultural associations, their work was in all cases supported heavily by state funds, and their advice was free to farmers. When the system was adopted in the kingdom of Bavaria in 1896, it was as an integral part of the state civil service; the extension workers were grandly titled Royal Agricultural Teachers (*Königliche Landwirtschafts-lehrer*) (Maier-Bode, 1910).

By the close of the nineteenth century, agricultural extension systems modeled to a considerable extent on the German *Wanderlehrer* had spread: to Denmark from 1870 onwards; to the Netherlands, where a few extension workers (*wandelleraren*) had been appointed by agricultural societies in the late 1840s and 1850s, but had then disappeared before being revived as a government system in the 1890s; to Italy, where the first itinerant agricultural teacher (*cattedra ambulate di agricoltura*) was appointed in 1886 at Rovigo, near the estuary of the River Po, with many others following in the next decade and funded largely by public donations, the church, and the banks; to Switzerland; to much of the Austro-Hungarian Empire; and to Russia.

Meanwhile, in France the first national, wholly state-funded agricultural extension service was established in 1879. The few itinerant agriculturalists appointed before 1848 (referred to earlier) had continued, but they served in only a very small minority of the country's *departements.* In 1874, the minister of public instruction in the reforming Third Republic issued a circular letter strongly commending the system and advocating its extension .This resulted in an additional thirty-three itinerant agricultural teachers being appointed by *departments* over the next five years, and a law passed in 1879 officially instituted the office of a department-level itinerant agricultural teacher (*professeur departmental d'agriculture*). This law was given practical effect by a decree in 1880 and an explanatory ministerial circular early in 1881 From then on, each *professeur* was a state-appointed civil servant. His duties included giving agricultural instruction to trainee primary school teachers. Mainly, however, under the responsibility of the Ministry of Agriculture, he was to be "nomadic" within his *departement,* "to keep farmers informed regarding modem discoveries and new inventions which could be applied economically and with advantage," "to be a popularize (*vulgarisateur*) of progress," "to carry enlightenment into the heart of the countryside." The number of these extension workers grew rapidly, and by the end of the 1880s the whole of France was being served (Jones, 1981).

The growth of agricultural education and extension work in continental Europe was to have a strong impact on the emergence of comparable activity in the United Kingdom. An official commission on technical education in the early 1880s included a detailed review of the European developments). At the end of the decade, a cluster of enactment's, which established county-based local government, created a board of agriculture, promoted technical (including agricultural) education, and allocated funds for the purpose, enabled agricultural extension work to be initiated. It was to be part of the services provided by the local government authorities. They either employed their own agricultural officer or more commonly sponsored lectures and travelling schools on agriculture (especially dairying) as part of the university extension system. This meant drawing on the staffs of the agricultural departments which were being created in new institutions of higher education. Government funds were available to support these activities, but funding also had to be provided by the local county authorities (Jones, 1994). By the turn of the century, such work existed throughout Great Britain.

This system and its underlying legislation, however, did not apply to Ireland (then entirely a part of the United Kingdom). There, agricultural extension work became established in 1900 as a result of the initiative of Horace Plunkett, well known for his advocacy of agricultural cooperation. An official committee in 1896, chaired by Plunkett, reviewed the developments in Europe and North America (Report, Recess Committee, 1896) and set out to adapt the various systems to suit Irish conditions. In 1900, a Department of Agriculture and Technical Instruction was established in Dublin, governed by a board of representative Irishmen. This initiated itinerant agricultural instruction, organized within each county as in Britain and similarly resourced partly from local and partly from central funds. A vague recollection existed of Lord Clarendon's "practical instructors" half a century earlier, and the title "itinerant instructors" was applied to the new extension workers, who were expected to provide information and advice, each to be "the guide, philosopher and friend of the existing farmers"

Many visitors and several official delegations from North America to Europe, particularly from the mid-nineteenth century onwards, reported back on the progress in agricultural research and education, including the itinerant teachers. In the United States and eastern Canada, agricultural societies had become common during the first half of the century and, usually supported by their state or provincial legislatures, some had at times sponsored itinerant lecturers in agriculture. However, two other developments after 1850 were of more significance to the evolution of agricultural extension in the United States. First was the Morrill Act of 1862, signed by President Lincoln during the Civil War, which was seminal in the creation of state colleges "of agriculture and the mechanic arts" in the northern United States; its land-grant provisions enabled the states to establish and fund their colleges. Second was the beginning at about the same time of the farmers' institute movement. These institutes organized one-or two-day (and later longer) meetings, which became popular after 1860, arranged by and for farmers.

Both developments had been widely discussed during the previous decade, and their growth over the next half century was closely interwoven. The visiting speakers at the institutes were largely professors at the state colleges of agriculture, and both depended on the formal support of their state legislatures and of farmers, especially through their agricultural societies Over the next forty years, these activities were influenced also by the university extension movement in Britain and the growing interest in adult self-improvement (inspired, for example, by the Chautauqua adult education institution in New York State). By 1890, when the second Morrill Act granted federal funds for the establishment of agricultural colleges in the remainder of the United States, the farmers' institutes had spread throughout and become a national institution with federal support and supervision, further stimulated by the formal establishment of experimental work at the state colleges of agriculture under the 1887 Hatch Act. A comparable development of farmers' institutes began in Ontario,

Canada, in 1885. These were financially supported by the provincial legislature and spread rapidly with lecturers mainly from the Ontario Agricultural College at Guelph (founded in 1874). A somewhat similar system began in Prince Edward Island.

Thus, by the end of the last century, a system of agricultural extension work had become well established in a large part of North America. In the United States, the colleges and their leading professors, including several notable proponents of more practical extension work, progressively took over the initiation and organization of the activity. This culminated in 1914 with the passage of the Smith-Lever Act, establishing the Cooperative Extension Service - a tripartite cooperation of federal, state, and local county governments, with the state college as the extension agency - "in order to aid in diffusing among the people of the United States useful and practical information on subjects relating to agriculture and home economics, and to encourage the application of the same."

In the Southern Hemisphere, extension work also became established along the wide coastal belts of southern and eastern Australia. Several agricultural ("show") societies were formed in the second half of the nineteenth century, although their effect was slight, but as the state administrations became more organized, departments of agriculture were established in the 1870s and 1880s with the aim of developing the potential of their territories. They recognized the importance of agricultural education, influenced by British, Irish, and some American examples whose activities were widely reported in the Australian press. Before the end of the century, under specific state legislation, the departments of agriculture had established agricultural colleges and experimental work in Victoria, South Australia, New South Wales, and Queensland.

Associated with this development was the official appointment in these states of the first itinerant agricultural instructors in the late 1880s. At the same time, because of the potential importance of milk products, travelling dairy schools were begun, while state exhibitions, especially the Centennial Exhibition in Melbourne in 1888, showed what was possible and gave considerable impetus to farming improvements. The few "government experts," some from the United Kingdom or the United States, grew in number during the 1890s and the first decade of this century, developing the range of the extension work. Its impact and that of the agricultural colleges in their early years was probably slight, but the basis had been laid for further development

Agricultural extension work had also started before 1900 in Japan. Following the Meiji Restoration in 1868, new administrative structures and various modernizing policies were adopted. Two agricultural colleges were established in the mid-1870s, staffed by Western (mainly European) teachers. At these colleges and government farms, experimental work was conducted and new practices were tested and developed. At the same time, agricultural fairs and exhibitions were begun, and progressive Japanese farmers gave talks and demonstrations at them. These led to the development of many agricultural societies from 1881 onwards, a "movement" formalized by legislation in 1899. In 1885, the government also initiated, at national and prefectural levels, a system of appointing experienced farmers as itinerant agricultural lecturers (because the Western "experts" knew little about rice husbandry). Supported by the work at government experiment stations, established from 1893, these farmers formed the basis of agricultural extension work. This activity, including the establishment of demonstration farms, was allocated in 1903 to the numerous agricultural societies which, with state funds, appointed agricultural technicians. In 1910, the 1899 law was strengthened; thereafter, farmers were required to belong to a village agricultural society which was linked to a national network and hierarchy of societies, and farmers were compelled to adopt the technical guidance and recommendations of the societies' extension workers - what became known as "forced extension" Ministry of Agriculture, 1993.

The development and organization of agricultural extension work was not entirely confined to temperate countries. In a variety of ways, it had also begun in tropical areas, especially in colonial territories. The European colonial powers looked to their overseas territories as a source of tropical agricultural products.

Despite a long connection with some of the colonial areas, the Europeans remained largely ignorant of many tropical agricultural plants. The solution was to establish experimental and demonstration "botanical gardens." The earliest was opened in 1821 at Peradeniya, Sri Lanka (Ceylon), and two others were established in the country later in the nineteenth century. Smaller ones were also created in several Caribbean islands and some West African territories. During the early years of this century, some of these developed considerably, although others were short-lived. Those which succeeded provided important sources of agricultural knowledge and innovation and formed the basis for an interest in agricultural societies and agricultural instruction. Some attempts were also undertaken to improve "native agriculture." This was often associated with the creation, as part of the administration, of departments of agriculture and the appointment of professional agriculturists as directors of agriculture.

A central department of agriculture was established in India after the 1866 Orissa famine, and the government of India soon after resolved to establish departments in each province. However, it was 1905 before a central government directive ordered every province to appoint a full time director of agriculture who should organize agricultural research and demonstration farms with staff who could advise farmers. The first British colony to appoint a director of agriculture was Zanzibar in 1896. Of more significance, however, was the creation in 1898 of the Imperial Department of Agriculture for the West Indies, with headquarters in Barbados. Before 1914, such departments of agriculture had been created in several African and Southeast Asian territories, as well as in several Caribbean islands. In Sri Lanka, a few agricultural instructors had been appointed about 1880 to work alongside government agents. When in 1904 the Ceylon Agricultural Society was formed to promote experimental work, it also began an agricultural extension service with the objective of reaching native cultivators Along with school gardens; the extension workers were considered an effective way of demonstrating improved cultivation practices to villagers. Similar developments also occurred in the Caribbean.

Modern Agricultural Extension

In the early years of this century, extension services were in their formative stage; they were relatively small in scale and limited in the scope of their work and contact with farmers, and their organization was often somewhat haphazard even though based on legislation. They were organized predominantly either by central or local governments, or by agricultural colleges, usually in close association with experiment stations, or by farmers' organizations (agricultural societies, cooperatives, farmers' unions, or chambers of agriculture), or combinations of these parent bodies. As the century has progressed, the organizations have matured. Changes have often occurred to their parent affiliations, government funding has become relatively more important, their objectives have become broader, especially in "the North," and the extension workers have become better trained and more professional. In addition, several other kinds of organizations have developed comparable work: agriculture-related commercial companies; agricultural commodity marketing boards, concerned to assure the supply and quality of their specific product; agricultural development projects, many of considerable territorial scale; and a variety of nongovernmental organizations (especially religious and charitable) involved in agricultural and rural development.

As agricultural extension organizations have grown and changed, they have invariably become more bureaucratic with distinct hierarchical structures. The work of dispersed extension workers had to be administered and controlled so that one or more levels of intermediary structure (for example, district, region) have been created between the field-level agents and their headquarters. Thus the management of extension activities has become a major preoccupation, and many organizations have been open to the criticism of being top heavy and top-down in their approach. However, with funding derived largely from national revenues (or international donors), senior managers have necessarily had to account for and justify their organization's activities. This has been equally pronounced in the North as in the South where, after colonial territories gained their independence, extension work has commonly been reinvented and staffed by nationals under the aegis of their new administrations (usually ministries of agriculture).

During the past quarter century, the work of extension services has often become more diversified. In the less developed countries, the main focus remains on agricultural (mainly food) production, but there has been a growing recognition of the need to reach, influence, and benefit the multitudes of small, resource-poor farmers. Strong efforts have been made in this direction, notably through the training and visit system. Among the commercial farmers of the North, a major problem has become surplus production, with farmers facing economic and policy pressures to restrict it. Associated with intensive production methods, many issues and problems regarding environmental deterioration and livestock welfare have also arisen. Thus these have become important aspects of extension work, particularly socio-economic guidance which focuses both on means by which farmers might maintain their income levels from their resources (for example, introduction of novel crops or livestock and involvement in various rural enterprises) and on the ways of assuring the longer term welfare of farmers and their families. Agricultural extension services are thus adding a strong social dimension to their activities.

Agricultural extension has now become recognized as an essential mechanism for delivering information and advice as an "input" into modem farming. Since commercial farmers can derive direct financial benefits from these inputs, there is a trend towards the privatization of the extension organizations, often as parastatal or quasi-governmental agencies, with farmers being required to pay for services which they had previously received free of charge. This trend is strong in the North, and there are examples of it beginning in the South.

The pace of change in the organization, functions, strategies, and approaches of agricultural extension is clearly accelerating.

The Future of Agricultural Extension

The need for agricultural and rural information and advisory services is likely to intensify in the foreseeable future. In much of the world, agriculture faces the challenge of keeping pace with rapidly increasing population with few reserves of potentially cultivable land. Farmers will have to become more efficient and specialized.

From government perspectives, whatever priority is given to production, extension will remain a key policy tool for promoting ecologically and socially sustainable farming practices.

Some of the most promising recent developments in extension methodology have occurred where the key agenda is environmental or is concerned with equity, for example in the need for the joint management of forests by professionals and local forest users and in integrated pest management. A consistent theme running through the innovative approaches being used, such as participatory rural appraisal is a fundamental change in what are the respective roles of extension agent and clients. The agent is no longer seen as the expert who has all the useful information and technical solutions; the clients' own knowledge and ingenuity, individually and collectively, are recognized as a major resource; solutions to local problems are to be developed in partnership between agent and clients. Since the scale at which extension support is required is thus often larger than the individual farm, extension workers need new skills of negotiation, conflict resolution, and the nurturing of emerging community organizations

The future is also likely to witness a reversal of recent trends towards bureaucratization within hierarchical extension services and a reduction in their levels of public funding. Moreover, a rapid increase can be expected in the use of information technology in support of extension. The forces for change in these areas will come from four main directions.

Economic and policy climate

With the collapse during the past decade of socialist forms of economic organization, the (dominant) role of the public sector in national economies has become questionable, with a strengthening trend to reduce levels of public spending. Thus government extension services and those which are largely publicly financed are, and will continue to be, under pressure to become more efficient, to reduce their expenditure and staff, and to pass on (some of) the costs of provision to their clients who directly benefit financially. This is particularly the case in countries where the farm population forms a small minority and agricultural production is in surplus. The case is weaker, but not absent, in less developed countries where farming households form a high proportion of the total population and where increasing food production is still important. Thus charging clients for services is likely to become more widespread, while governments will find it attractive to contract out the operation of services to the private or the voluntary sector.

Social context in rural areas

In the future, rural populations will undoubtedly be progressively better educated, while their exposure to the mass media will continue to reduce their isolation and detachment from information, ideas, and an awareness of their situation within a national and international context. However, this exposure will not reduce the need for extension. Rather, given the changing demands on agricultural producers from population growth, increasing urbanization, legislative changes, and market requirements, the more knowledgeable farming population will require different kinds of extension services. Social and economic trends within rural areas will therefore necessitate more highly trained, specialized, and technically competent workers, who also know where to obtain relevant information and problem solutions and various provision and organizational forms to replace monolithic government extension agencies. These agencies will need to recognize and serve different types of clients defined not in terms of "adopter categories" but of access to markets, degree of commercialization, and relative dependence on agriculture for family income and welfare.

Systems knowledge

Recognition of the locale-specific nature of farming systems and the agricultural information systems which support them is an important source of the pressure towards the debureaucratization and devolution of extension services. This recognition also implies that extension workers and farmers be jointly involved in the verification and adaptation of new technology, and thus that the extension workers respect farmers as experimenters, developers, and adapters of technology and devote more energy on communication within their local areas. The devolution of extension services to become local organizations is a reasonable corollary of this. Developments in mass media technology, already apparent over a decade ago will continue to support this localization of extension effort.

Information Technology

The continuing rapid development of telecommunications and computer-based information technology (IT) is probably the biggest factor for change in extension, one which will facilitate and reinforce other changes. There are many possibilities for the potential applications of the technology in agricultural extension (FAO, 1993; Zijp, 1994). IT will bring new information services to rural areas over which farmers, as users, will have much greater control than over current information channels. Even if every farmer does not have a computer terminal, these could become readily available at local information resource centres, with computers carrying expert systems to help farmers to make decisions. However, it will not make extension workers redundant. Rather, they will be able to concentrate on tasks and services where human interaction is essential - in helping farmers individually and in small groups to diagnose problems, to interpret data, and to apply their meaning.

The future will call for more able, more independent, more client-oriented extension workers. The emphasis will be on the quality of interaction between agent and client rather than on the movement of "messages" through a hierarchical system.

Flexibility and adaptability will be seen as virtues rather than aberrations. Paradoxically, these trends will bring us full circle to the early manifestations of modem extension in Europe. The itinerant agricultural teachers, unencumbered by large bureaucracies and tall hierarchies, will find their modem counterparts in the computer-carrying extension workers who are at ease helping farmers to identify the information they need in order to realize the potential of their farming operations. Looking back, we can regard the period from 1970 to 1995 as a necessary but expensive stage in the evolution of extension systems, after which extension agents were able to settle down to their main task - bringing together the expertise of farmers and the best available scientific knowledge to develop farms and local agricultural economies.

Extension goals

Goals lead the actions of individuals, groups, and organizations. While pointing towards a future state, they are influenced if not determined by past experiences. They reflect the interests of their stakeholders and differ, therefore, according to specific life situations, power positions, and development philosophies. The prominent features of a system, such as its organizational structure, the choice of clientele, its operational design, and the methods used, are directly influenced by its set of goals and must be evaluated in terms of their contribution to goal achievement.

Main actors within the extension system are the members of rural communities, extension and other development personnel, researchers, and staff of commercial or public service and support organizations. Empirical evidence shows a variety of forms in which interaction among these groups is institutionalized. The variety of forms suggests a similar variety of goals, and either could be used to classify extension approaches. In practice, however, one finds an almost inseparable mixture of goals inhibiting a clear-cut classification. It seems more appropriate, then, to use a broader category, namely, selectivity with regard to clientele, and treat the respective goals as a continuum. The two end points of this continuum would be marked as technology transfer and human resource development, suggesting either a rather narrow technical or a broader socio-economic view of development.

Technology transfer

Until the end of the eighteenth century, farming techniques developed gradually and steadily over centuries with few qualitative leaps. Colonialism and imperial expansion introduced innovations - the spread of maize, tobacco, potatoes are striking examples - but experimentation and dissemination of knowledge were basically at the local farm level. The rise of agricultural sciences has induced dramatic changes in this respect. Increasingly, new technology has been created outside the actual farming sector by public sector research organizations. More recently, private firms in industrialized economies find agricultural technology research and development a highly profitable business.

For decades the research-extension-farmer linkage, especially in developing countries, was based on a rather simple model. In order to achieve development, "modem" research results had to be transferred to the "traditional" farmer, and extension seemed to be the appropriate means to do so. The general faith in science and the commitment to modernization led to discrediting indigenous knowledge. Although this view is still held by many administrators, researchers, and extension agents, it is now being seriously questioned. Farming systems research and the "rediscovery" of farmers' knowledge have shown that "improved technology is a package of inputs and practices that usually comes from many sources" (UNDP, 1991, p. 2). The re-examination of the conventional view on agricultural knowledge cannot, however, result in questioning the important role of research as the source of new technology. For developing countries, one observes that the accelerated growth and spread of problems - such as the degradation of marginal land - surpass the problem-solving capacities of the local population. What is called for is a setting of new priorities and the building of knowledge systems based on problem solving rather than on information transfer.

Human Resource Development

The concept of human resource development is much broader than that of technology transfer, though both are closely interrelated. Increasing complexity not only of technology but also of the life situation of farmers even in remote areas demands new skills. With the help of these skills, rural women and men "acquire a better insight into the network of problems and recognize the alternative solutions available".

Traditionally, teaching the basic skills of literacy and numeracy has not been an extension activity. The limited success of literacy programms in poor countries has drawn attention to non formal education in which extension has an important part to play. Whereas, in most cases this would require a coordinated effort of different organizations of which extension is but one, human resource development may also be regarded as genuine extension content.

Extension may substitute over a certain period activities such as vocational education that are not yet in place, but more important will be the teaching of managerial and organizational skills that will enable farmers to increasingly solve their own problems. Human resource development thus aims at what may be called "critical competence." Extension clients know what to ask for, they can evaluate the appropriateness of technical information, they are responsible decision makers. Persons with this qualification exist in every rural community, and they will be the ones who actively seek further assistance. One important task of any extension system will therefore be to extend human resource development to underprivileged groups with less access to formal or vocational education - women farmers, rural youth, and generally small farmers in remote areas.

Alternative ways of organizing extension

The goals of extension may vary, as was shown, within the overall system as well as between different extension organizations. In addition, specific objectives may sometimes contradict each other. While smaller systems may come close to pursuing a consistent set of objectives or reconciling conflicting interests, large-scale organizations must work on a compromise basis. In this respect, Axon's principal observation is of particular importance: "The success of an agricultural extension programme tends to be directly related to the extent to which its approach fits the programme goals for which it was established"

The alternatives to organizing extension demand choices on various levels:

- Public versus private
  Government versus non-government
  Top-down (bureaucratic) versus bottom-up (participatory)
  Profit versus non-profit
  Free versus cost-recovery

General versus sector

Multipurpose versus single purpose

- Technology driven versus need oriented

In practice, extension organizations everywhere pursue the overall goals of technology transfer and human resource development, though the emphasis will differ. Within each organization there is a mix of objectives, and within countries there is often a mix of organizational patterns. When presenting an overview on the most important patterns, we will be using a well-established terminology (UNDP, 1991), though the grouping is different. We will differentiate between approaches that, at least in principle, target all persons in rural areas engaged in farming and those that purposely select clientele according to specific criteria.

General Clientele Approaches

Ministry-Based General Extension

Shortly before or after independence, organizing agricultural extension work under the wings of the ministry of agriculture seemed to be an ideal solution for many African and Asian governments. All options for reaching large numbers of clients and serving their needs in terms of quality information and assistance appeared to be open. The original colonial model combined research and extension within the same organization. All important aspects of small-holder agriculture - plant production, animal husbandry, home economics - could be attended to as the ministry established respective sections under its jurisdiction. The fact that the ministerial hierarchy followed the country's territorial subdivision allowed the systematic expansion of the system "down" to the village. The generalist nature of field extension staff functions corresponded to the set of problems faced by non-commercial growers. To cater to specific needs - in terms of technology or in terms of target groups - specialists could be employed. Thus clientele included in principle all persons engaged in agriculture. Commercial service and support organizations lacking, village-level extension staff could be expected to supplement information by rendering services necessary to apply it productively. A uniform and nationwide organizational pattern seemed to facilitate information flow - including the infusion of expatriate expertise - and corrective measures whenever weaknesses were identified. Public interest was to guide goal setting, programme formulation, and the implementation of fieldwork.

A review of the last thirty years of extension work in Africa and Asia shows that reality is quite far for failure are complex and manifold and cannot be removed from this vision. The reasons for failure are complex and manifold and cannot be reduced simply to incompetence or the ill-will of national governments.

One reason is the contradictory nature of goals. Public interest implies serving farmers and the urban population, securing subsistence production and promoting cash crops for export, reaching the masses of rural households and serving the needs of specific groups, extending assistance to high-potential and disadvantaged producers. In short, priorities will have to be set, and these are all too often pro urban in terms of price policy, favouring innovative individuals within the modem sector, neglecting poorer strata, and forgetting about women farmers.

In many ways, the hierarchical and highly bureaucratic way in which the services are organized hampers a full realization of their potential. Priority setting for research is rarely based on extension field evaluations because the system does not foster critical upward communication.

The way in which technical (and other) knowledge is transformed into field messages frequently leads to distorted and outdated information.

In the eyes of the ministry, extension has never been a purely educational activity. This is a legitimate view as long as the different functions to be performed by extension personnel are compatible and basically client oriented (such as helping to organize input supply). No educational activities may include anything from statistical data collection to attending to foreign visitors. Incompatible with and clearly detrimental to regular extension work are such activities as supervising credit repayment, policing disease control measures, organizing "voluntary" community work, and electioneering.

Ministry-based extension has been unable to reach a majority of its potential clientele for economic, socio-psychological, and technical reasons. Even dramatic quantitative increases in personnel - more staff closer to the farmer - have not produced manageable client-to-agent ratios. In recent years, the trend has even been negative. Financial constraints have produced a strong pressure to reduce staff, and the field level has been hit hardest. Those remaining have little if any material resources left to maintain mobility.

In addition, many extension workers select the more responsive section of their clientele. They may have to fulfill production plans, they may want to improve job satisfaction or status, or they may simply be prejudice against certain target groups. Lastly, extension often has little to offer in terms of messages to large sections of the rural population. Adequate and location-specific answers to a farmer's problem are often not available because it has not been a research concern or the solution has simply not reached the field.

Today's situation is aggravated by two additional aspects which refer to the internal structure of the service: management problems and lack of control from below. Ministry extension employs thousands of persons working under a wide variety of circumstances. Decision making and management are highly centralized and formalized. Extension fieldwork, on the other hand, demands location-specific, flexible, and often quick decisions and actions. Managing the "invisible" man or woman must be highly ineffective as long as he or she is expected to receive and execute orders.

All these problems are well known, and criticism has come both from within and outside the ministry. What has been lacking is organized feedback from clientele. Farmers may show their discontent by refusing to cooperate with extension, but they have virtually no way of influencing institutional reforms.

Training and Visit Extension (T&V)

In the strict sense of the word, T&V is not a separate approach but one way to organize ministry-based extension. The controversial debate on the merits of T&V tends to obscure the fact that it was originally meant to solve some very specific problems of conventional extension services.

Benor and Harrison's original paper - one of the most influential extension publications ever - critically evaluates the ministry-based extension system of the 1970s they found:

An inadequate internal organizational structure

Inefficiency of extension personnel

Inappropriateness or irrelevance of extension content

Dilution of extension impact

Whichever impact is reached serves "only a few favoured farmers in favoured areas rather than the bulk of the farming community"

When first being introduced, T&V seemed to be strikingly original and promising because it combined a set of rather convincing simple elements in a plausible way. Rather than trying to reach all farmers directly and thus pre-programming constant failure, the system concentrates on contact farmers expected to pass information on to fellow farmers with similar problems. To ensure regular field contacts, facilitate supervision and communication, and set clear and attainable objectives, fixed visits at regular intervals are prescribed. Similarly, regular sessions for extension workers to receive training and discuss administrative matters are held. Thus costly refresher courses are avoided, knowledge may be enhanced step-by-step, and up-to-date information can be fed into the system.

In addition, T&V operates under the assumption that its extension workers are exclusively engaged in educational activities and that a unified extension service exists. Agricultural research must not only be effective but also work in close collaboration with extension. Both external and internal evaluations are to be used to constantly modify and adapt the system to changing conditions.

Simple as the prescriptions seemed, implementation proved to be difficult. First, the contact farmer concept - implying a two-step flow of information from the extension worker to the contact, farmer and from there to other farmers - has frequently failed. Extension workers have been blamed for "wrong selection," but the root of the problem lies within the purely technical philosophy of T&V. Other aspects such as communication skills, leadership, and organizational capacities are neglected. In practice, T&V has been a top-down approach leaving little possibility for participation and initiative, both for farmers and village extension workers. Too little emphasis has been put on critical feedback based on self-evaluation. As a result, rigidity rather than flexibility characterizes local fieldwork.

Secondly, Benor's fear that extension services may "rapidly run out of anything to extend" characterizes many T&V field situations. The standardized messages passed on are often of little relevance to local conditions. Once T&V was extended to less favoured regions, it soon became clear that technology of the green revolution type showing quick and visible results is not available. Still, training sessions were held and visits made according to schedule, leaving behind disinterested farmers and demotivated extension workers.

The limited success of T&V in its present form as a nationwide extension system should not discredit the quality and appropriateness of many of its elements. Applied less rigidly and combined with the tools of human resource development as well as with the concept of participation, these elements may constitute a valuable base for reforming extension organizations, large or small (Nagel *et al.*, 1992).

The Integrated (Project) Approach

Integrated approaches aim at influencing the entire rural development process. Extension is only one though often crucial element in this strategy which targets the entire population in a given area but emphasizes work with disadvantaged groups. Integrated approaches are generally implemented in the form of large-scale and foreign-funded projects aiming at alleviating mass poverty in rural areas on the basis of "a simultaneous improvement in the utilization of natural resources and of human potential" (Rauch, 1993, p. 6). Measures to promote production are coupled with a strong emphasis on self-help. The underlying concept is typically multispectral.

Evaluations of more than a decade of integrated rural development (IRD) projects have revealed serious shortcomings in reaching the goal of mass poverty alleviation Sizeable numbers of the poor were not reached by project activities, nor were positive effects consolidated on a sustainable basis. Project deficiencies were in part management related and very often due to a serious under-estimation of the great complexity of multi-spectral programmers with ambitious goals. The disregard of the target group principle and of due consideration for framework conditions (economic and institutional) played an even more important role, as did the lack of compatible technical solutions.

Recent efforts to improve regional rural development (RRD) projects and enhance chances for a broad and sustainable impact are relevant for all general extension approaches. The key concept is the availability of locally adapted solutions established on a common basis. This requires not only participatory technology identification, test, and dissemination, but also an active role by the change agency in mediating between different institutions involved and their interests. A particular emphasis is laid on dealing with adverse framework conditions, explicitly taking them into account and attempting to influence them in favour of clients. Finally, in order to achieve these improvements, new efforts must be made to specify and operationalize (extension) objectives and concepts (sustainability, participation, gender-specific target-group approach, and poverty alleviation).

University-Based Extension. While the Cooperative Extension Service (CES) of the United States is still the only system in which the main extension function remains within the university, some developing countries, notably India, have integrated educational institutions into practical extension work. Within the United States of America, state universities have traditionally cooperated with local counties and the U.S. Department of Agriculture in doing extension besides education and research. Within the last 130 years, extension goals of the land-grant colleges have shifted from practical education to technology transfer and, more recently, to a much broader concept of human resource development.

With the emergence of strong private and other public sector research and development organizations and dramatic changes within the agricultural production sector, CES is facing new challenges with regard to coordination and cooperation. Apart from its traditional roles, *networking* will become a primary role (Bennet, 1996). In this model, industry as well as intermediate and end users of knowledge become part of the extension system.

While in most countries, the main contribution of educational institutions to extension will be the training of qualified, dedicated, and responsible personnel, some Indian agricultural universities have come close to the U.S. model without taking over the full load of extension work. In the field, they have taken over functions which are only inadequately performed by the ministry, thus supporting general extension work. Remarkable features are direct assessment of clients' needs, user-oriented research, quality training for state personnel, and a strong linkage between academic education and field practice. Models vary from state to state. The Punjab Agricultural University (PAU) has its own multidisciplinary extension team in each district, engaged in adaptive research, training, and consultancy. Backed up by extension specialists on campus, they are transmitters and receivers of experiences from researchers, farmers, and state extension workers. At PAU, a unique system of processing these experiences is practiced. Regular workshops are held which unite university and department staff from research and extension together with outstanding farmers. New findings and feedback are presented, evaluated, and published as a "Package of Practices" to be used by all extension staff for the next season (Nagel, 1980).

In the Philippines, which works with ministry-operated extension, university field contacts have been combined with practical development work. The University of the Philippines at Los Baños (UPLB) has its own "social laboratory" in rural areas. Transfer of ideas is not limited to production technology, but includes the testing of communication strategies as well as helping farmers to organize themselves. Experiences are channelled back into UPLB teaching and research.

Animation Rural

For a historically rather short period, the concept of Animation Rural (AR) gained importance in francophone African countries such as Senegal, Ivory Coast, and Madagascar;. Though the original approach is no longer pursued, some of its elements are now being reintroduced into rural development programmes.

Animation Rural was an answer to the authoritarian and often repressive nature of intervention before independence. Developed originally by the French Institute de Recherché et d' Application des Methods de Développement (IRAM), it shows many parallels to the Brazilian experiments of Paolo Freire.

Integration of rural areas into the national system was to be achieved by initiating a dialogue between rural communities (collectivities) and the state. In a dialectical way, increasing competence of villagers to express their own needs was to liberate them from colonial dependence. In order to initiate and perpetuate this process, AR relied on a large number of voluntary collaborators, so-called *animators.* Selected by the villagers themselves these *animators* had to be experienced and well-respected farmers but not traditional leaders. Training, supervision, and support of *animators* were organized by the Ministry of Rural Development. Their task was to initiate discussions within the community on local needs and objectives, thus empowering rural people for a dialogue with the state. At the same time they were to "interpret" government plans to the villagers and acquaint them with services available. The long-term perspective was a replacement of traditional institutions and the creation of "development cells" able to negotiate contracts with the state bureaucracy.

Sülzer and Payr maintain that AR "did not fail as a philosophy of extension... [Although]... it did not achieve a large-scale breakthrough on a national level." Lack of sustainable impact was due to internal as well as external factors. The objectives of AR were extremely difficult to operationalize and, as a result, the role of *animateurs* remained unclear. In addition, lack of rewards and selection mistakes contributed to the fact that many *animateurs* soon lost interest in their work. Farmers, as it turned out, were more interested in receiving qualified technical assistance; and even if *animateurs* had successfully initiated village projects, it was the "technicians" who reaped the benefits. Lastly, it is highly questionable whether the administration was seriously committed to creating a system which would curtail its own power.

What has remained is the philosophy of empowerment and many of the practical experiences. Many NGOs use the ideas of Animation Rural often without realizing their roots. The present discussion on participatory extension shows its lasting influence.

Extension to Selected Clientele

Commodity Based Extension

Next to the ministry-operated general approach, commodity-based extension run by government, parastatals, or private firms is the most frequent extension organization. Clients may be dispersed over a large area or closely connected, as in the case of large, centrally operated irrigation projects. Commodity-based extension is the predominant feature in many francophone countries of Africa but is also strong in other countries with commercial or export crops.

The original rationale was the generation of revenue as well as the assured supply of tropical products for the colonial powers. Today, goals are still clearly and intentionally production and profit oriented. All aspects of producing and marketing a particular crop are vertically integrated, spanning the whole range from research, advice, and material support given to farmers, to organizing marketing and even exports. Proponents of the approach argue that, by infusing modern technologies and monetary incentives into traditional farming, a cumulative chain of effects are triggered, thus contributing to overall development.

Advantages in terms of organizing the extension function seem obvious. One generally works with well-tested technologies. Objectives and targets can be clearly defined and the organizational structure kept simple. The focus on only one or two crops facilitates training of extension workers who are agents of the society or board concerned. Control of agents and farmers is easy, because they are judged in terms of defined targets.

A closer look at these advantages reveals that they are largely defined from the perspective of the commodity organization. This poses no problem as long as organizational and clients' goals are identical, as was the case for coffee, tea, or sisal boards in the private plantation sector. For small farmers, the situation may be quite different. The rigidity of the system leaves little room for incorporating farmers' needs. The border between control and coercion is often crossed, for example, when farmers are forced to plant commercial crops at the expense of traditional subsistence crops. Extension workers are regarded as successful once they have brought farmers to producing "what and how" the organization wants. The obvious advantage of guaranteed marketing does not automatically entail security for the agricultural producer. Farmers cannot react quickly to price fluctuations, and in some cases quality standards are arbitrarily set in order to increase personal or organizational profits. Many governments have used the approach to excessively extract revenue by dictating low farm-gate prices.

Strengths as well as limitations of the commodity approach lie in its narrow focus. It is useful in terms of technology transfer but leaves out important public interest issues (such as environmental protection), as well as target groups (such as noncommercial producers). A successful combination of general and commodity-based extension at the national level, as practiced in East Africa, demands clear policy goals and highly efficient management.

Extension as a Commercial Service

Commercial extension is a rather recent phenomenon and typical of either industrialized forms of agriculture or the most modem sector of an otherwise traditional agriculture. It may be either part of the sales strategy of input supply firms or a specialized consultancy service demanded by an agricultural producer. In both cases, the goal of the organization or the individual is profit earning, which in turn is tied very closely to customer satisfaction. Most directly this is the case for private consultants who will be retired only if their clients feel that expenses made have been profitable. Large input supply firms or rural banks that use their own extension workers as sales personnel must also have a long-term perspective with regard to the competitiveness of their products and services. Negative effects of incorrect application or use will be attributed to the product itself. The clients of commercial extension will also be profit oriented. Their objective is the optimal utilization of purchased inputs or contracted expertise.

The emergence of commercial extension has influenced the debate on who should bear the costs of extension. With escalating budget deficits, the idea of extension as a free public service is no longer being generally accepted. It is argued that those who can afford it should actually pay for advisory services. In the case of commercial input suppliers, the solution is very simple: the costs of extension are included in the product price, as are the costs for research or advertisement. Private consultancy, on the other hand, is costly and affordable only to either large-scale or highly specialized producers.

As a general trend, one observes that public extension in industrialized countries has been under pressure to introduce cost sharing or altogether commercialize advisory work. An approach which combines commercial and public elements is at present being introduced in some of the eastern states of Germany. For example, the Ministry of Agriculture in Brandenburg subsidizes consultancies once they have actually taken place. Farmers have the option either to organize themselves in "extension rings" and employ their own extension workers or to choose an extension consultant who is officially accredited by the ministry once he or she organizes at least twenty clients in an "extension association." In both cases, up to 80 per cent of extension costs within a certain limit are reimbursed to the farmer.

Privatization and cost sharing are propagated in the name of greater effectiveness and efficiency, but are largely motivated by financial constraints. It is obvious that the private sector will be active only in case of reasonable returns, and they will not be concerned with public interest issues:

> Because of the selective participation of the private sector, the provision of public good types of information will have to remain a public sector responsibility... public and non-profit organizations... will have to work together to satisfy the needs of those in "orphan" areas. (Umali & Schwartz, 1994).

Client-Based and Client-Controlled Extension

One way of dealing with the shortcomings of large extension systems has been to localize extension and utilize the self-help potential of rural groups. Often organized by outsiders, these decentralized approaches are in a better position to serve the needs of specific target groups, notably those in disadvantaged positions. Close contact with their clients and intimate knowledge of their life situations are essential for the planning of problem-oriented extension activities. Local personalities are identified who take over leader functions once the external (nongovernmental) organization withdraws. The principles of these organizations (awareness, empowerment, participation, self-help) are close to the philosophy of Animation Rural without the national dimension.

The impact of client-based approaches must be seen on two levels. Directly, they provide benefits to their clients. The diversity and large number of small projects forbid a general statement on their effectiveness in terms of human resource development. It appears, however, that their weakness lies more in the technical field (UNDP, 1991, p. 22). Besides, they can reach only a very limited number of people. Apart from this, they perform an important role as organizational innovators. They have proved that participation can work in practice and that many farmers are highly competent partners in technology development. Government extension services have been forced to rethink their top-down approach, to accept human resource development as an equally important extension goal, and to address the problems of rural women.

A rather unique approach has been practiced in Taiwan, where a large share of extension work is done through farmers' associations. Organized at provincial, county, and township levels, membership totaled 90 per cent of Taiwanese farmers. Extension education is done by agents employed by the farmers' associations at the township level and financed largely by the farmers themselves.

Unlike the small self-help groups discussed above, there are strong and institutionalized linkages with research and other services. The overall extension policy is defined by the government. On the other hand, the clientele is quite different: farms are highly modernized and extension advice is demand driven.

Present and Future Role of Extension Staff

Person-to-person communication has traditionally been the most important form of information transfer. Print media as well as radio and television were of a supplementary nature because they frequently lacked a target group or location specificity and information was not up-to-date. Revolutionary changes in communication technology have dramatically increased the speed and quality of information transfer and changed the role of extension workers in industrialized countries. Electronic communications systems may in part replace personal visits, and one of the major tasks of any agent will be to link her or his clients with other suppliers of information.

Notwithstanding the fact that their use is not restricted to industrialized countries, the fascination with modem communication means tends to obscure the fact that most extension personnel - 90 per cent of which are located in developing countries (FAO, 1990) - are working under extremely difficult and disadvantageous conditions. In fact, little has changed during the last two decades to remedy the basic ills, and the field agent is still the weakest link within the system.

There is a wide discrepancy between organizational goals and the potential of even well-trained and dedicated field staff. Early Indian community development activities covered close to fifty areas (from reclamation of wasteland to improved rural housing on a self-help basis), all to be administered by the local village-level worker (Dube, 1958, Fieldwork in most developing countries is characterized by conditions that foster low morale: lack of mobility, virtually no equipment, and extremely low salaries. For many extension workers, tapping additional income sources is a question of physical survival. Quality performance is further impeded by low educational qualifications and lack of advancement possibilities (Swanson, Farner, & Bahal, 1990,).

While working conditions of extension personnel have deteriorated, expectations with regard to their role are increasing. They are no longer to be simply transmitters of technical knowledge. They are to practice participatory methods, recognize and respect gender issues, identify indigenous needs and problem solutions, and serve as a link to the world outside the village, to name but a few of the present topics. The emerging role is closer to that of a "socio-economic community worker" (Blackburn & Flaherty, 1994,) than a technical expert, but their training is insufficient for either.

The situation sketched above is well known and documented. The sheer dimension of the problem surpasses, however, the capacities of poorer countries, notably in Africa. Foreign-funded projects have addressed the issue in a piecemeal fashion and have often drained no project areas of personnel. Staff reductions on a national level have not even secured the status quo. Neither approach has solved the basic dilemma: insufficiency of material and human resources to reach universally accepted societal goals. Having to count on their own resources for extension, many countries will not be in a position to implement technology transfer, much less the more demanding strategy of human resource development. Regardless of specific extension approaches, there is no alternative to a strong international commitment to strengthening and revitalizing extension personnel resources.

The conceptual foundation for extension impact

Two conceptual themes are relevant to extension impact. The first is the awareness-knowledge-adoption-productivity (AKAP) sequence. The second is the "growth gap" interrelationship between extension, schooling, and research.

The AKAP Sequence

It is convenient to visualize extension as achieving its ultimate economic impact by providing information and educational or training services to induce the following sequence:

A: Farmer awareness

K: Farmer knowledge, through testing and experimenting

A: Farmer adoption of technology or practices

P: Changes in farmers' productivity

Changes in farmer behaviour will be reflected in quantities of goods produced, the quantities of inputs used, and in their prices. These, in turn, can be measured as "economic surplus," which is the added value of goods produced from a given set of inputs made possible by the extension activities.

Studies of extension impacts have measured farmer awareness (and sources of awareness), knowledge (and testing of practices), adoption, and productivity. Not all studies have examined all parts of the sequence. Most have shown a statistical relationship between the quantity of extension services made available to farmers and increases in awareness, knowledge, adoption, and productivity.

While the AKAP sequence has a natural ordering, it is clear that real resources in the form of skills and activities by both extension staff and farmers are required to move along the sequence. Awareness is not knowledge. Knowledge requires awareness, experience, observation, and the critical ability to evaluate data and evidence. Knowledge leads to adoption, but adoption is not productivity.

Productivity depends not only on the adoption of technically efficient practices, but of allocatively efficient practices as well. Productivity also depends on the infrastructure of the community and on market institutions.

Extension services affect each part of the sequence. They can be seen as both substitutes for and complements to the acquired skills of their clientele farmers. Empirical evidence indicates that they are, on balance, net substitutes for farmers' skills as reflected in farmers' schooling. For example, extension services are typically not the only sources of information (awareness). Skilled farmers can seek information on their own. Farmers with few skills may not do so. Extension information then may have a higher impact on farmers with less schooling. It appears, however, that the awareness-knowledge part of the sequence is where extension services are strong substitutes for farmer schooling. Through organized frequent contact, they "teach" farmers, and this is more than simply informing farmers.[4]

The teaching versus informing distinction is also relevant to the "newness" of the information (that is, of the recommended practice or other technology) and of the nature of the practice or new technology. When technology is new (as for example with a recently released variety of rice) and is also "simple" to evaluate and adopt (where it is a matter of using new seed without altering other practices), information-awareness is relatively easily converted to knowledge and adoption. Farmers with few skills usually adopt such technology with a time lag. When the technological practice is more complex and requires substantial changes in activities and sometimes capital investment, teaching is required. Repeated messages clearly stated, followed up by field staff and often community organization, and are required to proceed through the AKAP sequence in this case.

Productivity Gaps and Extension

The AKAP sequencing is, as noted above, related to the flow of new technical information and to the existing state of undaunted technology. We can see this interrelationship more clearly in the context of productivity "gaps."

A: Actual yields

BP: Best practice yields

BPBI: Best practice, best infrastructure yields

BPBIRP: Best practice, best infrastructure, research potential yields

These yield levels in turn define three "gaps":

- *G(P):* A practices gap between the best practice (BP) yield and actual (A) farmers' yields
- *G(I):* An infrastructure-institutions gap between the best institutions, best practice (BPBI) yield and best practice (BP) yield
- *G(R):* A research gap between the research potential yield (BPBIRP) and the best practice, best institutions (BPBI) yield

These gaps provide a way to classify the contribution of extension activities and to show how research and extension are linked. A stylized sequence across technology types is depicted. This could also be visualized as a time sequence.

Extension programmes are designed to reduce both the practice gap, G(P), and the institutions gap, G(I). Extension programmes are not the only activities that reduce these gaps. Providing market information to farmers and developing organized farm groups reduce G(I). Information and teaching reduce G (P). Research programmes are generally required to reduce G(R), although extension programmes can facilitate the reduction of G(R) via facilitating the importing and local modification of improved technology developed elsewhere. Research programmes in most developing countries also modify and adopt imported technologies and germplasm.

Two of the gaps are closely linked. When G(R) is closed (that is, when the BPBI yields go up), G(P) is opened.[6] (This may happen with G(I) also, but to a lesser extent.) Further, it should be noted that the size of the gap is an index of the potential impact of research or extension. As extension succeeds in closing G(P), diminishing returns set in. Successful research opens up new potential by increasing G(P). The relative mix of teaching versus informing is also related to these gaps. When the BPBI yield level has been constant for some time, the G(P) gap is closed mostly by teaching. When BPBI is increased, as by "green revolution" rice and wheat varieties, information and testing advice play a larger role.

The pattern of gaps and yield levels across country groups is intended as a stylized pattern. It is roughly based on experience. For Type la countries, both G(I) and G(P) are depicted as large.[7] These are traditional economies with not much new technology being produced that is relevant to them.

As economies move to Type 1b, improvements in institutions allow BP (and A) to rise even without new technology; BPBI remains unchanged. Extension can contribute to reducing both G(I) and G(P), and these contributions are qualitatively different from those required in more advanced country groups. There is little new technology (few new practices) in these countries, farmers have little schooling, and infrastructure is poor. The teaching and organizing activities of extension dominate here.

As economies move to the Type 1c category, some new technology has been introduced (BPBI has risen), and the institutions gap has been further reduced. The practices gap, G(P), has been both opened (because BPBI increased) and closed because of continued teaching and because new practices now can be extended.

Four Generations of Extension in Asia

The development of extension services in modern Asia has differed from country to country. Despite the variations, it is possible to identify a general sequence of four periods or "generations"

- *Colonial agriculture*: Experimental stations were established in many Asian countries by the colonial powers. The focus of attention was usually on export crops such as rubber, tea, cotton and sugar. Technical advice was provided to plantation managers and large landowners. Assistance to small farmers who grew subsistence crops was rare, except in times of crisis.
- *Diverse top-down extension*: After independence, commodity-based extension services emerged from the remnants of the colonial system, with production targets established as part of five-year development plans. In addition, various schemes were initiated to meet the needs of small farmers, with support from foreign donors.
- *Unified top-down extension*: During the 1970s and '80s, the Training and Visit system (T&V) was introduced by the World Bank. Existing organizations were merged into a single national service. Regular messages were delivered to groups of farmers, promoting the adoption of "green revolution" technologies.
- *Diverse bottom-up extension*: When World Bank funding came to an end, the T&V system collapsed in many countries, leaving behind a patchwork of programmes and projects funded from

various other sources. The decline of central planning, combined with a growing concern for sustainability and equity, has resulted in participatory methods gradually replacing top-down approaches.

The fourth generation is well established in some countries, while it has only just begun in other places. While it seems likely that participatory approaches will continue to spread in the next few years, it is impossible to predict the long-term future of extension. Compared to 20 years ago, agricultural extension now receives considerably less support from donor agencies. Among academics working in this field, some have recently argued that agricultural extension needs to be reinvented as a professional practice. Other authors have abandoned the idea of extension as a distinct concept, and prefer to think in terms of "knowledge systems" in which farmers are seen as experts rather than adopters.

Four Paradigms of Agricultural Extension

Any particular extension system can be described both in terms of both how communication takes place and why it takes place. It is not the case that paternalistic systems are always persuasive, nor is it the case that participatory projects are necessarily educational. Instead there are four possible combinations, each of which represents a different extension paradigm, as follows:

- Technology Transfer (persuasive+paternalistic). This paradigm was prevalent in colonial times, and reappeared in the 1970s and 1980s when the Training and Visit system was established across Asia. Technology transfer involves a top-down approach that delivers specific recommendations to farmers about the practices they should adopt.
- Advisory work (persuasive+participatory). This paradigm can be seen today where government organizations or private consulting companies respond to farmers enquiries with technical prescriptions. It also takes the form of projects managed by donor agencies and NGOs that use participatory approaches to promote pre-determined packages of technology.
- Human Resource Development (educational+ paternalistic). This paradigm dominated the earliest days of extension in Europe and North America, when universities gave training to rural people who were too poor to attend full-time courses. It continues

today in the outreach activities of colleges around the world. Top-down teaching methods are employed, but students are expected to make their own decisions about how to use the knowledge they acquire.

- Facilitation for empowerment (educational +participatory). This paradigm involves methods such as experiential learning and farmer-to-farmer exchanges. Knowledge is gained through interactive processes and the participants are encouraged to make their own decisions. The best know examples in Asia are projects that use Farmer Field Schools (FFS) or participatory technology development (PTD).

It must be noted that there is some disagreement about whether or not the concept and name of extension really encompasses all four paradigms. Some experts believe that the term should be restricted to persuasive approaches, while others believe it should only be used for educational activities. Paulo Freire has argued that the terms 'extension' and 'participation' are contradictory. There are philosophical reasons behind these disagreements. From a practical point of view, however, communication processes that conform to each of these four paradigms are currently being organized under the name of extension in one part of the world or another. Pragmatically, if not ideologically, all of these activities are agricultural extension.

Extension Methods

Compact Area Group Approach (CAGA) is an innovative extension approach developed by Dr.K. Abdul Kareem, an agricultural extensionist of India to combat problems like coconut eryophyid mite in states like Kerala.

Creative Extension is a new branch of agricultural extension pioneered by the Krishi Vigyan Kendra, Kenner under Kerala Agricultural University. Creative Extension uses art form for communication of innovation to farmers and calls for inputs from people with histrionic talents.

Principles of Extension Education

Principles

A principle is a statement of policy to guide decision and action in a consistent manner (Mathews).

A principle is a universal truth that has been observed and found to be truth and a settled rule of action.

The principles of extension education are given hereunder:

1. Principle of cultural difference

1. People differ in thinking, living and culture.
2. Extension education methods should be in line with these differences.
3. It is difficult to recognize non-material culture than material culture.
4. The blue print of extension programme for one area may not suit to other areas.
5. Changes will have to be made in the programmes according to changing situations.

2. Principle of cultural change

1. Culture undergoes change due to extension.
2. Change occurs otherwise also.
3. Extension workers should gain the confidence of people.
4. Extension workers should organize result demonstration.
5. After increasing production, the extension workers should concentrate on marketing.
6. Extension workers have to change to meet the cultural changes among the people.

3. Principle of grass-root organization

1. Extension workers should pay attention to all the groups' needs and interests.
2. Imposed innovations have no relevance to groups.
3. People will accept the innovations only when they find those useful.

4. Principle of interests and needs

1. People and extension workers should work together.
2. Co-operation and help of each other needed for social upliftment.

5. Principle of interests and needs

1. People should voluntarily participate.
2. Work should start from interests and needs of people.
3. Fulfillment of needs create interests.
4. First concentrate on felt needs and then develop felt needs.

6. Principle of participation

1. Attachment will not develop by offering ready made things.
2. Participation develops leadership and increases confidence.
3. Involving leaders increases people's participation.

7. Principle of adaptability in the use of teaching methods

1. People differ in knowledge and understanding.
2. Method should vary accordingly.
3. Use of more than one method is beneficial.
4. If needed new methods must be devised to meet new situations.

8. Principle of leadership

1. Extension workers should utilize local leadership for increasing speed of work.
2. Identification, training and encouragement of leaders is necessary.

9. Principle of trained specialists

1. Agricultural and other sciences are developing speedily.
2. Maintaining competency in any of these sciences is a continuous process.
3. Without specialist's support the extension cannot strive.
4. SMS is responsible to solve the extension workers problems.
5. Subjects Matter Specialist (SMS) is a link between research and application of research.
6. SMS should have broad outlook and be well versed.

10. Principle of satisfaction

1. The extension programme should give satisfaction to the people.
2. People will not participate if they do not get satisfaction.

11. Principle of whole family approach

1. Extension work should reach all the family members.
2. Neglecting any member may result in rejection of innovations e.g. Hybrid maize in U.P.

12. Principle of evaluation

1. Determining the research results in unbiased way is necessary.
2. Intermittent review of progress is necessary.
3. Corrective measures are needed if the direction is found wrong
4. Behavioral changes should be measured.
5. Evaluation helps in improving the quality of work.

13. Principle of applied science and democracy

1. People have freedom to accept or reject the technology.
2. Applied agricultural Science is a two way process.

14. Principle of learning by doing

Some more other principles

- The citizen is the sovereign (Supreme) in the democracy.
- Home is the fundamental unit of civilization.
- Family is the first training group of the human race and
- Average farm is endowed with great resources and facilities.

❑❑❑

Chapter 2

Fundamentals of Agricultural Extension and Communication

Diffusion of Innovation

Perceived attributes

1. Relative advantage -HYV varieties	Orves older RCCTO	i. Economic gain
2. Compatibility	1. With values and beliefs	ii. Labour saving devices
3. Complexity	2. With previously	iii. Immediate award
4. Trial ability	3. With the need of adopters	iv. Crisis situation
5. Observability	4. Technology Clusters	v. Non Economical
		vi. Low initiatives cost
		vii. Innovation low risk

The innovation - decision process

i. Knowledge
ii. Persuasion
iii. Decision
iv. Implementation
v. Confirmation

Types of innovation decision

1. Optional Complete responsibility for the decision.
2. Collective Adopter has some degree of influence.
3. Authority Adopting individual has no influence in the decision.

Consequences of innovations

1. Desirable vs Undesirable
2. Direct vs indirect
3. Anticipated vs unanticipated

Elements in the diffusion of new idea

1. Innovation
2. Communication channel ICTS
3. Time-Short/Long
4. Social System

Types of diffusion research

1. Earliness of knowing about innovations
2. Rate of adoption of different innovations in a social system.
3. Innovativeness
4. Opinion leadership
5. Diffusion networks
6. Rate of adoption in different social systems.
7. Communication channel usage.
8. Consequences of innovation.

Criticisms of diffusion research

1. Pro innovation bias
2. Individual blame bias in the diffusion research.
3. The recall problem in diffusion research.
4. The issue of equality in the diffusion of innovations.

Steps in Innovation development process

1. Recognizing a problem or need.
2. Basic and applied research.
3. Development
4. Commercialization
5. Diffusion and adoption
6. Consequences

The innovation decision process (KPDIC)

1. Knowledge

Types of knowledge

- ❒ Awareness knowledge
- ❒ How to knowledge
- ❒ Principle knowledge

2. Persuasion

- ❒ Favorable
- ❒ Unfavorable attitude

3. Decision Stage

1. Adopt
2. Reject
   - Active
   - Passive

4. Implementation – Reinvention

- ❒ RI occurs at the implementation stage for many innovations and for many adopters.
- ❒ RI a higher degree of RI leades to a greater rate of adoption of an innovation.

- A higher degree of re-invention leads to a higher degree of sustainability of an innovation.

Why RI

- Complexity of innovation
- Adopters lack of detailed knowledge

5. Confirmation stage

- Dissonance
- Discontinuance

Variable determining the rate of adoption of innovations

i. Perceived attributes of innovations relai
ii. Types of Innovation decision OCA
iii. Communication Channels
   - Mass
   - Interpersonal
iv. Nature of social system
v. Extent of change agent's promotion efforts.

ADOPTER CATEGORIZATION/ADOPTER CATEGORIES AND INNOVATIVENESS

1. Characterization of adopter Categories
2. Socio economic Status
3. Age
4. Education
5. Literature
6. Utilities
7. Personality values
8. Communication behavior

Empathy	E.A.*	Greater empathy than later
Dogmatism	E.A.	Less dogmatic than are later adopters
Abstractions	E.A.	Greater ability to deal with abstractions than do later adopters

* E.A. = Early Adopter

Rationality	E.A.	Have greater rationality than do later adopters
Intelligence	E.A.	Have more intelligence than do later adopters
Attitude	E.A.	Have a more favorable attitude towards change than do later adopters
Risk	E.A.	Are better able to cope with uncertainly and risk than one later adopters
Attitude	E.A.	Have a more favorable attitude towards science than do later adopters
Fatalism	E.A.	Less fatalistic than are later adopters
Aspiration	E.A.	Have higher aspiration than do later adoption

Characteristics of adopter categories

- Social Economic characteristics
- Earlier adopters are no different from later adopter in age
- EA have more year of formal education than late adopter
- EA one more likely to be literate then are late adopters
- EA have greater degree of upward social mobility than later adopters
- Personality variables

Communication Behavior

- EA have more social participation than do late adopters.
- E.A. are more highly interconnected through interpersonal networks in their social system than are late adopters.
- Cosmopolitenesses E.A. are more consmoplitetion are later adopters.
- Change agent E.A. have more contact with charge agents than do late adopters.
- Exposure E.A. have greater exposure to mass media communication channels than do late adopters.
- E.A. has greater exposure to interpersonal communication channels than do late adopters.
- Innovation knowledge Earlier adopter have greater knowledge of innovations, than do late adopters.

- Earlier adopters have a higher degree of opinion leadership than do later adopters
- Innovativeness/need paradox and the strategy of least resistance.
- Audience segmentation and adopter categories
- Diffusion networks how many type of diffusion models of communication flows.
- The hypodermic needle model
- The two step flow model

Homophily and heteropily in communication networks

Measuring opinion leadership and network links

1. Econometric techniques
2. Informants ratings
3. Self designating techniques
4. Observation

Opinion leadership

Types

- Monomorphic
- Polymorphic

Characteristic of opinion leaders

1. External communication
2. Accessibility
3. Socio economic status
4. Innovativeness, opinion leadership and system norms
5. Opinion leadership organizations
   - Do opinion leader matter
   - Do opinion leaders matters
   - Communication network analysis.

The Change Agent

The sequence of change agent roles

1. To develop need for change
2. To establish an information exchange

3. To diagnose problems
4. To create an intent to change
5. To translate an intent into action
6. To stabilize adoption and prevent discontinuance.
7. To achieve a terminal relationship.

Factors in change agent success

1. Change agent effort
2. Client orientation
3. Compatibility with clients needs/sustainability
4. Change agent empathy.

Communication campaigns

Successful campaign strategies consisted of

1. Utilizing formative research.
2. Utilizing audience segmentation
3. Setting specific but reasonable campaign goals
4. Designing the campaign's mass media messages to trigger interpersonal network communication among members of the intended audience.

Homophily and change agent contact with clients

Socio-Economic status Change agent contact-immuness

1. Change agents contact with lower status clients.
2. Para professional aides
3. Change agent or edibility
4. Inauthentic professionalization of aides.

The use of opinion leaders

Client's evaluative ability Centralized and decentralized diffusion system characteristics

1. Degree of centralization.
2. Direction of diffusion.

3. Sources of innovations.
4. Who decides which into to diffuse.
5. Importance of the client's needs in driving the differences.
6. Amount of reinvention.

Advantages and Disadvantages of Decentralized Diffusion

Advantages

1. Technical expertise is difficult to bring to bear on decision and about which innovations to diffuse and to adopt.
2. Furthermore non exports in decentralized diffusion systems lack of understanding of diffusion strategies that should be utilized.
3. Sometimes national govt wants on innovation diffused for which the people do not feel a need.

This discussion suggests that

1. Decentralized diffusion systems are not appropriate under certain conditions.
2. Certain elements of centralized and decentralized diffusion system can be obtained to form a hybrid diffusion system. Innovation in organizations.

Stages in innovation process

1. Agenda setting
2. Matching
3. Redefining/Restructuring
4. Clarifying
5. Reutilizing

Consequences of innovations

1. Direct and indirect consequences.
2. Desirable and undesirable consequences.
3. Anticipated as unanticipated consequences.

Imperfections of the diffusion process of Rogers

1. Population is not homogeneous
2. Wind fall profits and other different rewards
3. Information distortion
4. Progressive farmer bias in the technology development.
5. Self reinforcement process
6. The consequences of rapid innovation in conditions of imperfect diffusion.

Diffusion : Can be defined as a process by which an innovation is communicated through certain channels over certain period of time among the members of a social system.

Communication	Diffusion
Communication covers all types of messages	Diffusion is concerned about the message of new ideas.
A degree of risk is not present in communication	A degree of risk on the part of receiver is present in diffusion
In communication, the focus is on changing the knowledge or attitude of the people	In diffusion to focus is on overt change
Communication is broader in sense elements SMCRF	Special type of communication elements of diffusion ICST

Diffusion research tradition

1. Anthropological research tradition
2. Early sociology
3. Rural sociology
4. Education
5. Public health and medical sociology
6. Marketing
7. Geography

Rural Development

1. Measurement
2. Perception real GNP
3. Per capita consumer expenditure
4. Physical quality of life index
5. HDI

Initiatives : Sewagram, shantiniketan etc.

Basic elements

1. Necessities of life
   i. Food, water, shelter, clothes
   ii. Primary healthcare and literacy
   iii. Security to life and property
2. Self respect
3. Freedom
   i. Political
   ii. Economical
   iii. Social servitude

Paradigms

1. Free world/modernization theory
2. Merit school/dependency theory
3. Big push/Rosenstein-Roderick theory
4. Human capital model of development
5. Gunnar Myrdal theory of spread and backward effect

Determinants

1. Technology
2. Capital
3. Human Resource
4. Natural Resource
5. Institutions and agencies

Typology of diffusion research

These are types of diffusion research

1. Earliness of knowing about innovation
2. Rate of adoption of different innovations in a social system
3. Innovativeness
4. Opinion leadership
5. Who interacts with whom?
6. Rate of adoption in different social system
7. Communication channel usage
8. Consequences of innovation

Criticism of diffusion research

1. Pro innovation bias
2. Individual blame bias
3. The recall problem
4. The issue of equality in diffusion of innovations
5. Focus on optional innovation diffusion

Innovation development and process

Stage - I Recognizing a problem/need
II Research
III Development
IV Commercialization
V Diffusion and adoption
VI Consequences

Adoption process

Willson and Gallup (1955)

Behavioral characteristics of farmers

1. Need
2. Awareness
3. Interest
4. Deliberation

5. Trial
6. Evaluation
7. Adoption

Difference

Diffusion process	Adoption process
According to Roger, "Diffusion process is the spread of a new idea from its source of invention or creation to its ultimate user or adopter	Adoption processes a mental process through which an individual passes from hearing about an innovation to final adoption
Diffusion is a process related to adoption of an innovation in an entire social system such as village or blocks etc	Adoption is a sequence of thoughts and actions which an individual goes through, before he finally adopts a new idea

Factors affecting the diffusion and adoption of new practices

1. Social
2. Cultural
3. Situational
4. Personal

A. Social factors

1. Locality groups
2. Family
3. Social cliques
4. Reference group
5. Formal groups
6. Status factors

B. Cultural factors

i. Culture
ii. Values and attitudes

C. Situational factors

i. Farm income
ii. Farm size
iii. Tenure status
iv. Source of farm information used
v. Standard of living

D. Personal factors

i. Age
ii. Education
iii. Psychological characteristics

What are the different types of decision how they are important in the decision process

1. Authority Fastest rate of adoption
2. Collective
3. Optional slowest rate of adoption
4. Contingent decision

Andragogy

Assumptions

Assumption about	Adult learner	Child learner
1. Self concept	Self directed individual	Depend on others
2. Experience	Have greater value and quality of experience	Less experience
3. Readiness to learn	Ready to learn according to their perception of developmental needs	Ready to learn when they are told by some authority
4. Orientation to	Problem of life centered learning	Subject centered
5. Motivation for learning	Motivated both by extrinsic and intrinsic rewards	Extrinsic motivation

Process of adult learning

1. Unfreezing
2. Diagnosis of problem
3. Setting objectives
4. Presentation of learning material
5. Refreezing

Principles and conditions

1. Learning involves active participation of learner.
2. Adult learning involves identification of needs by the learners.
3. In value the learners in planning of learning activities.
4. Catalytic role of teacher
5. Provide opportunities for learning through discovery.
6. Relate theory with practice.
7. Provide challenging and satisfying learning opportunities.
8. Adult learning is unique and the ability to learn varies widely.
9. Adult learning involves effective communication, cooperation's and interdependence.
10. Proper physical and psychological environment.

Differences between adult and child learning

Dimensions	**Child learning**	**Adult learning**
1. Goals and objective	To prepare the children to face life by integrating them with existing culture subject centered	To help individual to active their human potential
2. Contents	Contents chosen by those having delegated authority	Problem centered selecting contents based on learners need
3. Control	Teacher controlling the learning outcome	Teacher has very loss control over learning outcome
4. Method	Didactic	Facilitative
5. Role of teacher/Trainer	Clearly defined and considered as important	Not so important
6. Examination/Evaluation	Through public examination and tests arranged by teachers	Self assessment of learner unamphasis or easement

"Refers to perceived degree of expertise and trust worthiness accorded to a source by its audience at a given time".

Credibility refers to the perceived degree of expertise and trustworthiness accorded to a source by its audience at a given time.

Characteristics of credible communication

1. The individual knows	Objective
2. The communicate	Audience
3. Should be skillful	Message in Channels
4. Selecting	Organization
5. Tracing	The professional abilities and
6. Expressing	Limitations
7. Massage	

8. The individual prepares the plan of communication

Communication credibility : Credibility means trust worthiness and competence. Before the audience accepts any message it will judge whether the communicator and the communicator and the organization the individual represents, can be relied upon and is competent enough to give the information.

The scientists and extension agents having status, expertise, accomplistiment, authority and experience and perceives as highly credible by the farmers in communicating information on agriculture and rural development.

A communication to be credible should posses following characteristics'

1. The individual knows
   - The objectives - specifically defined
   - The audience - their needs, interest abilities, predispositions
   - The message - Its content, validity, usefulness importance
   - Channels
   - Organization and treatment of the message
   - The professional abilities and limitations
2. Communicator should be skillful in selecting, beating, expressing, message

3. The individual prepares
   - Plan of communication
   - Communication material

Different levels of credibility

Various sources of farm information enjoy different credibility among farmers eg scientist, demonstrations, block, extension agency progressive farmers have high credibility while agriculture.

Department officers, block officials, folder leaflet, newspaper enjoy low credibility among farmers. Further, this perceived credibility differs significantly amongst progressive and non-progressive farmers.

1. In a progressive village agri/scientist, radio, progressive farmers, blocked extension agency and demonstration were accorded descending order of credibileties.
2. In a non-progressive village demo progressive farmers, agri/ scientists, look first, second, third and fifth positions respectively.
3. Folders and newspaper were accorded least credibility in both types of villages.
4. The difference in credibility pattern of the progressive and non progressive villages should also be used to advantage by extension workers. Since the credibility of demonstrations in the progressible villages in low, we need not conduct demonstration in these villages on the other hand need for good demonstration is very high in non progressive villages.
5. The low credibility of written material may be due to the large scale and illiteracy amongst farmers. However they are effective in progressive villages.
6. Determinant of source credibility: - It is mostly the relative effectiveness of an information source as perceived by the clientele that determine the communication source credibility.
7. So by increasing relative effectiveness of information source use can increase the credibility of the same.

Important ingredients needed for credible message/for high credibility

1. Status
2. Expertise
3. Accomplishment
4. Authority
5. Experience
6. Resource base

On the part of message

1. Clear
2. Specific
3. Straightforward
4. Need before
5. Informative

Fidelity in communication

- Fidelity is the faithful performance of communication process by all its elements, communicator, message, channel and receiver
- Communication fidelity can be explained as the extent by which desirable changes in receiver's behavior as a result of communication.
- The basic step in affecting change in the communication in any field. More the information is sought more the adoption and high the contacts of farmers with communication sources higher is the adoption.
- Rate of adoption, extension agencies and use of extension teaching aid as primary source of information are better than community sources.

Communication Videlity

C.F. depends positively on six variables

1. Compatibility of language of change agent with receiver
2. Frequency of contact with the receiver
3. Usefulness of message

4. Different channels used and their use of frequency by extension agency
5. Socio-economic status of receiver
6. Past experience of the clientele with the message of extension agency.

All above relates to communications ability of change agent and his strategy of communication.

Noise and fidelity are two sides of the same coin. Eliminating noise increases fidelity, the production of noise reduces fidelity. The basic concern related to noise and fidelity is the isolation of those factors within determines the effectiveness of communication.

Role/Importance

1. Fidelity more - communication effective.
2. More fidelity - more adoption
3. More fidelity - High contact of the farmers with the source.
4. Faithful/reliable relation of the farmer with the source
5. Unbiased and faithful communication

Relationship with noise or how to increase fidelity

1. Reduces noise
2. Increases language compatibility
3. Increase frequency of contact
4. Make the message useful
5. Higher the noise lower the fidelity

Factors affecting fidelity in the communication

1. Compatibility of language of change agent with receiver.
2. Frequency of contact with the receiver.
3. Usefulness of message
4. Different channels used and their use of frequency by extension agency
5. Socio-economic status of receiver
6. Past experience of clientele with the message of extension agency.

FEEDBACK IN THE COMMUNICATION PROCESS

Feedback means carrying some significant responses of the audience back to the communicator.

Characteristics of feed back

i. Feedback is source oriented.

ii. Feedback varies in different communication situation.

iii. Feedback exerts control over future messages

iv. Feedback affects communication fidelity

v. Feedback maintains the stability and equilibrium of a communication system.

Other characteristics

Intention, specificity, description, usefulness, timeliness, clarity, validity, readiness.

Types of feed back

1. Self feedback
2. Listener feedback
3. Fact based feedback
4. Opinion based feedback

Functions of feedback

1. *Evaluative*: To evaluate what is right or wrong about a particular communication. If you give a speech to the class, your teacher will offer criticism and suggestions for improving your delivery.
2. *Simulative*: Feedback can also stimulate changes
3. *Reinforcing:* To reinforce, to give reward or punishment.

Use / benefits of feedback

Adequate and correct feedback provides communicator an opportunity to take corrective steps in communication work, help in identifying subsequent activities, and acts as a path finder for need based research.

Importance of feedback

If the problem does not relate to research, the extension agent shall find out whether the message has been relevant to the audience, or whether the channel, treatment, audio-visual aids have been appropriately used if not corrective steps should be taken without any loss of time. For a season bound programme. If nothing can be done in particular season the extension agent shall take appropriate steps in next season so that the mistaken are not repeated.

Concept of communication

- The term communication stems from the Latin word 'Communes' meaning common.
- Communication means the movement of knowledge to people in such ways that they act on that knowledge to achieve some useful results.
- Argil communication is defined a planned transfer of farm technologies from the research system to the farmer's system through extension system and media with a view to make desirable changes in respect of higher productivity, profitability and prosperity and also get feedback from the clients.
- Process of communication requires six steps
- Ideation Encoding transmission, receiving, recording and action

Elements

1. Communicator
2. Message
3. Channel
4. Audience

Level of communication

1. Intra-personal
2. Inter-personal
3. Organizational
4. Inter organizational

Form

1. Verbal
2. Non Verbal
   - Hepatics
   - Proximies
   - Oculesics
   - Vocalics
   - Chronemics

B Message flow
   - Upward
   - Crosswire
   - Downward
   - Horizontal

C Operation

Formal
   - Written
   - Oral

Informal
   - Gestures
   - Silence

Functions of communication

1. Informative
2. Persuasive / Influence
3. Instructive / Commend
4. Integrative

Origin of communication

1. Latin word communes
2. Meaning commonality
3. Exchange of facts feelings and ideas.
4. Create common understanding

Barriers

1. Semantic	Meaning Decoding
2. Social	Status Rate
3. Cultural	Norms Language
4. Psychological	Perception Self esteem Temporal
5. Technical	Space/distance Physical obstruction

Levels

A. Personal

- Interpersonal
- Intrapersonal

B. Organization

- Inter Organizational
- Intra organizational

Relationships

1. Action-reaction interdependence
2. Physical interdependence
3. Empathy
4. Interaction
5. Homophile
6. Heterophile

Elements

1. Source
2. Message
3. Channel
4. Receiver
5. Feedback

Extension Principles

1. Cultural difference
2. Grassroots organization
3. Indigenous knowledge
4. Learning by doing
5. Interest and needs
6. Whole family
7. Local leadership

Functions of Extension

1. Change knowledge, skill and attitude
2. Create better understanding
3. Bring better goal and action
4. Develop confidence

Approaches of Extension

1. Commodity specialized
2. General agricultural based
3. Training and visit
4. Participatory extension
5. Pro based
6. Farming system based
7. Cost sharing
8. General education
9. Bottom up

Types of Extension

1. Formative extension
2. Persuasive extension
3. Emancipating extension
4. Informative extension

Philosophy of Extension

1. Make people how to think
2. Brings voluntary change
3. Communication as intervention

Strategy

1. How do I get them?
2. Where I want them
3. How do I develop an offering?
4. Which my target clients want
5. How do I help people achieve
6. What they themselves want to achieve

Levels

1. Education
2. University education
3. Farmer training
4. Distance education

Service

1. Technology testing
2. Technology integration
3. Technology production
4. Technology dissemination

Principles of Learning

1. Self activity
2. Association
3. Dissociation
4. Transfer
5. Training
6. Readiness
7. Motivation
8. Attitude
9. Interest

Domains of Learning

1. Cognitive Domain
   - i. Knowledge
   - ii. Comprehension

iii. Application
iv. Analysis

2. Affective Domain
   i. Receiving
   ii. Responding
   iii. Valuing
   iv. Organization
3. Psycho-motor domain
   i. Speech behavior
   ii. Body movement

Modes of Learning

1. Concert experience
2. Clearing from feeling
3. Reflective observation
4. Learning from listening
5. Abstract conceptualization
6. Learning from thinking
7. Active experimentation
8. Learning by doing

Stages of competence

1. Unconscious incompetence
2. Conscious incompetence
3. Conscious competence
4. Unconscious competence

Learning Situation Situation

1. Learners
2. Subject matter
3. Teaching materials
4. Physical facilities
5. Teachers

Orientation

1. Field dependent
2. Field Independent

Learning styles

1. Visual learners
2. Auditory learners
3. Kinesthetic learners

"Learning is the process by which an individual through his own activity attains a change in his behavior. It is an active process on the part of the learner. The essential rule of an extension worker is to create effective "Learning situations".

Classification of Extension Teaching Methods

A. According to form

1. Written	1. Bulletins 2. Leaflets 3. Personal letters 4. Circular letters 5. Farm journals
2. Spoken	1. Meetings 2. Farm and Home visits 3. Office calls 4. Radio 5. Flag methods 6. Agricultural clinics
3. Visual	1. Result demo 2. Exhibits 3. Posters 4. Charts 5. Flash cards 6. Bulletin boards
4. Spoken and Visual	1. Method demo 2. Result demo

B. According to Use

A	Individual contact	1. Farm and home visit 2. Agricultural clinics 3. Personal letters 4. Flag method 5. Office calls
B	Group contact	1. Farm tours 2. Method demo 3. Result demo 4. Peripatetic team visit 5. Agricultural games 6. Media forum
C	Mass contact	1. Broadcast media 2. Screen media 3. Printer media 4. Others

3. Television
4. Movies
5. Puppets
6. Campaigns

Types of learning

1. Conditioning
2. Verbal learning
3. Motor learning
4. Perceptual learning
5. Attitude learning
6. Problem solving

Psychological factors in adult learning

1. Rapport and empathy
2. The four basic wishes - security response
3. Emotionality and rationality news experience

4. Suggestibility and suspiciousness
5. Imitativeness and assertiveness
6. Group loyalty and individualism
7. Adherence to condition and change ability
8. Creativity and pragmatism

PRA Participatory Rural Appraisal

Principles

1. Optimal ignorance
2. Seeking diversity
3. Offsetting bias
4. Triangulation
5. Listening and learning
6. Multidisciplinary team

Kind

1. PRA for Research
2. PRA for evaluation

Methods

1. Library
2. Direct observation
3. Semi-structured interview
4. Maps and models
5. Transect walk
6. Seasonal diagramming
7. Ranking and scoring
8. Venn diagrams
9. Historical profile
10. Futures possible
11. Livelihood analysis
12. Flow diagramming

Foundations

1. Behavioral/attitude
2. Methods of storing

Instructional aids

1. Audio - Visual aids

- Projected aids
  - i. Motion pictures
  - ii. Video
- None projected aids
  - i. Puppet show
  - ii. Talking doll

2. Selection

- Nature of audience
  - i. Size
  - ii. Education
  - iii. Ecosystem
  - iv. Location
  - v. Age
- Teaching objectives
  - i. Generate awareness
  - ii. Increase knowledge
  - iii. Change attitude
- Facilities available
  - a. Manpower
  - b. Fund
- Other factors
  - a. Urgency
  - b. Time
- Visual aids
- Projected aids
  - a. Slides
  - b. Filmstrips

None projected aids

1. Chalk Board
2. Bulletin Board
3. Picture and photographs
4. Flame graph
5. Flip chart, flab card
6. Poster diagram chart
7. Specimen, model, diagrama
8. Tran slide

Determinates

1. Technology
2. Communication channel
3. Innovation decision process
4. Level of innovativeness
5. Social system

Technology attributes

1. Relative advantage
2. Observability
3. Compatibility
4. Complexity
5. Trial ability

Communication channel

1. Interpersonal
2. Mass Media
3. Innovation decision process
4. Level of innovativeness
5. Social system
6. Social structure
7. Norms and culture

Three levels of TOT

1. Knowledge
2. Use
3. Commercialization

Programmes

- ATMA
- ATIC
- SREP
- NAIP
- Systems: - First line - ICAR
- SAUS
- VKS
- ARS

Frontline

Voluntary Agencies

1. Basic research
2. Strategic research
3. Technology generation
4. Technology testing
5. Technology integration
6. Technology production
7. Technology dissemination
8. Technology adoption

Effects

- Adoption
  1. New Ideas
  2. Replace present
- Rejection
  1. Passive
  2. Active

Barriers in Transfer of Technology

1. Social
2. Economic
3. Psychological
4. Cultural
5. Technical
6. Extension or/organizations
7. Political

Psychology

Applications

1. Assessment
2. Clinical
3. Counseling
4. Educational
5. Forensic
6. Health
7. Industrial
8. Legal

Area

1. Perception
2. Cognition
3. Emotion
4. Personality
5. Behavioral

Methods

1. Controlled experiment
2. Animal studies
3. Qualitative research
4. Observation in natural settings

5. Survey questionnaires
6. Longitudinal studies
7. Neuropsychological methods
8. Computational modeling

Schools

1. Behavioral
2. Structuralism
3. Cognitive
4. Gestalt psychology

Tool to develop need based, problem solving, area specific technology to ensure the better adoption of it

Farming System Research and Extension

FSRE: - Evolved in 1970s as

1. An alternative to conventional research and extension
2. Technology developed and disseminated as package of practice were not useful for resource poor, farmers and risk prone areas
3. Debase developed without farmers participation so lest adaptive
4. So make the technology to reach to farmer's research attention given to problems of different groups of farmers and their calculation is ensured at all levels of development to trail to adoption.

Characteristics of FSR/E

The FSR/E is a farmer based problem solving, comprehensive, interdisciplinary, complementary, interactive, dynamic and responsible approach to the society.

1. Farmer based
2. Problem solving
3. Comprehensive
4. Interdisciplinary
5. Complementary

6. Dynamic
7. Responsible to society
8. Location specific technology

Activities

The FSR/E involves a set of activities beginning from target and research area selection to problem diagnosis and development of relevant agenda.

Five basic activities of the FSR/E as follows

Step 1 Allocation of the target area - stratifying into subareas, according to critical physical biological and socio economic factors to get reasonable degree of homogeneity in conditions to ensure applicability of technologies Research site is selected because the technology will be developed according to location specificity.

Step 2 Interdisciplinary research team scouts new areas, review secondary data.

Step 3 Research teams review the problems and opportunities for on farm research programme.

Step 4 Three types of on form research are conducted viz

a. Research managed trials

b. Super imposed trials and

c. Farmer managed tests/trials

Step 5 Further investigation of technologies investigation before transfers to know how it can perform well in a broad and more intensive sale.

Evaluation

1. Unique and important aspect of the FSR/E methodology is its recognition that farmers operate different resource endowments, and therefore have different technogical needs.
2. All adaptive research/demo trials are carried out on farmers fields.
3. FSR/E represents major philosophical and methodological break with the traditional approach.

4. The most common problem that FSR/E addresses is the patron-client mentality of R and E personnel which blocks effective two way communication with farmers specially small-scale farmers.

Commonly encountered weaknesses of FSRL

1. FSR/E is marginal to most agricultural research organizations. Therefore there programmes are undervalued, receive lack priority and are generally ness funded.
2. FSR/E sometimes requires researchers to live in small towns or cities resulting in personal sacrifice to researchers and their families.
3. FSR/E is aimed at technology development and adoption, rather than research.
4. FSR/E does so at considerable professional risk. FSR/E is somewhat more costly than on statics research, because it requires adequate transportation, supplies and technical support.
5. FSR/E is especially vulnerable to budget cut.

Participatory Development Approaches

Participation includes peoples involvement in decision-making processes, in implementation programmes, their sharing in benefits of development programmes and their involvement in efforts to evaluate such programmes.

Steps in participatory approaches

1. Raport building
2. DIY?
3. Consultation with the villagers and outsiders
4. Secondary data collection
5. Training of trainers
6. Diagnosis

Different Participatory Approaches

RRA: Rapid rural appraisal

Planning for rural development required authentic information to provide the right salutions. The RRA is a process of collecting and analyzing information form field in a short period of time. It user multiple methods and involves people to collect and involves people to collect and validate information and does not rest on any single method.

Special features

1. Use of RRA demands respect for ITK the local people operate in the existing situation by virtue for their understanding of the environment.
2. Experts must learn from people in terms of their (people) preceptor in terms of their (people) precautionary things one not to become a learner and
3. It allows for natural learning that can take place through observation and interaction. It is an informal, flexible and creative way of gradual learning.
4. It requires experts to come in direct contact with people and area
5. It allows one to by different methods in order to get closer to the truth.

RRA methods

1. Use of existing information
2. Learning local knowledge
3. Direct observation
4. Key information
5. Group interview

PRA : Participatory rural appraisal

The PRA is a methodology for interacting with villagers and seeking their participation in putting forward their points of view about problems, analyzing them and utilizing the information to acquire learning. It takes into account the people's indigenous knowledge. It is learning through participation about livelihood system and interaction of various facts. It helps in joint evaluation of problems and opportunities.

Features of PRA

The PRA is concerned with collecting information through participation.

1. It helps farmers to communicate their perception in their own language and mode of communication.
2. It requires living and looking at life from close quarters using methods.
3. It encourages multidisciplinary investigation, asking multiple methods for cross checking to allow different perspectives.
4. Attitude of listening learning and respect for rural people.

PRA Methods

1. Diagram
2. Transact walk
3. Ranking and scoring
4. Do it yourself
5. Maps and models

Principles of PRA

1. Listening and learning
2. Offsetting biases
3. Utilization of precious community time
4. Seeking diversity
5. Triangulation/Crosschecking
6. Optimal ignorance and appropriate

Behavioral principles of PRA

1. Handing over the stick
2. Self critical awareness
3. Use your best judgments
4. From selecting to sharing

PLA Participatory learning and action.

In the world of PRA, the word "Appraised" has a much broader connotation than its dictionary meaning.

Category	RRA	PRA
Developed in	Late 1970's/1980's	Late 1980's 1990's
Key resource	Local peoples knowledge	Local peoples capabilities of analysis
Main Innovations	Methods	Change of behavior and attitude
Mode	Extractive	Facilitating
Mode of instruments	Verbal (Interview discussion)	Visual (Participate diagramming)
Ideal objectives	Learning from insiders by outsiders	Empowerment of local people
Outsider's role	Investigation	Initiator and catalyst
Insiders role	Respondent	Presenter, analysis planner
A model	For participatory intervention	For interaction
Who demand	Donor organization	Insider (Ideally)

PLA is an approach for learning and organizing participation of local communities and groups for interacting with them, understanding them and learning for them. It helps in initiating a participatory process, in sustaining it and in opening a vista of avenue for participation.

It provide an alternative frame work for data generation and analysis where the people generate different kinds of qualitative and quantitative data and information for their own analysis. In such participatory appraised both economic and non economic factors are important.

AEA Agro Ecosystem Analysis Conway et al. in the early 1980s developed this approach in Northern Pakistan within the scope of the Aga Khan in which it has been successfully applied at all levels of agrarian ecosystems.

Procedure

Study of sources, then field residence, resulting in a series of agro-ecosystem diagrams.

Time required 2-3 days per village.

Special features

An agro-ecosystem is created by the combination and interaction of ecological and socio economic process.

System properties

1. Productivity, 2. Stability, 3. Sustainability, 4. Equitable distribution of production & resources and inputs required to achieve it.

Tools / Techniques

Analysis of sources, direct observation, diagrams decision tree, identification of innovation potentials and needs by means of matrices.

Assessment

It is suitable during the orientation and/or planning phase of a project for capturing a many of them.

PALM

Participatory Learning Methods This approach is developed by MYRADA in Bangalore, India around INGO.

- Area in which it has been successfully applied is participatory planning of natural resource development at the village level and in urban neighborhood.
- Special features PALM use the key PRA concepts but emphasize participation by village residents and the function of the externals and catalyst and parties for self determination development.

Tools / Techniques

Geographical and historical transects participatory mapping, seasonality diagram my ranking and scoring, as well as other well known PRA/RRA techniques.

Assessment

With 25-30 externals, the PALM village studies involve a relative large team. This farm, PALM is tailored to conditions in India.

PAR Participatory Action Research

Developed in 1940 as an academic discipline of social psychology.

Participatory Action research was developed in social science circle in Latin America, and has been used since the early 1980 in the FAO's peoples participated.

Procedure

PAR is learning by dairy approach in which the investigator places his/her knowledge at the disposal of local groups.

The following is expected of an action researchers awareness of one been limitations and value orientations.

Special features

Studies of dependencies at local, national and international levels are intended to reveal solutions for disadvantaged groups to enable them to become more independent.

Tools / Techniques

Evaluation of official and locally available sources social stratification techniques, group interviews strategies for resulting conflicts.

Assessment

PRA has proved to be particularly useful for joint efforts to identify and come to terms with group conflicts.

National Agricultural Extension System (NAES)

In Indian context, there are four major organizational streams devoted to extension work for agriculture and rural development. Prasad these are :

1. The first time extension comprising mainly the Indian Council of Agricultural Research (ICAR) institution and the agricultural universities.
2. Extension system of ministry of agril and the state department of agriculture.
3. Extension system of ministry of rural development and the state development departments.
4. Development work by the voluntary organizations, business owners etc.

   NGOs

   - Non govt. organization
   - Cooperative and
   - Private organization
   - Often use a participatory approach in their extension programmes and often farmers an opportunity to participate in their programmes.

NGOs

Major types/ and roles

- Membership organizations
- Grass root support organizations
- Development support organizations

1. Membership organizations

- Educational roles
- Commercial and organizational role
- Management of common property
- Defending collective interest of members
- Other role

2. Organizations with grass roots support origins

- Churches
- Universities
- Business firms
- People with ideological motivation
- People with entrepreneurial support

3. Development support organizations

- Help membership and grass root through training and financial and material support.
- Help in developing country.
- Most of them based in industrialized countries.
- Financed partly by members and voluntary contribution and partly by ministry of development assistance.

Advantages and disadvantages

- Less bureaucratic more participatory
- They have a more motivational staff
- Success at reaching poor and landless
- Try to develop location specific solutions

- Result in fewer conflicting rules
- Give more attention to sustainable agril
- More competent at faciliting agril
- More likely to discover new methodologies

Disadvantages

- Staff often lacks competence in production technology.
- Ideology based solution rather than farmers problem.
- Small in size confined to local only
- Short term result given
- May be expensive programmes
- Can be used by politicians
- Leader may use the available resources

Importance of communications

1. Helping
2. HAD 5 M
3. Analyzing
4. Deciding
5. Informing
6. Monitoring

Methods of influencing human behavior

1. Compulsion/coercion
2. Exchange - goods/services
3. Advice
4. Openly influencing a farmer's knowledge and attitude
5. Manipulation
6. Providing means
7. Providing service
8. Changing the socio-economic structure

The farmer's decision-making process and how the extension agent can help.

A. Farmer's decision-making process

1. Perception of the problem assessing the present situation
2. Ascertain the desired situation
3. Assessing the problem
4. Formulating possible alternative solutions.
5. Evaluating alternatives
6. Choosing an alternatives
7. Implementing the choice
8. Evaluating
9. If the desired situation has not been achieved.

Limitations of adoption and diffusion research

1. Scientific knowledge
2. The desirableness of innovations
3. Advantages and disadvantages to the user
4. Content and interpretation of the message
5. Application of innovations
6. Feedback
7. The social effects of innovation
8. The effects of social system

Meaning and characteristics of attitude

Attitude refers to predisposition to perceive feel or behave towards specific objects/subjects in a particular manner. Attitude has three components

1. Cognitive
2. Affective
3. Behavioral component

Types of attitude

1. Social
2. Attitude towards specific persons
3. Attitude towards specific group community

Characteristics of attitude

1. Acquired by experience not born
2. Related to beliefs and social values
3. More or less permanent in nature
4. Attitude are changeable
5. It directs the behavior of individual
6. Emotional aspects combined with attitude
7. Socialization helps in attitude formation

Have subject relationship

B.C. Development of attitude

B.C. Factors affecting attitude change

A. Socio - cultural

1. Teaching and learning process
2. Religion
3. Social value and ideas
4. Economic situation
5. Family
6. Institution
7. Friend and colleges

B. Psychological determinants

1. Internal tension
2. Need and desire
3. Physical aspect
4. Habit
5. Ego defensive
6. Perceptional selective
7. Life goal

C. Functional determination

1. Motive role
2. Presentation in society
3. Observation
4. Personal value of individual
5. Knowledge

Methods of changing the "Attitude"

1. Learning theory
2. Consistency theory
3. Technique theory
    1. Observation
    2. Knowledge
    3. Presentation
    4. Motivation
    5. Personal value

Extension educational process

Steps as identified by J.P. Lagans

1. Situation and problems
2. Objective and solution
3. Teaching plan of work
4. Evaluation
5. Reconsideration

Social Structure of Rural Society

I	Family	1	Institution
II	Economic	2	Organisation
III	Political	3	Ecological Entities
IV	Religious	4	Groups
V	Education	5	Collectivities

Social strucutre is the term applied to the particular management of inter-related institutions and social pattern as well as the statuses and roles which each assume in the group. *T. Parson*

Elements of social structure

1. Social Institutions Family, Economic, Pol, Relig, Edu
2. Organizations
3. Ecological entities
4. Groups
5. Collectivities

Social institutions

"An institution is an organized system of social relationship which embodies certain common values and procedures and meets certain basic needs of society".

"An institution is a relatively permanent structure of social patterns of roles and relations that people enacts in certain sanctioned and unified ways for the purpose of satisfying basic social needs."

Formation of social structure

1. Individual
2. Family
3. Near group
4. Relative
5. Sub caste and gotra
6. Varna

Social structure develops social solidarity on the bases of

1. Marriage
2. Equality of language
3. Equal right and use of land
4. Kinship
5. Common responsibility
6. Common religion and beliefs
7. Use of common occupation

8. Mutual experience and action
9. Mutual help and security
10. Attachment to the social institutions

Functions of social structure

According to bottomore

1. It is a communication system.
2. Economic system relate to production and distribution
3. Socialization of new generations
4. Distribution of power and energy
5. Ritual system - control of society

Importance of social structure

1. Smooth running of society
2. Distribution of work
3. Labour distribution
4. In conducting social arrangement
5. Development of relation with other groups
6. In deciding the norms and behaviours of groups
7. Maintaining social control
8. Provide environment for individual development

Social institutions : "Institutions an organized system of social relationship which embodies certain common values and procedures and meets certain basic needs of society"

The Family : A family is a group of persons whose relation to one another are based upon consanguinity, who are therefore, kin to another.

Types of family

1. Nuclear
2. Joint - Matrilineal, patrilineal, matriarchal, patriarchal
3. Polyandroids
4. Polygamous

Functions of Rural Family

1. Emotional support to members
2. Sexual regulation
3. Reproduction
4. Social placement of members
5. Provide need and goal for the members
6. Other works

Characteristics of rural family

- Joint family system
- Greater inomogeneity
- Peasant household
- Greater discipline and interdependence
- Dominance of family ego
- Low status of women
- Religion and morality
- Welcome to guest
- Patriarchal
- Social control
- Ancestor workshop
- Social unit of society
- Economic cooperation
- Feeling of family respect
- Traditional influence
- Joint family system
- Main features/characteristics
- Demerits
- Factors affecting change in JFS
- Families
- Affected social fields of familism
- Factors making familism
- Caste system

- Caste is such a group based on birth, which directs its members to follow certain restrictions in regards to eating, drinking, marriage occupation and social interaction.

Characteristics

- Sentimental division of society
- Hierarchy
- Lack of un-restricted choice of occupation
- Restriction on marriage
- Restriction on feeling and intercourse
- Social division
- Religious disabilities

Origin of caste system

1. Traditional theory - Vedic
2. Religious theory
3. Occupational theory
4. Economic theory
5. Political theory
6. Geographical theory
7. Multiple theory

Functions of caste in India

1. Psychological security
2. Determine social status
3. Social security
4. Efficiency in occupation
5. Choosing of mat
6. Control our behavior
7. Religious security
8. Cultural security
9. Maintain purity of blood
10. Unity in diversity
11. Division of labour
12. Educational function

Demerits of caste system

1. Factors responsible for stability
2. Factors disintegrating caste system

Organization

"Those classes of human relationship structures where in people purposefully associate in systematically arranged units to promote and achieve some common purposes or interests that are not specifically expressed in the institution. Each member has a formal status and role".

Essential characteristics of organizations

1. Clearly defined limits
2. Formal membership, status and role
3. Self contained administrative structure
4. Operative principles, procedures and goals
5. Provision for control, authority and decision making
6. Creation, stimulated by dissatisfaction and periods of crisis.

Classification of organizations

1. On the basis of political structure within which they are created
    - Prescribed
    - Voluntary
2. On the basis of motives of participation
3. On the basis of organizational operation
4. On the basis of admission to membership

Development of organizations

1. Period of cultivation
2. Period of normal function
3. Period of decline and passing

The social structure of rural society

Ecological entities :

"Spatial pattern of symbiotic relationships. Associating in varying degrees and forms of inter-related organization with others in society, human being distributes themselves in relation to land areas in such a way as to create land-based relationships or units called ecological entities."

Natural, unplanned ecological entities

1. The region
2. The community
3. Neighborhood
4. Planned ecological entities

Groups: Factors affecting Group integration or cohesiveness

1. Homogeneity
2. Size
3. Physical mobility
4. Effectiveness

Classification of groups

a. Primary and secondary
b. Formal and informal groups
c. Gemeinschaft and geselischaft
d. Groups based on the quality of relationship
e. Group based on social class
f. Groups based on size
g. Groups based on personal feeling of belonging
h. Groups based on compulsion of participation
i. Reference groups

Collectivities

Collectivity is a number of persons whose behavior is specifically polarized around a temporary centre of attraction that leads to interaction and unification which exists only as long as the centre of attracts exists.

Characteristics

- Composition
- Four of attention or polarization
- Structure
- Duration of existence
- Standard of behavior
- Permanent structure may result from collectivities.

Social conditions that produce collectivity forms of collective behavior

1. Crowds
2. Audiences
3. Publics
4. Social movements

Culture : "Culture is that complex whole, which include knowledge, art, belief, moral, atoms and any other capabilities and habits acquired by man or a member of its society".

Characteristics of culture

1. Product of human behavior
2. Transmitted through learning
3. Gratify human needs
4. Trends towards integration of society as national culture
5. Utilize cultural technology

Types of culture

1. Ethnocentrism
2. Xenocentrism
3. Temporocentrism
4. Cultural relativism

Types of culture

1. Culture traits and complex
2. Sub culture
3. Counter culture
4. Idea culture
5. Real and ideal culture

Elements of culture

1. Symbols
2. Language
3. Value
4. Norms
5. Technology and material

Norms

"Norms are the rules of conduct or social expectations for the behavior those people should or should not behave in various situations".

They are both

- Perspective what they should do
- Prescriptive what they should not do.

Definition : "Norms are controls; it is through them that the human society regulates the behavior of its members in such a way that they perform activities fulfilling social needs - even sometimes, at the expense of organic needs.

Types of norms

1. Folk ways
2. Mores

Folk ways are customs or conventions

They provide rules for the conduct the violation of folkways bring only mild censure. They are learned interactions with others and are passed down from generation to generation, folkways changed or culture change.

Characters of Folk ways

1. Strong
2. Folkways are base of the culture
3. They are habits of individuals and custom of the group
4. Traditionally handed down

Mores

They are more important than folkways an reaction to their violation are more serious violation are more serious inspires intense reactions and sometimes of punishment mores prohibit something, that state are called taboos.

Definition : "It is recognized, that when the folkways have added to them conception of group welfare, standards of right and wrong, they are converted into mores".

Characteristics of mores

i. Always true
ii. Always correct
iii. Conservative - not easy to change

Types of more

- Morals Direct for proper behavior
- Taboos - which

Social and technological change processes

Social change process

1. Process
2. Evaluation
3. Progress
4. Revolution

5. Adaptation
6. Disorganization
7. Cyclic

- *Process* : The process means continuous change taken place in a definite manner through the operation of forces presented from the first within the institution cooperation, completion, integration and disorganization.
- *Evaluation* : In this process developmental direction, movement and results are not limited. Evaluation is a movement from the simple to complex can, be and has been serious disputed. It is the integration of matter and concomitant dissipation of motion during matter passes from indefinite incoherent homogeneity to a definite coherent heterogeneity.
- *Progress* : It is change but it changes in desired or approved direction - not in any direction. We can say growth of life in respect of these qualities to which human being can attach or can rationally attach values which may faster human welfare.
- *Revolution* : It is more or less sudden change, usually accomplished by forcible overthrow of existing political order and leading to establishment of new forms of social and legal controls. It is a transvaluation of attitude and value towards the existing social structure - security tension, freedom tension and stratification tension.
- *Adaption process* : For the setting in particular environment an individual has to adopt some procedure, and methods of that environment therefore, change in the habit occur.
- *Disorganization process* : It occurs when there is change in equilibrium of forces so that many former expectations no longer apply and many forms-social control no longer function offensive. It is the process by which group relation are broken. It is a decrease of the influence of existing social rules of behavior upon individual members of the group (Thomas). Uncertainty of roles and status, conflict of mores and institution.

The cyclic process : The cycle of development runs through char change and response to challenge there are three conditions of the cycle.

i. Response to challenge
ii. Time of trouble
iii. Down fall

I. Institution

1. Family
2. Economics
3. Political
4. Religious
5. Educational

II. Organization

III. Ecological Entities

1. Region
2. Community
3. Neighborhood

IV. Groups

1. Primary/Secondary
2. Formal/Informal
3. Gememchaft/geselchaft

V. Collectivities

1. Crowd
2. Audiences
3. Publics
4. Social Movements

Technological change process

Social change process stages

1. Invention
2. Diffusion
3. Consequences

Social change processes

1. Process
2. Evaluation
3. Progress
4. Revolution
5. Adaptation
6. Disorganization
7. The cyclic process

Social change process stages

1. Invention
2. Diffusion
3. Consequences

Social structure

Control tools

- Folkways
- Mores
- Taboos
- Norms
- Laws

Stratification

- Wealth
- Ancestry
- Functional utility
- Religion
- Biological characteristics

Social interaction process

Interaction

(A) Competition	Absolute/Relative Personal/Impersonal
(B) Cooperation	1. Primary 2. Secondary 3. Tertiary
(C) Conflict	1. Compromise 2.Conversion 3. Tolerance 4. Arbitration 5. Truce 6. Displacement
(D) Accommodation	
(E) Assimilation	

Physical structure

1. Natural Resources
2. Man Made Resources
3. Human Resources

Classification of groups

1. Group mass public crowd.
2. Primary/secondary/tertiary
3. Formal/informal
4. Gemeinschalft geselicaft
5. Small, medium large
6. Aggressive passive
7. Cohesive/In cohesive
8. Democratic/Autocratic
9. Voluntary/involuntary
10. Political/social/religious/caste
11. Functional/dysfunctional
12. Statutory/non-statutory

Group dynamics

Social group: - "Social group are aggregate or categories of people who have consciousness of membership and of interaction" Honten and Hant

OR

"Where two or more person came together and influence one another, they may be said to contribute a social group." - Ogburns and Ninkoff.

Group integration or cohesiveness

1. Homogeneity of the group
2. Size of group
3. Physical mobility
4. Effectiveness
5. Quality
6. Social class
7. Size

A. Classification of groups / Types of groups

- Primary and secondary groups
- Formal and informal groups
- Gaminchaft and geselschaft

B. Groups based on social classes

- Horizontal
- Vertical

C. Groups based on size

- Small
- Medium
- Large

Ingredients of group

- Interdependency among members and
- A common identification There is combination create some degree of interpersonal bond.

Another dimension of understanding group is defining the boundry who is inside and who is outside.

Thus group may be defined a two or more people interacting in which existence of all the existence of a group as a group, is necessary for the needs of the individual group members to be satisfied.

Stage of group development

From the late 60's it was believed that group passed through a standard sequence of five stages

1. Forming
2. Storming
3. Norming
4. Performing
5. Adjourning

1st Stage: - Much uncertainty

2nd Storming: - Intergroup conflict

3rd Norming: - Close relationship and cohesiveness

4th Performing: - Group development, when the group is finally functional

5th Adjourning : - The final stage group development for temporary groups, characterized by concern with wrapping up activities rather than task performance.

- The punctured equilibrium model studies have shown that groups do not develop in a universal sequence of stages. But the timing of when group form and change the way they work is highly consistent specifically its, been found that
- The first meeting sets the groups direction
- The first phase of group activity is one of inertia
- A transition takes place at the end of the first phase which occurs exactly when group has used up half its allotted time
- The transition initiate major changes
- A second phase of inertia follows the transition
- The group's loot meeting is characterized by markedly accelerated activity. These are shown

The punctured - equilibrium model

While there are many kind of group. We are primarily interested in group, at work or groups in an organization called the work group, psychological groups since us dealing with it in psychological terms and define this as follows.

"A Psychological group is any number of people who

1. Interact with one another
2. Are psychologically aware of one another?
3. Perceive themselves as a group

Diagram

A crowd is not a group since there no mutual awareness and interaction. Perhaps the most important rationale for group formation is that when a person goes to work in a factory he looks for and needs companionship, understanding, sympathy, appreciation, and even affection besides doing the job, which is most cases has become monotonous, fatiguing and mind slogging and he needs the warmth of recognition that its is 'me', a person and not a 'cold machine'.

Groups can be 'formal' or 'informal'. Formal group a 'designated work group' defined by a organizations structure while an informal group is that group formal, informal

Dynamic factors refer to how the group are organized the group of training in membership and leadership skill, the tasks given to group its prior history of success or failure etc most of these factors are dynamic and keep changing. There is considerable amount of literature highlighting a body of knowledge and a related of skills concerned with actually helping group to change, grow and become more effectively achieving the following.

1. Sensitivity of members to each other, their feelings start up, needs and aspirations
2. Paying attention to each other.
3. Open free and frank communication with each others.
4. Other paying respect to each other.
5. Empathizing with another
6. Leaders become more aware of difference of opinions and explore all sides of issues and alternatives. They mere sensitive to group names, pressures, and feelings.

7. In increased commitment to a spirit of inquiry and a diagnostic approach.

8. Commitment to the value of open and now communication where ever appropriate.

An important aspect of group dynamics is the importance of roles that individuals play within groups. Those roles can be helping a hindering. It is not that a particular role by itself or hindering to a group performance. But that a high performance group member plays a role that is a given situation contribution to maximizing productivity of the group. The key is whether or not is appropriate to the situation.

SS HR/LT	**S2 HT/HR**
Committing	Persuading
Facilities involvement	Questioning
Synthesizing Summarizing	Guiding encouraging
Gaining commitment	Developing alternative
Problems solving	Advocating
Depending	Manipulative
Agreeing with every thing	Topic jumped
Avoiding decision closure through area	Statements as question
Seeking, sympathy	Selective interpretation
Expressing facility	Gate keeping
Resignation, helplessness	

Helping and hindering roles in groups

SA LR/LT	**S1 HT/LR**
Attending	Establishing
Listening	Getting started
Showing interest	Clarifying purpose
Taking notes	Defining goals
Monitoring	Maintaining direction
Avoidance	Aggressive
With drawing psychologically	Sanitizing
Withdrawing physically	Attacking personally
Reflecting boredom	Dominating
Escaping the group	Name salvage

R4	R3	R2	R1
Able and willing confident insecure	Able unwilling or confident	Unable but unwilling insecure	unable and unwilling

Each style S1 to S4, represents and behavioral

Competency, within watch competency is two categories of behavior one helping and one hindering. There categories are further divided into indicators of the types of behavioiur an individual crowd engage in under each category.

Observation Guidelines to Process Analyze Group Behavior

According to Bfeifffer and Jonex (1972) the main indicators for analyzing group dynamics are

1. Participation
2. Influence
3. Styles of influence
4. Decision making procedure
5. Task functions
7. Group atmosphere
8. Membership
9. Feeling
10. Norms

1. Participation

1. Verbal participation
2. High participation
3. Low participation
4. Shift in participation
5. Treatment of silent people

2. Influence

This is different from participation. Some participants may speak very little, yet they capture the attention of whole group styles of influence. Inlufence can be positive or negative it can enlist the cooperation of other an alienate them.

The different styles of influence

1. Auto cratic
2. Peacemaker
3. Laisez fair
4. Democratic

Indicators of group dynamics

1. Participation
2. Influence
3. Style of influence
4. Decision making procedure
5. Task junctions
6. Group atmosphere
7. Membership
8. Feelings
9. Norms

Introduction

The success or failure of managers depends on their leadership qualities. They can be successful leaders by helping subordinates to find solutions to their problems. Managers are involved with bringing together resources, developing strategies, organizing and controlling activities in order to achieve objectives. At the same time managers, as leaders, have to select the goals and objectives of an organization, decide what is to be done and motivate people to do it. Thus, leadership is that function of management which is largely involved with establishing goals and motivating people to help achieve them. Leaders set goals and help subordinates find the right path to achieve these goals.

A person may be an effective manager – a good planner, and an organized administrator – but lack the motivational skills of a leader. Another may be an effective leader – skilled at inspiring enthusiasm and devotion – but lack the managerial skills to channel the energy he/she arouses in others. Given the challenges of dynamic engagement in today's business world, most organizations today are putting a premium on managers who also possess leadership skills.

Definition and Meaning of Leadership

- Leadership is the use of non-coercive influence to shape the group or organization's goals, and motivate behavior towards the achievement of those goals.
- It is a process in which one individual exerts influence over others.
- Leadership involves other people - employees or followers - who by the degree of their willingness to accept direction, help to define the leader's status.
- It involves authority and responsibility, in terms of deciding the way ahead and being held responsible for the success or failure in achieving the agreed objectives.
- Leadership involves an unequal distribution of power between leaders and group members. Group members are not powerless; they can and do shape group activities in a number of ways. Still, the leader will usually have more power.

Key elements of leadership

It has been observed that every group that attains its goals or performs efficiently has a skilled leader. A leader's skill comprises of four major elements:

1. The ability to use power effectively and in a responsible manner,
2. The ability to understand the fact that people are motivated by different forces at different times and in different situations,
3. The ability to inspire and
4. The ability to behave in a manner
5. That will develop a harmonious work culture.

Leadership Theories

Trait theory of leadership

According to this theory, leaders are born, not made. Many researchers have tried to identify the physical, mental, and personality traits of various leaders. leaders do not possess all the traits mentioned in these theories, whereas many non-leaders possess many of them. Moreover, the trait approach does not give one an estimate of how much of any given trait a person should possess. Different studies do not agree about which traits are leadership traits, or how they are related to leadership behavior. Most of these traits are really patterns of behavior.

Behavioral theories

When it became evident that effective leaders did not seem to have a particular set of distinguishing traits, researchers tried to study the behavioral aspects of effective leaders. In other words, rather than try to figure out who effective leaders are, researchers tried to determine what effective leaders do - how they delegate tasks, how they communicate with and try to motivate their followers or employees, how they carry out their tasks, and so on.

Situational or contingency theories

The use of the trait and behavioral approaches to leadership showed that effective leadership depended on many variables, such as organizational culture and the nature of tasks. No one trait was common to all effective leaders. No one style was effective in all situations. Researchers, therefore, began trying to identify those factors in each situation that influenced the effectiveness of a particular leadership style. They started looking at and studying different situations in the belief that leaders are the products of given situations. A large number of studies have been made on the premise that leadership is strongly affected by the situations in which the leader emerges, and in which he or she operates. Taken together, the theories resulting from this type of study constitute the contingency approach to leadership.

Situational or contingency approaches obviously are of great relevance to managerial theory and practice. They are important for practicing managers, who must consider the situation when they design an environment for performance. The contingency theories focus on the following factors.

1. Task requirements
2. Peers' expectations and behavior
3. Organizational culture and policies

There are four popular situational theories of leadership: (1) Fiedler's contingency approach to leadership (2) The path-goal theory, (3) The Vroom-Yetton model and (4) Hersey and Blanchard's situational leadership model.

Path-Goal theory

Practice I	Challenging the process of change
Commitment 1	Search out for challenging opportunities
Commitment 2	Experiment with new ideas and take risk
Practice II	Inspiring as shared vision
Commitment 3	Develop a vision for future
Commitment 4	Enlist others to act
Practice III	Enabling other to act
Commitment 5	Footer collaboration by promoting cooperative goals and building track.
Commitment 6	Strengthen people by giving power away, providing choice, developing competence assigning Centrica looks and offering visible tasks
Practice IV	Modeling the way
Commitment 7	Set the example by behaving in way that are constant with shared values.
Commitment 8	Achieve small win that promote constant progress and build commitment
Practice V	Encouraging the heart
Commitment 9	Recognize individual contribution to the success of every project
Commitment 10	Celebrate team's success regularly

This theory was developed largely by Robert J. House and Terence R. Mitchell. The path-goal theory of leadership attempts to explain how a leader can help his subordinates to accomplish the goals of the organization by indicating the best path and removing obstacles to the goals.

The path-goal theory indicates that effective leadership is dependent on, firstly, clearly defining, for subordinates, the paths to goal attainment; and, secondly, the degree to which the leader is able to improve the chances that the subordinates will achieve their goals. In other words, the path-goal theory suggests that the leaders should set clear and specific goals for subordinates. They should help the subordinates find the best way of doing things and remove the impediments that hinder them from realizing the set goals.

Emerging pattern of leadership

1. Emergence of democratic leadership
2. Importance of education
3. Specialization in leadership
4. Increasing representation of the youths
5. Dominance of middle class
6. Mobility in leadership
7. Rise of collective leadership
8. Leadership combined with politics
9. Decline of the impact of land ownership
10. Emphasis on bureaucratic leadership
    - i. Problem
    - ii. Diffusion
    - iii. Education
    - iv. Cooperative
    - v. Coordinate
    - vi. Self help
    - vii. Motivation
    - viii. Sources
    - ix. Target

Need of leadership

1. To know the need and problems of rural people.
2. Decide the target of work
3. To educate follower in easy way about the programs
4. In arranging the sources for his group
5. To maintain co-relation amongst the followers
6. In developing self help and natural cooperation.
7. In diffusion of new technology intimation.
8. In motivating the people towards adopt of technology
9. In getting the cooperation in developed activities
10. To increase the numbers of teachers

Role and functions of leadership

1. Spokesman
2. Harmonizer
3. Planner
4. Executive
5. Symbol of group
6. Stimulatar
7. Recorder
8. Evaluation
9. Initiates
10. Orientor
11. Facilitation
12. Encouragement
13. Summons
14. Fact speak or
15. Fact give or
16. Compromising
17. Expeditor
18. Example
19. Arbitrator
20. Ideologist
21. Fathers figure
22. Scape goat
23. Controller
24. Executive
25. Policy maker

Methods of identifying leaders

1. Discussion
2. Workshop
3. Careful observation
4. Socio-metric

Theories/approaches of rural leadership

1. Traditional and genetic theory
2. Super natural origin
3. Margin uniqueness
4. The great man theory
5. Focalization of Spectre energy
6. Situational theory
7. Trait theory
8. Behavioral theory
9. Contingency theory

Diffusion is the process by which

1. An innovation
2. Is communicated through certain channels
3. Overtime
4. Among the members of a social system

Factors determining effectiveness of leader

1. Need fulfillment
2. Prestige
3. Value membership
4. Cooperative relationship
5. Heightened interaction
6. Clarity of goal

7. Small unit
8. Homogeneity
9. Outside events
10. Increased position
11. Attacks from environment

Styles of leadership

- Autocratic
- Democratic
- Theory
- Charismatic approach
- Free-rein leader
- Leadership as continuum
- Path goal approach to leadership effectiveness

Contingency Theory of Leadership

The theory holds that people become leaders not only because of the attributes of their personalities but also because of various situational factors and the interactions between leaders and group members.

1. Position power
2. Task structure
3. Leader member relations

Bases of power of leadership

1. Reward power
2. Coercive power
3. Legitimate power
4. Reference power

Experiential learning cycle

The ELC is a step by step model of adults learns new skills and gain knowledge. A graphic presentation of ELC is given below

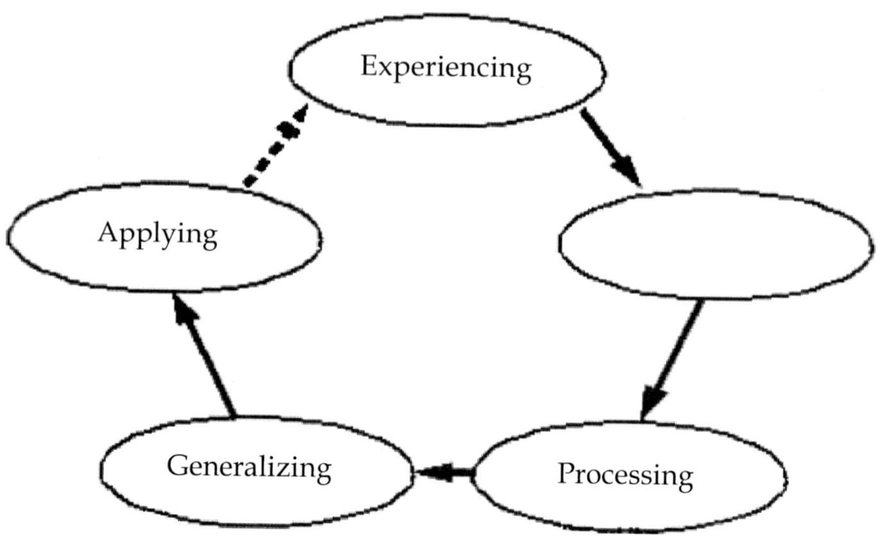

Capacity building for farmers groups

Capacity Building is the process of upgrading or strengthening group capabilities to manage work effectively to become self reliant and sustainable.

Approaches/techniques of capacity building

1. Exposure visit
2. Group Management
3. Identification and understanding resources, problem, solution, root cause of cause effect analysis
4. Participatory planning
5. Training of volunteer, leaders
6. Democratic decision making
7. Tie up with other development programmes.
8. Income generating activities
9. Analysis of success and difficulties
10. Arranges contacts with people
11. Formation of groups
12. Intergroup linkage

Steps in capacity building

1. Enabling
2. Empowering
3. Decision making
4. Self reliance

The role of nongovernmental organizations in extension

John Farrington is an agricultural economist and Director of the Rural Resources and Poverty Research Programme at the Overseas Development Institute in London.

In recent years, many observers have suggested that agricultural and rural development strategies would benefit from increased collaboration between government research and extension organizations and nongovernmental development organizations, hereafter called GOs and NGOs, respectively.

These advocates of closer NGO-GO collaboration have tended to underemphasize:

1. The wide range of interaction that currently exists, not all of it collaborative; much involves pressure by one side or the other.
2. The limitations facing efforts to work together.
3. The preconditions for successful collaboration; in particular, the prior informal contacts necessary to build up mutual trust.
4. Limitations as well as successes of NGO action.
5. The extent to which certain functions relating to, for example, "public goods" will remain more cost-effectively performed by the public sector than by NGOs. Analysis of how GOs might work with NGOs must be accompanied by continuing attention to ways of improving public sector management, an area in which structural adjustment reforms have not had the success expected.

NGO Characteristics

NGOs are defined here as nonmember ship development-oriented organizations. Our concern here is with the stronger of the south-based NGOs that provide services either directly to the rural poor or to grass-roots membership organizations, and with the local branches of international NGOs that enjoy varying degrees of autonomy. They are therefore distinct from (but, as discussed below, often linked with) formal and informal membership organizations such as farmers' associations. But even within this definition, there exists wide diversity of origins and philosophy. Some NGOs were set up by left-leaning professionals or academics in opposition to the politics of government or its support for or indifference to the prevailing patterns of corruption, patronage, or authoritarianism. Some are based on religious principles, others on a broadly humanitarian ethos, and yet others were set up as quasi-consultancy concerns in response to recent donor-funding initiatives. Some NGOs reject existing social and political structures and see themselves as engines for radical change; others focus on more gradual change through development of human resources (usually through group formation) to meet their own needs or to make claims on government services; yet others focus more simply on the provision of services (e.g., advice, input supply) largely within existing structures. Their ideological orientations also differ widely in relation to agricultural technology: many are concerned with low external input agriculture, 1 others pursue fundamentally organic approaches, 2 and, especially in the case of Andean societies, some are concerned to strengthen or reinstate traditional agricultural practices which formed the basis of social organization (CAAP, 1991).

Of crucial importance when considering NGO-GO links is that NGOs are *independent:* they are not mandated to collaborate with research and extension services in the way that government departments might be. They will therefore collaborate only if GOs have something useful to offer.

NGO Strengths

1. The majority of NGOs is small and horizontally structured with short lines of communication and is therefore capable of responding flexibly and rapidly to clients' needs and to changing circumstances. They are also characterized by a work ethic conducive to generating sustainable processes and impacts.

2. NGOs' concern with the rural poor means that they often maintain a field presence in remote locations, where it is difficult to keep government staff in post.

3. One of NGOs' main concerns has been to identify the needs of the rural poor in sustainable agricultural development. They have therefore pioneered a wide range of participatory methods for diagnosis[3] and, in some contexts, have developed and introduced systems approaches for testing new technology, for example in Chile (Sotomayor, 1991). In some cases, these approaches have extended beyond fanning systems into processing and marketing, as with soya in Bangladesh (Buckland & Graham, 1990), sesame in the Gambia (Gilbert, 1990), and cocoa in Bolivia (Trujillo, 1991).

4. NGOs' rapport with farmers has allowed them to draw on local knowledge systems in the design of technology options and to strengthen such systems by ensuring that the technologies developed are reintegrated into them (Chaguma & Gumbo, 1993).

5. NGOs have also developed innovative dissemination methods, relying on farmer-to-farmer contact, whether on a group or individual basis (e.g., Sollows, Thongpan, & Leelapatra, 1993).

6. In some cases, NGOs have *developed* new technologies such as soya production in Bangladesh (Buckland & Graham, 1990) or management practices such as the sloping agricultural land technology in the Philippines (Watson & Laquihon, 1993), but more often they have sought to *adapt* existing technologies, such as PRADAN's efforts in India to scale down technologies developed by government for mushroom and raw silk production and so make them accessible to small-scale farmers (Vasimalai, 1993).

7. Undoubtedly, one of the main strengths of NGOs has been their work in group formation. This has been in response to perceived needs at several levels: (1) To meet the technical requirements of certain types of innovation. Thus, Action for World Solidarity in India worked with grass-roots organizations to achieve simultaneous action in an integrated pest management programme (Satish & Vardhan, 1993). In the Gambia and Ethiopia, NGOs helped farmers to organize local informal seed

production in ways to avoid undesirable cross-pollination (Henderson & Singh, 1990). (2) To manage "lumpy" assets. In Bangladesh, NGOs have helped to organize landless labourers to acquire and operate water-pumping technology (Mustafa, Rahman, & Sattar, 1993). (3) To manage common property resources. Many examples exist of formal and informal associations, often supported by NGOs, which manage irrigation water. In other cases, NGOs have supported group efforts in soil and water conservation, whether on private land or on a micro-watershed basis involving both private and common land (Fernandez, 1993a). They have also helped in managing common grazing and forest land in a sustainable fashion in relation both to technology and the creation of a capacity to make demands on government over, for example, access issues (Femandez, 1993b).

8. In recent years, many observers have suggested that agricultural and rural development strategies would benefit from increased collaboration between government research and extension organizations and nongovernmental development organizations, hereafter called GOs and NGOs, respectively (Can-oil, 1992; de Janvry *et al.*, 1989; Jordan, 1989; Kurten, 1987). Donors in particular have begun to call for more NGO involvement in programmes that have traditionally been implemented through the public sector, and there has been a recent upsurge of donor interest in direct-funding south-based NGOs (World Bank, 1991a, 1991b; Farnsworth, 1991; Babington & Riddell, 1994).

These advocates of closer NGO-GO collaboration have tended to underemphasize:

1. The wide range of interaction that currently exists, not all of it collaborative; much involves pressure by one side or the other.
2. The limitations facing efforts to work together.
3. The preconditions for successful collaboration; in particular, the prior informal contacts necessary to build up mutual trust.
4. Limitations as well as successes of NGO action.

5. The extent to which certain functions relating to, for example, "public goods" will remain more cost-effectively performed by the public sector than by NGOs. Analysis of how GOs might work with NGOs must be accompanied by continuing attention to ways of improving public sector management, an area in which structural adjustment reforms have not had the success expected.

This chapter draws on a recent major study of the role of NGOs in sustainable agricultural development and the potential for collaborative links with GOs (Farrington & Bebbington, 1993; Babington & Thiele, 1993; Farrington & Lewis, 1993; Wellard & Copestake, 1993). The following will be reviewed: the characteristics of NGOs, their strengths and weaknesses in relation to agricultural technology, and the practical ways in which they and public sector extension services might collaborate more fully in the future.

NGO Weaknesses

1. NGOs' small size means that their projects rarely address the structural factors that underlie rural poverty. Small size, independence, and differences in philosophy also militate against learning from each other's experience and against the creation of effective forums, whether at national or provincial levels.
2. Some "fashionable" locations have become so densely populated by a diversity of NGOs that problems have arisen not merely of competition for the same clientele, but of some undermining the activities of others (Ayers, 1992).
3. NGOs have limited capacities for agricultural technology development and dissemination and limited awareness of how to create effective demand-pull on government services.
4. Some NGOs are more accountable to external funding agencies than to the clientele they claim to serve. Donor pressure to achieve short-term impacts, combined with a lack of cross-learning, has led in some cases to the promotion of inappropriate technology, such as protected horticultural systems in the Bolivian Andes (Kohl, 1991).

5. Many NGOs place great emphasis on voluntarism. Whilst such concepts as "volunteer extension workers" have great intuitive appeal and reflect widely commended values, they are sometimes promoted at the expense of financially sustainable alternatives. This was evident in SIDA's farm-level forestry project in North Vietnam, for instance, where the scope for supporting an emerging private nursery sector in the provision of technical advice was ignored, and complex and largely voluntary advisory services at the village level were promoted instead (author's observation, April 1994).

Examples of potentially replicable NGO-GO interaction

The examples that follow illustrate the types of NGO-GO configurations that offer potential for replication and adaptation in three areas: providing technical advice and feedback, training, and working with groups.

NGO-GO Configurations for Providing Technical Advice and Feedback

This chapter argues that the extreme institutional position in which all extension services are provided by the public sector is likely to be inefficient. At the other extreme, only in very specific circumstances can government hand over large parts of the extension function entirely to NGOs. This has been done in Chile, where government has contracted private technology companies to cater to the larger commercial farmers, and NGOs for small subsistence-oriented farmers.

However, similar attempts in India have been largely unsuccessful. A proposal in the Eighth Plan to hand over the entire range of technology transfer and training activities to NGOs in parts of Gujarat, Rajasthan, Orissa, Kerala, and West Bengal, with some technical support from the state agricultural universities and departments of agriculture, has generated only a lukewarm response from NGOs. One reason for this is the NGO concern that many of the technical recommendations from GOs that they would be expected to disseminate are not relevant to small-scale farmers. Another reason is that mechanisms for bottom-up feedback in existing technologies and for the articulation of demands for new technologies remain weak.

An attempt by the secretary of state for agriculture in Rajasthan to hand over responsibility for extension to NGOs in Udaipur District-renowned for its high density of NGOs - provoked a reaction that is likely to be typical of NGOs in many countries: namely, that by doing so the state is abrogating what is properly its responsibility to ensure a regular supply of technologies relevant to small-scale farmers.

Models developed elsewhere to provide a division of tasks more closely corresponding to the respective comparative advantages of the two sides have been more successful. One of the best-known models is in eastern Bolivia, where public sector extension services have long been characterized by chronic weakness (Thiele, Davies, & Farrington, 1988). Under a new strategy devised in 1989, Centre for Tropical Agricultural Research (CIAT) established a coordination unit - the Technology Transfer Department (DTT) - whose role was not to work directly with farmers, but with various intermediate users of technologies who had their own local extension services. NGOs are one of the most important types of intermediate users.

The DTT has subject-matter specialists and zonal specialists whose work is supported by a communications section. The subject-matter specialists (SMSs) are in regular contact with their corresponding CIAT researcher and collaborate on some research work. They package research information for delivery to intermediate users and are mandated to transmit feedback on farmer needs to the researcher.

Frequently, technologies developed in the experimental centers are still not ready for transfer. SMSs therefore carry out on-farm adaptive trials, in addition to ensuring that extensionists pass on the appropriate messages to farmers. Other dudes of the SMSs include preparing technical bulletins for extensionists, enhancing feedback and advice to extensionists (for instance, on the establishment of demonstration plots), and explaining how to give talks to farmers.

Informal collaborative arrangements rely heavily on the initiative of GO staff to feed lessons back into the next round of the research and extension agenda. In Bolivia, feedback was encouraged through a range of instruments, including NGO representation on the research planning committee of the local research station, and consultation with a number of zonal substations, part of whose function was to assemble users' views on the technologies being made available.

A different type of formal arrangement being developed in Udaipur District of India is a quarterly forum hosted by a "hybrid" NGO-GO institution - a government Farm Science Centre located in an NGO - in which interested NGOs and GOs participate. In essence, it is intended to promote familiarization by allowing cross-visits to be set up to each other's programmes, to allow training courses to be designed to meet NGOs' requirements, and to allow NGOs to "feed back" on currently available technologies and to voice further technology needs that are not currently being met. The forum is also intended to develop lessons for future interaction from efforts to monitor the process and impacts of current NGO-GO interaction within the district. The forum is beginning to make progress in some of these areas, but a major difficulty remains in the form of pre-programmed "targets" that the Department of Agricultural Extension has to meet each season in terms of, for example, the number of demonstrations of a given type. The forum has had the added benefit of facilitating substantive interaction among GOs (departments of agricultural extension, of animal production and health, and of horticulture) which otherwise meet on a monthly basis, but only for administrative purposes, and among the large number of NGOs in the district.

What remains to be addressed is the scope for similar formal links between NGOs, the membership organizations they are working with, and the GOs (such as the revenue and forest departments) which control large areas of land suitable for grazing or fodder collection. Although the central government has approved wider access by villagers (often with the assistance of NGOs) to such land under "joint management" arrangements, parts of the land have been encroached upon by wealthier farmers, and local-level officials find it difficult to rectify this so as to make the land accessible to village groups.

NGO-GO Configurations in Training

Some of the farmer training conducted by GOs is linked more strongly with GOs' programmes and targets than with farmers' needs. Much training is given in a classroom environment, without the practical content necessary to engage farmers' interests. NGOs have sought to work with GOs to address these shortcomings in several contexts:

1. In Gujarat, India, the Aga Khan Rural Support Project (AKRSP) identified village training needs through discussions with farmer groups (Shah & Mane, 1993). Initially, AKRSP organized government provision of this training, but the courses were formal in style (lectures in a classroom), and farmers' evaluations showed that they had learned little of practical value from them. In response, AKRSP developed an alternative needs-based training and dissemination methodology which it tested over several areas. Government staff were then brought in to observe, participate in, and finally adopt the methodology. Successful adoption was reinforced by informal networks and exchange of experience at workshops and consultations. AKRSP, along with Myrada (Femandez, 1993a), has also been instrumental in training GO staff in participatory methods.

2. In a different context, the International Institute for Rural Reconstruction in the Philippines brought together resource people from NGOs and GOs at a one-week workshop, the objective of which was to produce a completed *Agroforestry Resources Training Manual.* The manual is now widely used (Gonsalves & Miclatteves, 1993).

3. Clearly, there are also many instances in which NGO staff benefit from the skills which GO staff are able to impart; training in grafting techniques, for instance, has been found useful by a number of NGOs (Giordano, Satish, & Farrington, 1993).

NGO-GO Configurations in Group Formation

Substantial scope exists for GOs to benefit from NGOs' group-organizing skills. In India, for instance, recent modifications to the training and visit extension system now require village-level extension workers to interact with groups of approximately twenty farmers instead of with individual "contact farmers." However, extension workers are not trained in group formation skills, and groups that they form are unlikely - if they survive at all - to become interested in anything other than the testing of immediately available technology. The examples cited above illustrate how NGOs can effectively organize groups around integrated pest management, soil and water management, and the management of common property resources and capital assets.

What extension services can do to further collaborate with NGOs

The foregoing has several implications for extension services which aim to develop closer links with NGOs:

1. Explicit recognition of the wide diversity of NGO types will be necessary. Not all Many NGOs seek to support the establishment and growth of membership organizations capable of meeting their technology requirements over the longer term either from their own resources or by creating demands on government services or by a combination of both. Thus in seed supply, Cromwell and Wiggins (1993), for instance, quote numerous examples of ways that NGOs have supported local groups to produce seed, including vegetable and soya bean seed production in Bangladesh, and the multiplication of planting material for potatoes in the Ecuadorian Andes. In other countries (e.g., Nepal, the Gambia) local seed production initiatives have arisen plant breeding focused more fully on the needs of the rural poor, and the facilitation of linkages among the various agencies concerned with seed production and distribution. Some of these efforts see viable commercial arrangements as an essential feature of long-term sustainability. Thus PRADAN, in India, in an effort spanning several years to support the introduction of chrome-leather tanning by a local group, encouraged links with commercial lending organizations and private leather traders, not least because the latter could give accurate feedback on product

quality. In a more complex example of experimenting over several years with several types of women's groups for poultry production, the Bangladesh Rural Advancement Committee finally devised a multitier structure embracing rearing, local feed production, and health care by women parapets. These women drew on government for the necessary vaccines, earned a living by charging for injections, and provided elements of the extension function by giving advice on management and nutrition to those who paid for vaccinations will be willing to enter into a collaborative "service delivery" relationship with government, and those that do will do so only if GOs have something to offer appropriate to their clients' needs. Extension services therefore need to search for links with NGOs from a position of confidence that the research-extension system already has, or at least has the capacity to generate, something useful to NGOs and their clients.

2. Close interaction will be impossible if extension departments expect NGOs merely to assist in fulfilling preset targets such as the achievement of a given number of demonstrations of a given kind each season. GOs will have to bring an open agenda into the relationship, where possible keeping some resources "unallocated" in order to be able to respond to needs as they are articulated by NGOs.

3. Very specific efforts will have to be made to convey both feedback on existing technologies and NGOs' requirements for new technologies to researchers. In many GOs, reward systems provide no incentive among either researchers or extensionists to respond to feedback.

4. GO and NGO staff can jointly participate in training courses (ideally led by a joint team) in the action-oriented methods such as participatory rural appraisal favored by NGOs. The relevance of these to individual NGO staff will vary, but their capacity to enhance awareness of farmers' perspectives is important.

5. Depending on their philosophy, NGOs are concerned to develop local capacities for experimentation which build solely on farmers'

indigenous knowledge or on this and relevant "outside" ideas. This strategy may contribute to rural advancement in its own right, and the capacity it creates may prove a useful independent source of innovations in the absence of usable technologies from government. Alternatively, where GOs are willing and able to work with the poor, it will be a useful complement to what extension service can offer.

Re-invention

Degree to which an innovation is changed or modified by a user in the process its adoption and implementation.

History Until about 70's reinvention was not thought to occur or was considered a very infrequent behavior and considered a "noise" in diffusion research. Adopters were thus considered as passive adopter.

Characteristics

1. RI occurs at the implementation stage for many innovations and for many adopters.
2. Higher degree of RI leads to a faster rate of adoption of an innovation attitude.
3. A Higher degree of re-invention leads to a higher degree of sustainability of an innovation - sustainability is defined as the degree to which an innovation continues to be used over time after a diffusion program ends.

Why does Re-Invention occur?

Some reasons for re-invention lie in the innovation itself, while others invalue the individual or organization adopting the new idea.

- Innovations that are relatively more complex and difficult to understand are more likely to be reinvented
- Re-invention in such cases may be simplification of the innovation or ever represents a misunderstanding of it.

- Complexity of innovation.
- Lack of knowledge about innovation.
- Re-invention can occur owing to an adopters lack of detailed knowledge about the innovation, such as when there is little direct contact between adopter and change agents and previous adopters.
- An innovation that is general concept or a tool with many possible applications is more likely to be reinvented.
- When an innovation is implemented in order to solve a wide range of users problems, reinvention is more likely to occur. A basic reason for reinvention is that.
- Local pride of ownership of innovating may also be a cause of re-invention.
- Re-invention may occur because a change agency influences its clients to modify or adopt an innovation.
- Re-invention occurs when a innovation must be adopted to the structure of organization that is adopting it.
- Re-invention may be more frequent later in the diffusion process for an innovation as later adopters profit from the experience gained by earlier adopters.

Re-Invention Reason

A. Reasons related to innovation itself

1. Complexity
2. Wider applicability
3. Multiple problems based
4. RI more frequent in later diffusion process

B. Reasons related to adopter organization or person

1. Lack of knowledge or (detailed knowledge)
2. Local pride of ownership

3. Influence of change agency to modify innovation
4. Compulsion/necessity of adopting with organizational climate

Rogers identifies several additional characteristics dominant in the innovator type:

1. Venturesome, desire for the rash, the daring, and the risky,
2. Control of substantial financial resources to absorb possible loss from an unprofitable Innovation.
3. The ability to understand and apply complex technical knowledge, and
4. The ability to cope with a high degree of uncertainty about an innovation.

Characteristics Rogers identified in the early adopters

- Integrated part of the local social system,
- Greatest degree of opinion leadership in most systems,
- Serve as role model for other members or society,
- Respected by peers, and
- Successful.

Characteristics Rogers identified in the early majority

1. Interact frequently with peers,
2. Seldom hold positions of opinion leadership,
3. One-third of the members of a system, making the early majority the largest category.
4. Deliberate before adopting a new idea.

Characteristics Rogers identified in the late majority

- One-third of the members of a system,
- Pressure from peers,

- Economic necessity,
- Skeptical, and
- Cautious

Characteristics Rogers identified in the laggards

- Possess no opinion leadership,
- Isolates,
- Point of reference in the past,
- Suspicious of innovations,
- Innovation-decision process is lengthy, and
- Resources are limited.

Although additional names and titles for the adopters of an innovation have been used in other research studies, Everett Rogers labels for the five adopter categories are the preferred or Standard for the industry. Moreover, the specific characteristics that Rogers' identifies for each Adopter category is of significance to advertisers interested in creating an integrated marketing Plan targeting a specific audience.

The adoption process

In his book Diffusion of Innovations, Rogers defines the diffusion process as one "which is the Spread of a new idea from its source of invention or creation to its ultimate users or adopters".

Rogers differentiates the adoption process from the diffusion process in that the diffusion process Occurs within society, as a group process; whereas, the adoption process is pertains to an Individual. Rogers defines "the adoption process as the mental process through which an Individual passes from first hearing about an innovation to final adoption".

Five stages of adoption

Rogers breaks the adoption process down into five stages. Although, more or fewer stages may exist, Rogers says that "at the present time there seem to be five main functions". The five stages are:

1. Awareness,
2. Interest,
3. Evaluation,
4. Trial, and
5. Adoption

In the awareness stage "the individual is exposed to the innovation but lacks complete Information about it". At the interest or information stage "the individual becomes interested in the new idea and seeks additional information about it". At the evaluation stage the "individual mentally applies the innovation to his present and anticipated future situation, and then decides whether or not to try it". During the trial stage "the individual makes full use of the innovation".

At the adoption stage "the individual decides to continue the full use of the innovation". Why is the Adoption Process of any relevance to advertisers? The purpose of marketing and Advertising is to increase sells, which hopefully results in increased profits. It is through Analyzing and understanding the adoption process that social scientists, marketers and advertisers are able to develop a fully integrated marketing and communication plan focused at a predetermined stage of the adoption process.

Rejection and discontinuance

Of course, as Rogers's points out, an innovation may be rejected during any stage of the adoption process. Rogers defines rejection as a decision not to adopt an innovation. Rejection is not to be confused from discontinuance. Discontinuance is a rejection that occurs after adoption of the Innovation. Rogers synopses many of the significant research findings on discontinuance. Many "discountenances occur over a relatively short time period" and few of the "discountenances were caused by supersedence of a superior innovation replacing a previously adopted idea". One of the most significant findings was done by Johnson and Vandan Ban (1959):

The relatively later adopters had twice as many discountenances as the earlier adopters. Previous researchers had assumed that later adopters were relatively less innovative because they did not adopt or were relatively slow to adopt innovations. This evidence suggests the later adopters may adopt, but then discontinue at a later point in time.

Rogers identifies two types of discontinuance

1. Disenchantment discontinuance - a decision to reject an idea as a result of dissatisfaction with it's performance, and
2. Replacement discontinuance - a decision to reject an idea in order to adopt a better idea.

Rogers defines the innovation-decision process as the "process through which an individual (or Other decision making unit such as a group, society, economy, or country) passes through the Innovation-decision process".

There are five stages in the Innovation-Decision Process:

1. Knowledge of innovation,
2. To forming an attitude toward the innovation,
3. To a decision to adopt or reject,
4. To implementation of the new idea,
5. To confirmation of this decision.

It should be noted that prior conditions affect the innovation-decision process. Prior conditions such as:

1. Previous practice,
2. Felt needs/problems,
3. Innovativeness, and
4. Norms of the social systems.

The first stage of the innovation-decision process entails seeking one or more of three types of Knowledge about the innovation. Rogers describes these as:

1. Awareness knowledge is information that an innovation exists.
2. How-to-knowledge consists of the information necessary to use an innovation properly, and
3. Principles knowledge consists of information dealing with the functioning principles
4. Underlying how the innovation works.

Rogers states that awareness and knowledge of an innovation can be made most efficiently through mass media. It will be interesting in twenty years or so, to ascertain if mass media will still be considered the most efficient means to create product awareness and knowledge.

The following chart identifies seven characteristics consistently found in 'early knower's.' These characteristics should be taken into consideration when targeting the early or late knower's segment of the population.

- Earlier knower of an innovation has more formal education than laterknowers.
- Earlier knowers of an innovation have higher socioeconomic status than late knower's.
- Earlier knower of an innovation have more exposure to mass media channels of communication than later knowers.Earlier knower of an innovation have more exposure to interpersonal channels than later knowers.Earlier knower of an innovation have more change agent contact than later knowers.Earlier knower of an innovation have more social participation than later knower's.
- Earlier knower's of an innovation have more cosmopolite than laterknowers.

The knowledge stage of the innovation-decision process is of great value to advertisers because at this vulnerable stage of the innovation-decision process, advertisers are able to create an impressionable impact on their target audience. Advertisers should focus their efforts on creating awareness and knowledge when promoting a new product or innovation.

Consequences of Innovations

Before concluding our discussion on the innovation-decision process, it is important to consider the consequences or changes that occur to an individual or to a social system as a result of the adoption or rejection of an innovation. Rogers identifies three consequences or changes:

1. Desirable versus undesirable consequences
2. Direct versus indirect consequences, and
3. Anticipated versus unanticipated consequences

For the most part, the world of advertising is concerned with the diffusion of innovation process in terms of how such research studies can facilitate product adoption and therefore market segmentation. But it should be mentioned that additional research exists on the diffusion of innovation theory in other scientific disciplines, such as economic development and in the technological sector.

Five Stages of the Diffusion Process

In his book Inventive Activity, Diffusion, and Stages of Economic Growth, Stanislav Gomulka identifies five stages of technological growth that any economy in the world can be divided. They are:

1. By and large balanced growth at a low rate,
2. Transition phase of gradually increasing the characteristic rate of growth.(Gomulka states that there are four characteristic rates of growth),
3. High level of roughly semi-balanced growth,

4. Transition phase of gradually decreasing characteristic rate of growth,
5. By and large balanced growth at a relatively low rate, possibly close to the rate of growth of the country's population.

The first stage

In this phase of development the technological sector as well as the level of technology are in their "embryonic" stages. Both the share and rate of growth are low. The society is frequently faced with a great scarcity of primary commodities, living on a low level of subsistence, and with a lack of medical facilities (Gomulka 1971). Although the birth rate is high, the expansion rate of the technological sector is almost zero. There are two channels for diffusion to a less developed country. The first channel for diffusion is the exchange of knowledge and the second channel for diffusion is innovations from other countries. For either one of these channels to be utilized, so that the diffusion rate of a less developed country expands, is dependent upon two factors:

1. The degree of openness and receptivity from the underdeveloped country, and
2. The rate of growth of exports.

The 'degree of openness and receptivity' of an underdeveloped country is influenced by three main conditions

1. Transportation sectors within the country,
2. Communication sectors within the country, and
3. The general education levels of the population.

All of these conditions tend to be low in underdeveloped nations. Low levels of transportation, communication, and education produces a low degree of openness, making the less developed country almost closed to the diffusion process. Gomulka summarizes that the above hypothesis of a low rate of growth during the first stage can be due to the following three main reasons:

1. Low rate of growth of the total population,
2. Very limited growth of the technological sector,
3. Relatively little communication with more advanced countries.

The second stage

The rate of growth of the total population gradually begins to accelerate due to increased knowledge and achievement of a certain level of technology. The increased knowledge and technology achievement is due to improvements in food production and medical facilities. "New expanded supply and demand by society necessitates a larger absorption of foreign-made innovations and augmentation of the technological sector." (Gomulka 1971).

The third stage

The third stage is a continuation of the growth rate at the end of the second stage. Growth is brought about by the high rate of growth of the technological sector and/or by massive diffusion.

The fourth stage

The rapid rate of the third stage decreases due to an exhaustion of one or more of the growth rate variables.

The fifth stage

The fifth stage is congruent with the growth of the country when it:

a. Is already a part of the technologically leading area of the world,
b. Expansion of the economy follows growth of the country's population. The growth of the country's population is the independent variable, or principal determinant of the former.

Just as the adoption process relates to market segmentation, Gomulka presents an example of how the diffusion process can be applied to the economy and technological levels of less developed countries.

The New learning about innovation

Mark Dodgson and John Bessant in their book "Effective Innovation Policy: A New Approach"recognize that 'success' in innovation is not simply a matter of moving a resource from A to B, but "the capability on the part of the recipient to do something useful with that resource", in other words, to innovate effectively.

Dodgson and Bessant acknowledge that innovation is not an "instantaneous event, but a time based process involving several stages". They have identified these stages as:

1. Initial recognition of opportunity or need,
2. Search,
3. Comparison,
4. Selection,
5. Acquisition,
6. Implementation, and
7. Long-term use (involving learning and development).

The diffusion of innovation process consists of four main elements: the innovation, Communication through certain channels, over time, and among the members of a social system. Research concerning the diffusion of innovation process has increased significantly the past Several decades due to its' versatility. A universality or similarity found amongst the various research studies on the diffusion of innovation process is that the adoption process or the rate of diffusion can be charted on an S-shaped curve.

Of vast importance to those in the advertising field is the innovation-decision process. Rogers defines the innovation-decision process as the process through which an individual passes from first knowledge of an innovation to forming an attitude toward the innovation, to a decision to adopt or reject, to implementation and use of the new idea, and to confirmation of this decision.

The diffusion of innovation process can be tracked on a micro level as is the case of an individual who is a targeted member of an audience, or traced at the macro level when considering economic development or technological advances. In either instance, during the course of the twentieth century the diffusion of innovation theory has proven to be versatile, universal, but most important relevant.

Diffusion of innovations theory

Diffusion is the process by which an innovation is communicated through certain channels over time among the members of a social system. Diffusion is a special type of communication concerned with the spread of messages that are perceived as new ideas.An innovation, simply put, is "an idea perceived as new by the individual."An innovation is an idea, practice, or object that is perceived as new by an individual or other unit of adoption. The characteristics of an innovation, as perceived by the members of a social system, determine its rate of adoption. The four main elements in the diffusion of new ideas are:

1. The innovation
2. Communication channels
3. Time
4. The social system (context)

1. The innovation

Why do certain innovations spread more quickly than others? The innovation, to spread and be adopted should show:

The characteristics which determine an innovation's rate of adoption are:

1. Relative advantage
2. Compatibility
3. Complexity
4. Trial ability
5. Observability to those people within the social system.

2. Communication

Communication is the process by which participants create and share information with one another in order to reach a mutual understanding. A communication channel is the means by which messages get from one individual to another. Mass media channels are more effective in creating knowledge of innovations, whereas interpersonal channels are more effective in forming and changing attitudes toward a new idea, and thus in influencing the decision to adopt or reject a new idea. Most individuals evaluate an innovation, not on the basis of scientific research by experts, but through the subjective evaluations of near-peers who have adopted the innovation.

3. Time

The time dimension is involved in diffusion in three ways.

First, time is involved in the **innovation-decision** process. The **innovation decision** process is the mental process through which an individual (or other decision making unit) passes from first knowledge of an innovation to forming an attitude toward the innovation, to a decision to adopt or reject, to implementation of the new idea, and to confirmation of this decision. An individual seeks information at various stages in the innovation-decision process in order to decrease uncertainty about an innovation's expected consequences.

4. Step process

1. *Knowledge* – person becomes aware of an innovation and has some idea of how it Functions
2. *Persuasion* – person forms a favorable or unfavorable attitude toward the innovation
3. *Decision* – person engages in activities that lead to a choice to adopt or reject the innovation
4. *Implementation* – person puts an innovation into use
5. *Confirmation* – person evaluates the results of an innovation-decision already made

The second way in which time is involved in diffusion is in the **innovativeness** of an individual or other unit of adoption. **Innovativeness** is the degree to which an individual or other unit of adoption is relatively earlier in adopting new ideas than other members of a social system. There are five adopter categories, or classifications of the members of a social system on the basis on their innovativeness:

1. Innovators - 2.5%
2. Early adopters - 13.5%
3. Early majority - 34%
4. Late majority - 34%
5. Laggards - 16%

The third way in which time is involved in diffusion is in **rate of adoption**. The rate of adoption is the relative speed with which an innovation is adopted by members of a social system. The rate of adoption is usually measured as the number of members of the system that adopt the innovation in a given time period. As shown previously, an innovation's rate of adoption is influenced by the five perceived attributes of an innovation. – (Time/Infected Population)

5. The social system

The fourth main element in the diffusion of new ideas is the social system. A social system is defined as asset of interrelated units that are engaged in joint problem-solving to accomplish a common goal. The members or units of a social system may be individuals, informal groups, organizations, and/or subsystems. The social system constitutes a boundary within which an innovation diffuses. How the system's social structure affects diffusion has been studied. A second area of research involved how norms affect diffusion. Norms are the established behavior patterns for the members of asocial system. A third area of research has had to do with opinion leadership, the degree to which an individual is able to influence informally other individuals' attitudes or overt behavior in a desired way with relative frequency. A change agent is an individual who attempts to influence clients' innovation-decisions in a direction that is deemed desirable by a change agency. A final crucial concept in understanding the nature of the diffusion process is the critical mass, which occurs at the point at which enough individuals have adopted an innovation that the innovation's further rate of adoption becomes self-sustaining (the shaded area The concept of the critical mass implies that outreach activities should be concentrated on getting the use of the innovation to the point of critical mass. These efforts should be focused on the early adopters, the 13.5 percent of the individuals in the system to adopt an innovation after the innovators have introduced the new idea into the system. Early adopters are often opinion leaders, and serve as Role-models for many other members of the social system. Early adopters are instrumental in getting an innovation to the point of critical mass, and hence, in the successful diffusion of an innovation.

Conceptual Model

Diffusion of innovation model.

Source: Rogers (1995)

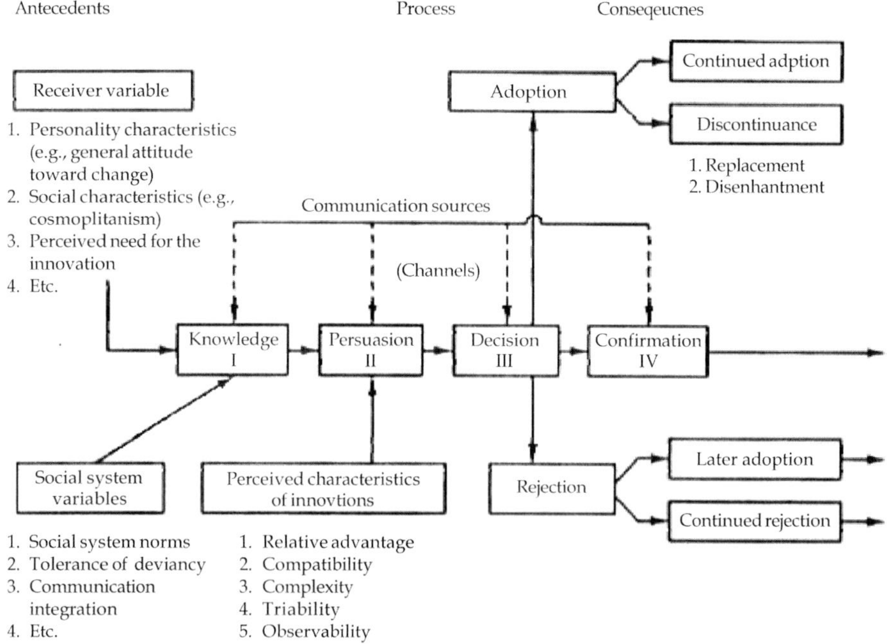

Technology Adoption and Diffusion

Introduction

Had technological change and innovation proceeded at today's rate in fifteenth century Europe when printing technology was introduced, one can only speculate on the economic and political effects that its adoption and diffusion might have had on that era and subsequent history. And, if Internet technology is the "printing" technology of today, its potential effects on modern society might be compared to those of that centuries-old innovation; that is, momentous, but difficult to foresee. Fortunately, we now know more about the adoption/diffusion process.

This review examines that process and the social and other factors influencing the diffusion of Internet/World Wide Web technology. Attributes of Internet technology that differ from those of traditional instructional technologies and that modify the adoption and diffusion process are discussed, as are characteristics of the potential adopters and strategies that contribute to successful technology adoption and integration within an organization.

Within this context, "adoption" refers to the stage in which a technology is selected for use by an individual or an organization. "Innovation" is similarly used with the nuance of a new or "innovative" technology being adopted. "Diffusion" refers to the stage in which the technology spreads to general use and application. "Integration" connotes a sense of acceptance, and perhaps transparency, within the user environment.

Typically, past adoptions of a new technology for education have signaled a confidence in its potential to alleviate a particular problem or to make a job easier or more efficient. Rarely has bringing about new social and functional conditions been a consideration. Internet and World Wide Web technology, however, may provide a means of creating totally new learning environments, and it may be to that end that adoption is initiated. In such instances, "innovation" and "adoption" may be seen as virtually synonymous elements of the adoption/diffusion process.

Technology adoption / Diffusion : Two views

Since early in this century, various "new" educational technologies have been touted as the revolutionary pedagogical wave of the future. Classroom films programmed learning devices, language laboratories, educational television, and computer-assisted instruction and, more recently, interactive videodisc technology have been adopted and integrated into the curriculum with varying degrees of success. Each technology was widely perceived as meeting a need, and each gained a measure of initial commitment of resources from a high level administrative or legislative entity. Their adoption and diffusion process generally followed what has been termed the "traditional model," a "top-down" process in which administrative "mandate" introduced the technology and administrative perceptions, decisions and strategies drove adoption and diffusion. Successful adoption was highly dependent on the degree, stability and wisdom of administrative sponsorship.

None of these technologies, however, has been generally available for individual or private use due to cost, scope or application. This deterred a "grass roots" technology adoption cycle as it was nearly impossible to generate movement from the bottom up by influencing faculty peers and administrators with demonstrations of successful applications.

Today's educational generation, however, sees personal computers, the Internet and the World Wide Web as technology's new wave. Proponents of distributed learning environments and distance learning on the World Wide Web forecast dramatic innovation at all levels and in all areas of education. And although this enthusiasm is reminiscent of that of past innovators, there are significant differences in the nature of this technology revolution in education and that of earlier ones with corresponding implications for adoption and diffusion.

Unlike earlier technologies which were thrust upon the education community, Internet technology is individually available to faculty and students who can use their own systems to serve their own purposes. The impetus for the innovation frequently grows from individual users of the technology, and as their communication and influence moves laterally through their contacts, a body of support can grow and exert "pressure" on the institutional administration to commit to adoption of the technology. There is, therefore, a high potential for a "bottom-up" or "grass roots" adoption process to succeed.

Indeed, Everett Rogers (1986), considered by many the "guru" of adoption/diffusion research since publishing Diffusion of Innovations (now in its fourth edition) in 1960, reveals three important ways in which the adoption of interactive communications differs from that of previous innovations. 1) A critical mass of adopters is needed to convince the "mainstream" teachers of the technology's efficacy. 2) Regular and frequent use is necessary to ensure success of the diffusion effort. 3) Internet technology is a tool that can be applied in different ways and for different purposes and is part of a dynamic process that may involve change, modification and reinvention by individual adopters.

Internet technology actually embodies a number of technologies—e-mail, databases, chat rooms, and information and education resources, among others. Additionally, the Internet exhibits many elements that constitute a culture or community—language, symbols, rituals, interaction, and other elements of communication. It thus essentially becomes an environment into which users enter (December, 1993; North, 1995). "Visionary" innovation and "pragmatic" application can begin with grass-roots enthusiasts who enter this environment. Viewed as a culture or community, however, the Internet can be perceived as a threatening competitor to the established norms of an existing culture or community, such as an academic department or some other institutional entity.

Adoption / Diffusion theories

The "top-down" and "bottom-up" models of adoption/diffusion provide a directional perspective to the process. Another theory dichotomy relates to the scale of innovation efforts by distinguishing between **macro-level theories** and **micro-level theories.** Macro-level theories focus on the institution and systemic change initiatives. Innovation typically involves broad aspects of curriculum and instruction and might encompass a wide range of technologies and practices. Micro-level theories, on the other hand, focus on the individual adopters and a specific innovation or product rather than on large-scale change.

Rogers (1995) recently presented four additional adoption/diffusion theories

Innovation Decision Process theory. Potential adopters of a technology progress over time through five stages in the diffusion process. First, they must learn about the innovation **(knowledge);** second, they must be persuaded of the value of the innovation

(persuasion); they then must decide to adopt it **(decision);** the innovation must then be implemented **(implementation);** and finally, the decision must be reaffirmed or rejected **(confirmation).** The focus is on the user or adopter.

Individual Innovativeness theory. Individuals who are risk takers or otherwise innovative will adopt an innovation earlier in the continuum of adoption/diffusion.

Rate of Adoption theory. Diffusion takes place over time with innovations going through a slow, gradual growth period, followed by dramatic and rapid growth, and then a gradual stabilization and finally a decline.

Perceived Attributes theory. There are five attributes upon which an innovation is judged: that it can be tried out **(trial ability),** that results can be observed **(observability),** that it has an advantage over other innovations or the present circumstance **(relative advantage),** that it is not overly complex to learn or use **(complexity),** that it fits in or is compatible with the circumstances into which it will be adopted **(compatibility).**

Each of the above can be considered in the context of either a top-down or a bottom-up adoption/diffusion process and in either macro-level or micro-level reforms. But there is one other adoption/diffusion theory dichotomy that is relevant to the discussion of Internet innovation. The distinction is between a **determinist (developer-based)** focus and an **instrumentalist (adopter-based)** one.

Determinists regard technology as the primary cause of social change. The process is seen as a series of revolutionary advances that are thought to be out of direct human control. Consequently, focus is on an innovation's technical characteristics. Successful adoption/diffusion is the assumed result of an innovation's technological superiority. The innovation's developer is viewed as the primary change agent.

For instrumentalists the process is evolutionary, and the causes of change are in social conditions and in human aspirations for change and improvement. Thus their focus is on the user (adopter) of a technology and its value as a tool to bring about desired change. Human control over the innovation is a key issue, and it is considered essential to understand the social context in which it will be used and the function that it will serve.

Comparative Approaches to Technology Adoption and Applications Development

Internet and World Wide Web technology offer opportunities for creating new and significantly different educational applications. But technological functionality is not the sole force currently driving its rapid adoption. Rather, the hope held by many in the educational community that new ways of teaching and learning that the technology makes possible appears to be increasing the rate of adoption. Concomitantly, many instructional technologists and educators now reject technical superiority alone as a sufficient basis for the successful diffusion of Internet and Web-based innovations.

David Jaffee (1998), in his analysis of resistance to asynchronous learning networks (ALN)—which would include many Internet/Web based learning applications—reminds us that classroom teaching is an established practice and cultural tradition of the teaching faculty. It centralizes power and influence with the instructor and serves as a focal point of professional identity. Jaffee suggests that institutionalization of the classroom teaching model is a major factor in reluctance to adopt ALN technology. He also observed that faculty exhibit less opposition to the use of televideo for transmission of classroom lectures, perhaps because the classroom setting is maintained between transmission and reception sites. The virtual classrooms afforded by ALNs and Web based learning modes, on the other hand, can be viewed as a threat in this context. He notes, however, that where professional identity is based on classroom presentation and student "reaction" to it, a disinterested and disengaged student audience might adversely affect that identity. Given that circumstance, Jaffee suggests that the ALN (Internet/Web) based virtual classroom concept might be conceived, or at least presented and promoted, as professionally enhancing.

Thus the focus of the process has shifted to the potential adopter and the organization into which the technology will be integrated. An adopter based, instrumentalist approach incorporating both macro- and micro-level perspectives now appears to be the most widely used to promote the adoption and diffusion of Internet technology.

Ernest Burkman's (1987) user-oriented development approach exemplifies those currently favored for the adoption and diffusion of instructional technology generally and Internet technology in particular. It consists of 5 adopter-focused steps: 1) potential adopter identification, 2) measurement of their relevant perceptions, 3) user (adopter)-friendly product design and development, 4) informing the potential user (adopter) of the product, and 5) support after adoption. An alternative model developed at the University of Minnesota's Telecommunication Center recommended a complete analysis of educational need and user characteristics along with the identification of a new educational technology's relevant and appropriate features and factors (Stockdill and Morehouse, 1992).

Tessmer (1991) stresses the need to analyze the environment in which the potential adopter is expected to use the technology. This process includes identifying the relevant physical and use characteristics of both the instructional situation and the support system. The approach is intended to ensure actual, correct and continual product use.

An adoption analysis approach (Farquhar and Surry, 1994) considers the process from the broader perspective of both user-perception and organization attributes, resulting in a plan for carrying out the adoption of technology that is rooted in an organizational context and addresses issues of concern to the intended user. Product and application design and development are also significantly influenced by this approach.

No single approach or process may be sufficient to ensure successful innovation adoption. But clearly, Internet and Web-based technology is individual-user based in application, and the adoption/ diffusion process should start at that level. It should focus on the potential adopters and address their characteristics in the context of the environment in which they will be using the technology.

Differentiation of technology adopters

The traditional adoption/diffusion continuum recognizes five categories of participants:

1. Innovators who tend to be experimentalists and "techies" interested in technology itself;
2. Early adopters who may be technically sophisticated and interested in technology for solving professional and academic problems;
3. Early majority who are pragmatists and constitute the first part of the mainstream;

4. Late majority who are less comfortable with technology and are the skeptical second half of the mainstream;
5. Laggards who may never adopt technology and may be antagonistic and critical of its use by others. The distribution of these groups within an adopter population typically follows the familiar bell-shaped curve.

Moore (1991) sees these groups as significantly different "markets" in the "selling" of an innovation to faculty adopters. He suggests that the transition from the early adopters to the early majority—one that is essential to an innovation's success—offers particular potential for breakdown because the differences between the two groups are so striking.

Table 1 : Adapted from Geoghegan, 1994.

Early Adopters	Early Majority
• Technology focused	• Not technically focused
• Proponents of revolutionary change	• Proponents of evolutionary change
• Visionary users	• Pragmatic users
• Project oriented	• Process oriented
• Willing to take risks	• Averse to taking risks
• Willing to experiment	• Look for proven applications
• Individually self-sufficient	• May require support
• Tend to communicate horizontally (focused across disciplines)	• Tend to communicate vertically (focused within a discipline)

Need-based diffusion strategies

Addressing the needs implied by the early adopter-early majority differences when designing diffusion strategies can greatly enhance the likelihood that a technology will be successfully integrated into the curriculum by groups beyond the innovators and early adopters (Geoghegan, 1994).

Need for recognition and process involvement. The chances of successfully "selling" an innovation to the pragmatic early majority will significantly increase if their differences are addressed in terms of their perceptions and needs. They should be recognized as a distinct group within the community and made a part of the planning and policy making process. Attempts to "convert" them to the point of view of the innovators and early adopters are likely to be futile, not to mention almost certainly disastrous to impose the technology on them otherwise. Diffusion of the innovation to the late majority and laggards is more likely to occur through this early majority involvement since the vertical lines of communication between the three groups are more direct than with the innovators and early adopters.

Need for vertical support structure to overcome technophobia. When technology adoption begins from the grass roots, innovators and early adopters, with their strong technology orientation, may be able to get by on their own initiative. Narrowly focused technical support staff may not pose a threat or discouragement to them and their needs for initial training and support may be relatively easy to accommodate. Members of the early majority, however, tend to have no interest in the technology per se and some may exhibit a form of technophobia. Their introduction to the technology should be related to their perceived program and process needs. Since they tend to focus vertically within a discipline, training and support provided by staff who enjoy discipline/content credibility will likely be best received. Correspondingly, such training and support will be more transferable to the late majority and laggards.

Need for well-defined purpose or reason. The very existence of a technology may be reason enough for innovators and early adopters to pursue it. Their bent for experimentation and their innate interest in technology may dispose them to adopt it and be content with "finding a problem to fit the solution". Members of the early majority (and the others by extension), however, tend to derive their purposes from problems related to their disciplines. If the innovation can be demonstrated as an effective, efficient and easily applied solution to those focused needs, it is more likely to be adopted and integrated into the program.

Need for ease of use and low risk of failure. The early majority's aversion to risk quite naturally translates into a need for ease of use and early success if they are to adopt and diffuse the technology. The overlap with support and training requirements is obvious.

Need for institutional/administrative advocacy and commitment. In the top-down adoption effort, institutional sponsorship and support is a given. The innovation may be mandated and grant moneys or other funds are committed. Without advocacy and resource commitment by the institution's "policy setters" and "holders of the purse strings", other issues become moot as the process is likely doomed to stalemate, if not to an early demise. But innovation that occurs from the bottom-up also requires institutional attention and an administration as an entity (except for some possible rare exceptions) tends to emulate the early majority rather than the innovators and early adopters. And even when an institution initiates an innovation from the top, their perspective tends to be a pragmatic one based on a problem or need that a given technology promises to alleviate. It may relate to staffing, financing, scheduling, teaching, distance or communication. In any case, the mindset is similar to that of the early majority and, as always, there is a need for advocacy to occur if the conditions and activities that can promote adoption by the early and late majorities and laggards are to prevail.

Meeting these needs is an essential part of any successful diffusion strategy. From their work at the University of Colorado, Wilson, Ryder, McCahan and Sherry (1996) derived several principles that apply particularly to situations in which students and faculty are introduced to networked learning environments.

First-time success. No one enjoys frustration or failure. An innovation is most likely to be accepted and integrated by the early and late majorities if success is experienced initially and subsequently built upon. E-mail is typically introduced early on because of its ease of use, and its success is almost guaranteed. It also extends the peer network, both within and outside the institution, thereby magnifying its impact on adoption and diffusion.

On-going peer support. Complementing the experience of initial success, there should be ample "hand-holding" along the way of integration as other Internet applications are introduced. Live peer support not only serves as assistance and encouragement; it contributes to the person-to-person communication that promotes diffusion throughout an educational community. In addition to a training cadre of recognized peers, a network of on-line mentors can expand the potential of the support structure to promote the exchange of innovative techniques.

Real task activities The early and late majorities are pragmatists who see technology in terms of real problem and task solutions. Activities designed to introduce and teach the technology should address those needs. As pointed out earlier, institutional administrations tend to emulate this pragmatic perspective. Internet access to information and resources, and its use for intra and inter-institutional communication can address many administrative needs in addition to those of the faculty, as well as establish a well-defined and recognizable need for adopting the technology.

Ownership and identity on the Internet Encouraging and enabling faculty and students to "create an active presence" on the Internet is important. Participating in listservs, creating a personal home page, publishing electronic papers all contribute to the electronic world-community and help ease "cultural assimilation." As with using e-mail to ensure early success with the technology, this "presence" extends the peer network impact on its adoption and diffusion. Beyond that, it also creates a professional identity and a credibility standing similar to that derived from traditional publication.

Variety of incentives. Attempts to impose a technology through explicit mandates and requirements, as in the top-down scenario, are not likely to be effective. This is particularly true with Internet and Web technology because the technology is so generally available to anyone who has a mind to adopt it. Policies and procedures promoting the technology should grow naturally from its application, and incentives for using it likewise should be tied to its practical use. Adoption and diffusion is more likely to occur where incentives and policies encourage a natural acceptance and use of the new technology.

Technology innovation in the educational community has often been hindered by the lack of a reward structure. Written publication has long been held as evidence of scholarly work that is worthy of recognition through promotions or tenure. In contrast, time consuming effort directed to pragmatic problem solving, instructional materials design and development or innovative classroom teaching has rarely received similar recognition. Integrating a technology like the Internet into one's teaching is time consuming and "effort intensive," usurping time and energy that otherwise could be devoted to more traditional—and more rewarded—endeavors. If innovative behavior is to be sustained, there must be a recognized and acknowledged system of rewards parallel to, and equal to, that associated with "traditional" academic pursuits.

Issues, Implications and Questions

From an innovation integration perspective, a collaborative, inter-institutional application of Internet technology for distance learning, or other purposes, raises issues with serious implications. Whereas the adoption/diffusion lines of communication for previous technologies tended to be confined within a single institution and vertical in nature, Internet and World Wide Web technology inherently extend those communication lines externally and broadly in a horizontal fashion. The aggregate of grass root innovators and early adopters at multiple institutions magnifies the impact that these "players" have on the process. What effect does this have on the adoption and integration at any of the individual institutions? How might that body of innovators be organized to best advantage? Is there a NIH (Not-Invented-Here) syndrome that must be overcome? If so, what strategies might be employed to do so?

The multi-institutional mass of innovators and early adopters provides opportunities for sharing trainers and training opportunities, and for expanding the scope and magnifying the impact of adoption efforts. Local, live peer support can be augmented by distributed "hand holding," but how might that be coordinated and integrated? And is distributed peer support effective?

Early success with the innovation, and ownership of and identity with the technology can be promoted through a variety of activities such as e-mail, listservs and chat rooms. Shared development of resources and papers can be collaborative efforts in which individuals at different sites can participate at their own level at any given time. Are some of these activities and tasks more likely to positively affect adoption/diffusion than others? How might they most effectively be introduced? Opportunities abound to develop resources and course material for alternative delivery on the Internet and World Wide Web, not to mention alternative, electronic publication of papers and research. What is the current extent of professional recognition for development and publication of this sort? How might professional recognition be fostered?

Advocacy of innovation and commitment to integration on the part of the institution's administration is essential to success. The inter-institutional nature of Internet and World Wide Web application, however, introduces a need for inter-institutional collaboration at the administrative policy/procedure setting level. Are there models of administrative collaboration that might be adapted to the integration of Internet technology?

Other issues that impact Internet and World Wide Web adoption need to be addressed, including, to mention a few, protection of intellectual property delivered on the Internet, copyright protection of electronically published materials, liability with regard to certain content and resources such as those of a medical nature, and licensure where "license boundaries" are crossed.

Diffusion of Innovations is a theory that seeks to explain how, why, and at what rate new ideas and technology spread through cultures. Everett Rogers, a professor of rural sociology, popularized the theory in his 1962 book Diffusion of Innovations. He said diffusion is the process by which an innovation is communicated through certain channels over time among the members of a social system. The origins of the diffusion of innovations theory are varied and span across multiple disciplines.

History

The concept was first studied by the French sociologist Gabriel Tarde (1890) and by German and Austrian anthropologists such as Friedrich Ratzel and Leo Frobenius. Its basic epidemiological or internal-influence form was formulated by H. Earl Pemberton who provided examples of institutional diffusion such as postage stamps and compulsory school laws.

In 1962 Everett Rogers, a professor of rural sociology published Diffusion of Innovations. In the book, Rogers synthesized research from over 508 diffusion studies and produced a theory for the adoption of innovations among individuals and organizations.

The book proposed 4 main elements that influence the spread of a new idea: the innovation, communication channels, time, and a social system. That is, diffusion is the process by which an innovation is communicated through certain channels over time among the members of a social system. Individuals progress through 5 stages: knowledge, persuasion, decision, implementation, and confirmation. If the innovation is adopted, it spreads via various communication channels. During communication, the idea is rarely evaluated from a scientific standpoint; rather, subjective perceptions of the innovation influence diffusion. The process occurs over time. Finally, social systems determine diffusion, norms on diffusion, roles of opinion leaders and change agents, types of innovation decisions, and innovation consequences. To use Rogers' model in health requires us to assume that the innovation in classical diffusion theory is equivalent to scientific research findings in the context of practice, an assumption that has not been rigorously tested.

The origins of the diffusion of innovations theory are varied and span across multiple disciplines. Rogers identifies six main traditions that impacted diffusion research: anthropology, early sociology, rural sociology, education, industrial, and medical sociology. The diffusion of innovation theory has been largely influenced by the work of rural sociologists.

Elements

The key elements in diffusion research are:

Element	Definition
Innovation	Rogers defines an innovation as "an idea, practice, or object that is perceived as new by an individual or other unit of adoption".
Communication channels	A communication channel is "the means by which messages get from one individual to another".
Time	"The innovation-decision period is the length of time required to pass through the innovation-decision process". "Rate of adoption is the relative speed with which an innovation is adopted by members of a social system".
Social system	"A social system is defined as a set of interrelated units that are engaged in joint problem solving to accomplish a common goal".

Decisions

Two factors determine what type a particular decision is :

- Whether the decision is made freely and implemented voluntarily,
- Who makes the decision?

Based on these considerations, three types of innovation-decisions have been identified within diffusion of innovations.

Type	Definition
Optional Innovation-Decision	This decision is made by an individual who is in some way distinguished from others in a social system.
Collective Innovation-Decision	This decision is made collectively by all individuals of a social system.

Authority Innovation-Decision	This decision is made for the entire social system by few individuals in positions of influence or power.

Mechanism

Diffusion of an innovation occurs through a five–step process. This process is a type of decision-making. It occurs through a series of communication channels over a period of time among the members of a similar social system. Ryan and Gross first indicated the identification of adoption as a process in 1943 (Rogers 1962, p. 79). Rogers categorizes the five stages (steps) as: awareness, interest, evaluation, trial, and adoption. An individual might reject an innovation at any time during or after the adoption process. In later editions of the Diffusion of Innovations Rogers changes the terminology of the five stages to: knowledge, persuasion, decision, implementation, and confirmation. However the descriptions of the categories have remained similar throughout the editions.

Five stages of the adoption process

Stage	Definition
Knowledge	In this stage the individual is first exposed to an innovation but lacks information about the innovation. During this stage of the process the individual has not been inspired to find more information about the innovation.
Persuasion	In this stage the individual is interested in the innovation and actively seeks information/detail about the innovation.
Decision	In this stage the individual takes the concept of the change and weighs the advantages/disadvantages of using the innovation and decides whether to adopt or reject the innovation. Due to the individualistic nature of this stage Rogers notes that it is the most difficult stage to acquire empirical evidence (Rogers 1964, p. 83).
Implementation	In this stage the individual employs the innovation to a varying degree depending on the situation. During this stage the individual determines the usefulness of the innovation and may search for further information about it.
Confirmation	Although the name of this stage may be misleading, in this stage the individual finalizes his/her decision to continue using the innovation and may use the innovation to its fullest potential.

Rates of adoption

The rate of adoption is defined as the relative speed with which members of a social system adopt an innovation. It is usually measured by the length of time required for a certain percentage of the members of a social system to adopt an innovation (Rogers 1962, p. 134). The rates of adoption for innovations are determined by an individual's adopter category. In general, individuals who first adopt an innovation require a shorter adoption period (adoption process) than late adopters.

Within the rate of adoption there is a point at which an innovation reaches critical mass. This is a point in time within the adoption curve that enough individuals have adopted an innovation in order that the continued adoption of the innovation is self-sustaining. In describing how an innovation reaches critical mass, Rogers outlines several strategies in order to help an innovation reach this stage. These strategies are: have an innovation adopted by a highly respected individual within a social network, creating an instinctive desire for a specific innovation. Inject an innovation into a group of individuals who would readily use an innovation, and provide positive reactions and benefits for early adopters of an innovation.

Rogers' 5 Factors

Rogers defines several intrinsic characteristics of innovations that influence an individual's decision to adopt or reject an innovation.

Factor	Definition
Relative Advantage	How improved an innovation is over the previous generation.
Compatibility	The level of compatibility that an innovation has to be assimilated into an individual's life.
Complexity or Simplicity	If the innovation is too difficult to use an individual will not likely adopt it.
Trialability	How easily an innovation may be experimented with as it is being adopted. If a user has a hard time using and trying an innovation this individual will be less likely to adopt it.
Observability	The extent that an innovation is visible to others. An innovation that is more visible will drive communication among the individual's peers and personal networks and will in turn create more positive or negative reactions.

Adopter categories

Rogers defines an adopter category as a classification of individuals within a social system on the basis of innovativeness. In the book Diffusion of Innovations, Rogers suggests a total of five categories of adopters in order to standardize the usage of adopter categories in diffusion research. The adoption of an innovation follows an S curve when plotted over a length of time.[10] The categories of adopters are: innovators, early adopters, early majority, late majority, and laggards (Rogers 1962, p. 150)

Adopter category	Definition
Innovators	Innovators are the first individuals to adopt an innovation. Innovators are willing to take risks, youngest in age, have the highest social class, have great financial lucidity, very social and have closest contact to scientific sources and interaction with other innovators. Risk tolerance has them adopting technologies which may ultimately fail. Financial resources help absorb these failures.
Early Adopters	This is the second fastest category of individuals who adopt an innovation. These individuals have the highest degree of opinion leadership among the other adopter categories. Early adopters are typically younger in age, have a higher social status, have more financial lucidity, advanced education, and are more socially forward than late adopters. More discrete in adoption choices than innovators. Realize judicious choice of adoption will help them maintain central communication position
Early Majority	Individuals in this category adopt an innovation after a varying degree of time. This time of adoption is significantly longer than the innovators and early adopters. Early Majority tend to be slower in the adoption process, have above average social status, contact with early adopters, and seldom hold positions of opinion leadership in a system (Rogers 1962 5th ed,
Late Majority	Individuals in this category will adopt an innovation after the average member of the society. These individuals approach an innovation with a high degree of skepticism and after the majority of society has adopted the innovation. Late Majority are typically skeptical about an innovation, have below average social status, very little financial lucidity, in contact with others in late majority and early majority, very little opinion leadership.

Laggards	Individuals in this category are the last to adopt an innovation. Unlike some of the previous categories, individuals in this category show little to no opinion leadership. These individuals typically have an aversion to change-agents and tend to be advanced in age. Laggards typically tend to be focused on "traditions", likely to have lowest social status, lowest financial fluidity, be oldest of all other adopters, in contact with only family and close friends, very little to no opinion leadership.

Heterophily and communication channels

Lazarsfeld and Merton first called attention to the principles of homophily and its opposite, heterophilyUsing their definition, Rogers defines homophily as "the degree to which pairs of individuals who interact are similar in certain attributes, such as beliefs, education, social status, and the like".[11] When given the choice, individuals usually choose to interact with someone similar to him or herself.[12] Furthermore, homophiles' individuals engage in more effective communication because their similarities lead to greater knowledge gain as well as attitude or behavior change.[12] However, most participants in the diffusion of innovations are heterophilous, meaning they speak different languages, so to speak.[12] The problem is that diffusion requires a certain degree of heterophony; if two individuals are identical, no diffusion occurs because no new information can be exchanged. Therefore, an ideal situation would involve two individuals who are homophilous in every way, except in knowledge of the innovation.[12]

Opinion leaders within a social system

Throughout the diffusion process there is evidence that not all individuals exert an equal amount of influence over all individuals. In this sense there are Opinion Leaders, leaders who are influential in spreading either positive or negative information about an innovation. Rogers relies on the ideas of Katz & Lazarsfeld and the two-step flow theory in developing his ideas on the influence of Opinion Leaders in the diffusion process [13] Opinion Leaders have the most influence during the evaluation stage of the innovation-decision process and late adopters (Rogers 1964, p. 219). In addition opinion leaders have a set of characteristics that set them apart from their followers and other individuals. Opinion Leaders typically have greater exposure to the mass media, more cosmopolitan, greater contact with change agents, more social experience and exposure, higher socioeconomic status, and are more innovative.

Organizations

Innovations are often adopted by organizations through two types of innovation-decisions: collective innovation decisions and authority innovation decisions. The collection-innovation decision occurs when the adoption of an innovation has been made by a consensus among the members of an organization. The authority-innovation decision occurs when the adoption of an innovation has been made by very few individuals with high positions of power within an organization (Rogers 2005, p. 403). Unlike the optional innovation decision process, these innovation-decision processes only occur within an organization or hierarchical group. Within the innovation decision process in an organization there are certain individuals termed "champions" who stand behind an innovation and break through any opposition that the innovation may have caused. The champion within the diffusion of innovation theory plays a very similar role as to the champion used within the efficiency business model Six Sigma. The innovation process within an organization contains five stages that are slightly similar to the innovation-decision process that individuals undertake. These stages are: agenda-setting, matching, redefining/restructuring, clarifying, reutilizing.

Consequences of adoption

There are both positive and negative outcomes when an individual or organization chooses to adopt a particular innovation. Rogers states that this is an area that needs further research because of the biased positive attitude that is associated with the adoption of a new innovation (Rogers 2005, p. 470). In the Diffusion of Innovation, Rogers lists three categories for consequences: desirable vs. undesirable, direct vs. indirect, and anticipated vs. unanticipated.

In her article, "Integrating Models of Diffusion of Innovations," Barbara Wagner details two categories for consequences: public vs. private and benefits vs. costs.

Public vs. Private

Public consequences refer to the impact of an innovation on those other than the actor, while private consequences refer to the impact on the actor itself Public consequences usually involve collective actors, such as countries, states, organizations, or social movements. The results are usually concerned with issues of societal well-being. Private consequences usually involve individuals or small collective entities, such as a community The innovations are usually concerned with the improvement of quality of life or the reform of organizational or social structures.

Benefits vs. Costs

The benefits of an innovation obviously refer to the positive consequences, while the costs refer to the negative. Costs may be monetary or nonmonetary, direct or indirect Direct costs are usually related to financial uncertainty and the economic state of the actor. Indirect costs are more difficult to identify.] An example would be the need to buy a new kind of fertilizer to use innovative seeds. Indirect costs may also be social, such as social conflict caused by innovation[15]

Mathematical treatment

Main article: Logistic function

The diffusion of an innovation typically follows an S shaped curve which often resembles a logistic function.

International Institute for Applied Systems Analysis (IIASA)

Several papers on the relationship between technology and the economy were written by researchers at the International Institute for Applied Systems Analysis (IIASA). The pertinent papers deal with energy substitution and the role of work in the economy as well as with the long economic cycle. Using the logistic function, these researchers were able to provide new insight into market penetration, saturation and forecasting the diffusion of various innovations, infrastructures and energy source substitutions.[16] Cesare Marchetti published on Concretive waves and on diffusion of innovations.[

A mathematical discussion of diffusion and substition models can be found in th Grübler (1990).

Diffusion data

Diffusion curves for radio, television, VCR, cable, flush toilets, clothes washer, refrigerator, home ownership, air conditioning, dishwater, electrified households, telephone, cordless phones, cellular phone, per capita airline miles, personal computers, internet are available from link on footnote.[19]

Diffusion curves for infrastructures (canals, railroads, highways, pipelines, airlines) and replacement of sailing ships by steam, then steam by diesel and replacement of steam locomotives by diesel are available from a link on the footnote.[18]

Criticism

Much of the evidence for the diffusion of innovations gathered by Rogers comes from agricultural methods and medical practice.

Various computer models have been developed in order to simulate the diffusion of innovations. Veneris developed a systems dynamics computer model which takes into account various diffusion patterns modeled via differential equations.

There are a number of criticisms of the model which make it less than useful for managers. First, technologies are not static. There is continual innovation in order to attract new adopters all along the S-curve. The S-curve does not just 'happen'. Instead, the s-curve can be seen as being made up of a series of 'bell curves' of different sections of a population adopting different versions of a generic innovation.

Rogers has placed the contributions and criticisms of diffusion research into four categories: pro-innovation bias, individual-blame bias, recall problem, and issues of equality.[22]

Electronic communication social networks

Prior to the introduction of the Internet, it was argued that social networks had a crucial role in the diffusion of innovation particularly Tacit knowledge in the book The IRG Solution - hierarchical incompetence and how to overcome it. The book argued that the widespread adoption of computer networks of individuals would lead to the much better diffusion of innovations, and with greater understanding of their possible shortcomings, and the identification of needed innovations that would not have otherwise occurred - the Relevance paradox.

Definition

In many cases in which action or decision is required, it is obvious what information relevant to the matter at hand may be lacking: a military attack may not have maps so reconnaissance is undertaken, an engineering project may not have ground condition details, and these will be ascertained, a public health program will require a survey of which illnesses are prevalent, and so on.

However, in many significant instances across a wide range of areas, even when relevant information is readily available, the decision makers are not aware of its relevance because they lack the information which would make its relevance clear. As a result, they do not attempt to look for it.[2] These decision makers will seek only the information and advice they believe is the minimum amount required as opposed to what they actually need to fully meet their goals.

An analogy would be that of a short-sighted person who is unaware of the condition. They would not perceive any benefit in getting the glasses they need to improve their sight, until they tried wearing the glasses. Such a situation could be resolved by a third party, aware of its relevance, recommending an eye test.

The Relevance Paradox is cited as a cause of the increase in diseases in developing countries even while more money is being spent on them: "Relevance paradoxes occur because of implementation of projects without awareness of the social or individual tacit knowledge within a target community. The understanding of the individual and the social tacit knowledge in a given community, which is a function of knowledge emergence, is the foundation of effectiveness in leadership practice." [3]

Tacit knowledge (as opposed to formal or explicit knowledge) is knowledge that is difficult to transfer to another person by means of writing it down or verbalizing it. For example, stating to someone that London is in the United Kingdom is a piece of explicit knowledge that can be written down, transmitted, and understood by a recipient. However, the ability to speak a language, use algebra,[1] or design and use complex equipment requires all sorts of knowledge that is not always known explicitly, even by expert practitioners, and which is difficult to explicitly transfer to users.

While tacit knowledge appears to be simple, it has far reaching consequences and is not widely understood.

Definition

With tacit knowledge, people are not often aware of the knowledge they possess or how it can be valuable to others. Effective transfer of tacit knowledge generally requires extensive personal contact and trust. Another example of tacit knowledge is the ability to ride a bicycle. The formal knowledge of how to ride a bicycle is that in order to balance, if the bike falls to the left, one steers to the left. To turn right the rider first steers to the left, and then when the bike falls right, the rider steers to the right.[2] Knowing this formally, however, is no help in riding a bicycle, and few riders are in fact aware of this.

Tacit knowledge is not easily shared. It involves learning and skill, but not in a way that can be written down. Tacit knowledge consists often of habits and culture that we do not recognize in ourselves. In the field of knowledge management, the concept of tacit knowledge refers to a knowledge possessed only by an individual and difficult to communicate to others via words and symbols. Knowledge that is easy to communicate is called explicit knowledge.

Tacit knowledge has been described as "know-how" - as opposed to "know-what" (facts), "know-why" (science), or "know-who" (networking). It involves learning and skill but not in a way that can be written down. The process of transforming tacit knowledge into explicit or specifiable knowledge is known as codification, articulation, or specification. The tacit aspects of knowledge are those that cannot be codified, but can only be transmitted via training or gained through personal experience.

A chief practice of technological development is the codification of tacit knowledge into explicit programmed operations so that processes previously requiring skilled employees can be automated for greater efficiency and consistency at lower cost. Such codification involves mechanically replicating the performance of persons who possess relevant tacit knowledge; in doing so, however, the ability of the skilled practitioner to innovate and adapt to unforeseen circumstances based on the tacit "feel" of the situation is often lost. The technical remedy is to attempt to substitute brute-force methods capitalizing on the computing power of a system, such as those that enable a supercomputer programmed to "play" chess against a grandmaster whose tacit knowledge of the game is broad and deep.

The conflicts demonstrated in the previous two paragraphs are reflected in Ikujiro Nonaka's model of organizational knowledge creation, in which he proposes that tacit knowledge can be converted to explicit knowledge. In that model tacit knowledge is presented variously as uncodifiable ("tacit aspects of knowledge are those that cannot be codified") and codifiable ("transforming tacit knowledge into explicit knowledge is known as codification"). This ambiguity is common in the knowledge management literature. Nonaka diverted from Polanyi's original view of 'tacit knowing' that lacks empirical or conceptual foundation, as is discussed in detail in a later article by Ikujiro Nonaka and Georg von Krogh.[3]

Examples

One example of tacit knowledge is the notion of language itself—it is not possible to learn a language just by being taught the rules of grammar—a native speaker picks it up at a young age almost entirely unaware of the formal grammar which they may be taught later. Another example is how to ride a bike; this can only be learned through personal experimentation.

Collins showed[1] that a particular laser (The pate laser) was designed in America and the idea, with specific assistance from the designers, was gradually propagated to various other universities world-wide. However, in the early days, even when specific instructions were sent, other labs failed to replicate the laser, it only being made to work in each case following a visit to or from the originating lab or very close contact and dialogue. It became clear that while the originators could clearly make the laser work, they did not know exactly what it was that they were doing to make it work, and so could not articulate or specify it by means of monologue articles and specifications. But a cooperative process of dialogue enabled the tacit knowledge to be transferred.

Another example is the Bessemer steel process — Bessemer sold a patent for his advanced steel making process and was sued by the purchasers who couldn't get it to work. In the end Bessemer set up his own steel company because he knew how to do it, even though he could not convey it to his patent users. Bessemer's company became one of the largest in the world and changed the face of steel making.[4]

As apprentices learn the craft of their masters through observation, imitation, and practice, so do employees of a firm learn new skills through on-the-job training. When Matsushita started developing its automatic home bread-making machine in 1985, an early problem was how to mechanize the dough-kneading process, a process that takes a master baker year of practice to perfect. To learn this tacit knowledge, a member of the software development team, Ikuko

Tanaka, decided to volunteer herself as an apprentice to the head baker of the Osaka International Hotel, who was reputed to produce the area's best bread. After a period of imitation and practice, one day she observed that the baker was not only stretching but also twisting the dough in a particular fashion ("twisting stretch"), which turned out to be the secret for making tasty bread. The Matsushita home bakery team drew together eleven members from completely different specializations and cultures: product planning, mechanical engineering, control systems, and software development. The "twisting stretch" motion was finally materialized in a prototype after a year of iterative experimentation by the engineers and team members working closely together, combining their explicit knowledge. For example, the engineers added ribs to the inside of the dough case in order to hold the dough better as it is being churned. Another team member suggested a method (later patented) to add yeast at a later stage in the process, thereby preventing the yeast from over-fermenting in high temperatures.

Knowledge Management

According to Parsaye, there are three major approaches to the capture of tacit knowledge from groups and individuals. They are

- Interviewing experts.
- Learning by being told.
- Learning by observation.

Interviewing experts can be done in the form of structured interviewing or by recording organizational stories. Structured interviewing of experts in a particular subject is the most commonly used technique to capture pertinent, tacit knowledge. An example of a structured interview would be an exit interview. Learning by being told can be done by interviewing or by task analysis. Either way, an expert teaches the novice the processes of a task. Task analysis is the process of determining the actual task or policy by breaking it down and analyzing what needs to be done to complete the task. Learning by observation can be done by presenting the expert with a sample problem, scenario, or case study and then observing the process used to solve the problem. Some other techniques for capturing tacit knowledge are:

Action learning

All of these approaches should be recorded in order to transfer the tacit knowledge into reusable explicit knowledge.

Professor Ikujiro Nonaka has proposed the SECI (Socialization, Externalization, Combination, Internalization) model, one of the most widely cited theories in knowledge management, to present the spiraling knowledge processes of interaction between explicit knowledge and tacit knowledge (Nonaka & Takeuchi 1995).

- The IRG Solution - hierarchical incompetence and how to overcome it argued that tacit knowledge was essentially a property of social networks and that much tacit knowledge was held in, and communicated by this informal lateral communication between network members.

Knowledge management (KM) comprises a range of strategies and practices used in an organization to identify, create, represent, distribute, and enable adoption of insights and experiences. Such insights and experiences comprise knowledge, either embodied in individuals or embedded in organizations as processes or practices.

An established discipline since 1991 (see Nonaka 1991), KM includes courses taught in the fields of business administration, information systems, management, and library and information sciences (Alavi & Leidner 1999). More recently, other fields have started contributing to KM research; these include information and media, computer science, public health, and public policy.

Many large companies and non-profit organizations have resources dedicated to internal KM efforts, often as a part of their business strategy, information technology, or human resource management departments (Addicott, McGivern & Ferlie 2006). Several consulting companies also exist that provide strategy and advice regarding KM to these organizations.

Knowledge management efforts typically focus on organizational objectives such as improved performance, competitive advantage, innovation, the sharing of lessons learned, integration and continuous improvement of the organization. KM efforts overlap with organizational learning, and may be distinguished from that by a greater focus on the management of knowledge as a strategic asset and a focus on encouraging the sharing of knowledge.

History

KM efforts have a long history, to include on-the-job discussions, formal apprenticeship, discussion forums, corporate libraries, professional training and mentoring programs. More recently, with increased use of computers in the second half of the 20th century, specific adaptations of technologies such as knowledge bases, expert systems, knowledge repositories, group decision support systems, intranets, and computer-supported cooperative work have been introduced to further enhance such efforts.[1]

In 1999, the term personal knowledge management was introduced which refers to the management of knowledge at the individual level (Wright 2005).

In terms of the enterprise, early collections of case studies recognized the importance of knowledge management dimensions of strategy, process, and measurement (Morey, Mayberry & Thuraisingham 2002). Key lessons learned included: people and the cultural norms which influence their behaviors are the most critical resources for successful knowledge creation, dissemination, and application; cognitive, social, and organizational learning processes are essential to the success of a knowledge management strategy; and measurement, benchmarking, and incentives are essential to accelerate the learning process and to drive cultural change. In short, knowledge management programs can yield impressive benefits to individuals and organizations if they are purposeful, concrete, and action-oriented.

More recently with the advent of the Web 2.0, the concept of Knowledge Management has evolved towards a vision more based on people participation and emergence. This line of evolution is termed Enterprise 2.0 (McAfee 2006). However, there is an ongoing debate and discussions (Lakhani & McAfee 2007) as to whether Enterprise 2.0 is just a fad that does not bring anything new or useful or whether it is, indeed, the future of knowledge management (Davenport 2008).

Research

KM emerged as a scientific discipline in the earlier 1990s. It was initially supported solely by practitioners, when Skandia hired Leif Edvinsson of Sweden as the world's first Chief Knowledge Officer

(CKO). Hubert Saint-Onge (formerly of CIBC, Canada), started investigating various sides of KM long before that. The objective of CKOs is to manage and maximize the intangible assets of their organizations. Gradually, CKOs became interested in not only practical but also theoretical aspects of KM, and the new research field was formed. The KM ideas taken up by academics, such as Ikujiro Nonaka (Hitotsubashi University), Hirotaka Takeuchi (Hitotsubashi University), Thomas H. Davenport (Babson College) and Baruch Lev (New York University). In 2001, Thomas A. Stewart, former editor at FORTUNE Magazine and subsequently the editor of Harvard Business Review, published a cover story highlighting the importance of intellectual capital of organizations. Since its establishment, the KM discipline has been gradually moving towards academic maturity. First, there is a trend towards higher cooperation among academics; particularly, there has been a drop in single-authored publications. Second, the role of practitioners has changed. Their contribution to academic research has been dramatically declining from 30% of overall contributions up to 2002, to only 10% by 2009 (Serenko et al. 2010).

A broad range of thoughts on the KM discipline exists with no unanimous agreement; approaches vary by author and school. As the discipline matures, academic debates have increased regarding both the theory and practice of KM, to include the following perspectives [

- **Techno-centric** with a focus on technology, ideally those that enhance knowledge sharing and creation.
- **Organizational** with a focus on how an organization can be designed to facilitate knowledge processes best.
- **Ecological** with a focus on the interaction of people, identity, knowledge, and environmental factors as a complex adaptive system akin to a natural ecosystem.

Regardless of the school of thought, core components of KM include people, processes, technology (or) culture, structure, technology, depending on the specific perspective (Spender & Scherer 2007). Different KM schools of thought include various lenses through which KM can be viewed and explained, to include:

- Community of practice (Wenger, McDermott & Synder 2001) social network analysis
- Intellectual capital (Bontis & Choo 2002)

- Information theory (McInerney 2002)
- Complexity science
- Constructivism (Nanjappa & Grant 2003)

The practical relevance of academic research in KM has been questioned (Ferguson 2005) with action research suggested as having more relevance (Andresen) and the need to translate the findings presented in academic journals to a practice (Booker, Bontis & Serenko 2008).

Dimensions

Different frameworks for distinguishing between different 'types of' knowledge exist. One proposed framework for categorizing the dimensions of knowledge distinguishes between tacit knowledge and explicit knowledge. Tacit knowledge represents internalized knowledge that an individual may not be consciously aware of, such as how he or she accomplishes particular tasks. At the opposite end of the spectrum, explicit knowledge represents knowledge that the individual holds consciously in mental focus, in a form that can easily be communicated to others.[9] (Alavi & Leidner 2001). Similarly, Hayes and Walsham (2003) describe content and relational perspectives of knowledge and knowledge management as two fundamentally different epistemological perspectives. The content perspective suggest that knowledge is easily stored because it may be codified, while the relational perspective recognizes the contextual and relational aspects of knowledge which can make knowledge difficult to share outside of the specific location where the knowledge is developed.

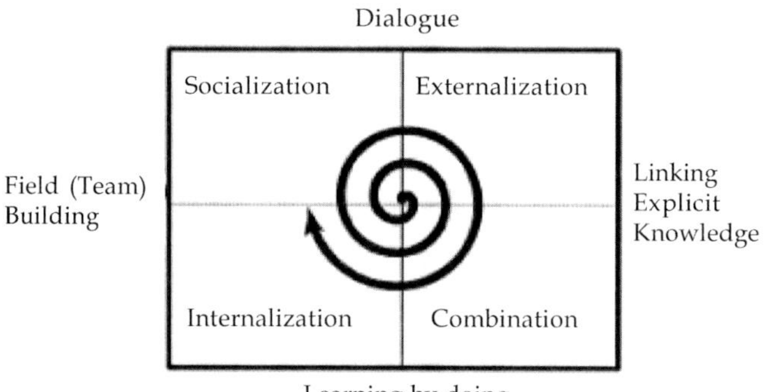

The knowledge spiral as described by Nonaka & Takeuchi

Early research suggested that a successful KM effort needs to convert internalized tacit knowledge into explicit knowledge in order to share it, but the same effort must also permit individuals to internalize and make personally meaningful any codified knowledge retrieved from the KM effort. Subsequent research into KM suggested that a distinction between tacit knowledge and explicit knowledge represented an oversimplification and that the notion of explicit knowledge is self-contradictory. Specifically, for knowledge to be made explicit, it must be translated into information (i.e., symbols outside of our heads) (Serenko & Bontis 2004). Later on, Ikujiro Nonaka proposed a model (SECI for Socialization, Externalization, Combination, Internalization) which considers a spiraling knowledge process interaction between explicit knowledge and tacit knowledge (Nonaka & Takeuchi 1995). In this model, knowledge follows a cycle in which implicit knowledge is 'extracted' to become explicit knowledge, and explicit knowledge is 're-internalized' into implicit knowledge. More recently, together with Georg von Krogh, Nonaka returned to his earlier work in an attempt to move the debate about knowledge conversion forwards (Nonaka & von Krogh 2009).

A second proposed framework for categorizing the dimensions of knowledge distinguishes between embedded knowledge of a system outside of a human individual (e.g., an information system may have knowledge embedded into its design) and embodied knowledge representing a learned capability of a human body's nervous and endocrine systems (Sensky 2002).

A third proposed framework for categorizing the dimensions of knowledge distinguishes between the exploratory creation of "new knowledge" (i.e., innovation) vs. the transfer or exploitation of "established knowledge" within a group, organization, or community. Collaborative environments such as communities of practice or the use of social computing tools can be used for both knowledge creation and transfer.[11]

Strategies

Knowledge may be accessed at three stages: before, during, or after KM-related activities. Different organizations have tried various knowledge capture incentives, including making content submission mandatory and incorporating rewards into performance measurement plans. Considerable controversy exists over whether incentives work or not in this field and no consensus has emerged.

One strategy to KM involves actively managing knowledge (push strategy). In such an instance, individuals strive to explicitly encode their knowledge into a shared knowledge repository, such as a database, as well as retrieving knowledge they need that other individuals have provided to the repository.[12] This is also commonly known as the Codification approach to KM.

Another strategy to KM involves individuals making knowledge requests of experts associated with a particular subject on an ad hoc basis (pull strategy). In such an instance, expert individual(s) can provide their insights to the particular person or people needing this (Snowden 2002). This is also commonly known as the Personalization approach to KM.

Other knowledge management strategies and instruments for companies include:

- Rewards (as a means of motivating for knowledge sharing)
- Storytelling (as a means of transferring tacit knowledge)
- Cross-project learning
- After action reviews
- Knowledge mapping (a map of knowledge repositories within a company accessible by all)
- Communities of practice
- Expert directories (to enable knowledge seeker to reach to the experts)
- Best practice transfer
- Knowledge fairs
- Competence management (systematic evaluation and planning of competences of individual organization members)
- Proximity & architecture (the physical situation of employees can be either conducive or obstructive to knowledge sharing)
- Master-apprentice relationship
- Collaborative technologies (groupware, etc.)
- Knowledge repositories (databases, bookmarking engines, etc.)

- Measuring and reporting intellectual capital (a way of making explicit knowledge for companies)
- Knowledge brokers (some organizational members take on responsibility for a specific "field" and act as first reference on whom to talk about a specific subject)
- Social software (wikis, social bookmarking, blogs, etc.)
- Inter-project knowledge transfer

Motivations

A number of claims exist as to the motivations leading organizations to undertake a KM effort. Typical considerations driving a KM effort include:

- Making available increased knowledge content in the development and provision of products and services
- Achieving shorter new product development cycles
- Facilitating and managing innovation and organizational learning
- Leveraging the expertise of people across the organization
- Increasing network connectivity between internal and external individuals
- Managing business environments and allowing employees to obtain relevant insights and ideas appropriate to their work
- Solving intractable or wicked problems
- Managing intellectual capital and intellectual assets in the workforce (such as the expertise and know-how possessed by key individuals)

Debate exists whether KM is more than a passing fad, though increasing amount of research in this field may hopefully help to answer this question, as well as create consensus on what elements of KM help determine the success or failure of such efforts (Wilson 2002).[14]

Technologies

Early KM technologies included online corporate yellow pages as expertise locators and document management systems. Combined with the early development of collaborative technologies (in particular Lotus Notes), KM technologies expanded in the mid-1990s. Subsequent KM efforts leveraged semantic technologies for search and retrieval and the development of e-learning tools for communities of practice[(Capozzi 2007). Knowledge management systems can thus be categorized as falling into one or more of the following groups: Groupware, document management systems, expert systems, semantic networks, relational and object oriented databases, simulation tools, and artificial intelligence (Gupta & Sharma 2004).

More recently, development of social computing tools (such as bookmarks, blogs, and wikis) have allowed more unstructured, self-governing or ecosystem approaches to the transfer, capture and creation of knowledge, including the development of new forms of communities, networks, or matrixed organizations. However such tools for the most part are still based on text and code, and thus represent explicit knowledge transfer. These tools face challenges in distilling meaningful re-usable knowledge and ensuring that their content is transmissible through diverse channels (Andrus 2005).

Software tools in knowledge management are a collection of technologies and are not necessarily acquired as a single software solution. Furthermore, these knowledge management software tools have the advantage of using the organization existing information technology infrastructure. Organizations and business decision makers spend a great deal of resources and make significant investments in the latest technology, systems and infrastructure to support knowledge management. It is imperative that these investments are validated properly, made wisely and that the most appropriate technologies and software tools are selected or combined to facilitate knowledge management.

Knowledge management has also become a cornerstone in emerging business strategies such as Service Lifecycle Management (SLM) with companies increasingly turning to software vendors to enhance their efficiency in industries including, but not limited to, the aviation industry.

Knowledge Managers

"Knowledge manager" is a role and designation that has gained popularity over the past decade. The role has evolved drastically from that of one involving the creation and maintenance of knowledge repositories to one that involves influencing the culture of an organization toward improved knowledge sharing, reuse, learning, collaboration and innovation. Knowledge management functions are associated with different departments in different organizations. It may be combined with Quality, Sales, HR, Innovation, Operations etc. and is likely to be determined by the KM motivation of that particular organization.

Knowledge managers have varied backgrounds ranging from Information Sciences to Business Management. An effective knowledge manager is likely to be someone who has a versatile skills portfolio and is comfortable with the concepts of organizational behavior/culture, processes, branding & marketing and collaborative technology.

Knowledge Management System

Knowledge Management System (KM System) refers to a (generally generated via or through to an IT based program/ department or section) system for managing knowledge in organizations for supporting creation, capture, storage and dissemination of information. It can comprise a part (neither necessary nor sufficient) of a Knowledge Management initiative.

The idea of a KM system is to enable employees to have ready access to the organization's documented base of facts, sources of information, and solutions. For example a typical claim justifying the creation of a KM system might run something like this: an engineer could know the metallurgical composition of an alloy that reduces sound in gear systems. Sharing this information organization wide can lead to more effective engine design and it could also lead to ideas for new or improved equipment.

A KM system could be any of the following:

1. Document based i.e. any technology that permits creation/ management/sharing of formatted documents such as Lotus Notes, SharePoint, web, distributed databases etc.
2. Ontology/Taxonomy based: these are similar to document technologies in the sense that a system of terminologies (i.e.

ontology) are used to summarize the document e.g. Author, Subj, Organization etc. as in DAML & other XML based ontology's

3. Based on AI technologies which use a customized representation scheme to represent the problem domain.
4. Provide network maps of the organization showing the flow of communication between entities and individuals
5. Increasingly social computing tools are being deployed to provide a more organic approach to creation of a KM system.

KMS systems deal with information (although Knowledge Management as a discipline may extend beyond the information centric aspect of any system) so they are a class of information system and may build on, or utilize other information sources. Distinguishing features of a KMS can include:

1. *Purpose*: a KMS will have an explicit Knowledge Management objective of some type such as collaboration, sharing good practice or the like.
2. *Context*: One perspective on KMS would see knowledge is information that is meaningfully organized, accumulated and embedded in a context of creation and application.
3. *Processes*: KMS are developed to support and enhance knowledge-intensive processes, tasks or projects of e.g., creation, construction, identification, capturing, acquisition, selection, valuation, organization, linking, structuring, formalization, visualization, transfer, distribution, retention, maintenance, refinement, revision, evolution, accessing, retrieval and last but not least the application of knowledge, also called the knowledge life cycle.
4. *Participants*: Users can play the roles of active, involved participants in knowledge networks and communities fostered by KMS, although this is not necessarily the case. KMS designs are held to reflect that knowledge is developed collectively and that the "distribution" of knowledge leads to its continuous change, reconstruction and application in different contexts, by different participants with differing backgrounds and experiences.
5. *Instruments*: KMS support KM instruments, e.g., the capture, creation and sharing of the modifiable aspects of experience, the

creation of corporate knowledge directories, taxonomies or ontologism, expertise locators, skill management systems, collaborative filtering and handling of interests used to connect people, the creation and fostering of communities or knowledge networks.

A KMS offers integrated services to deploy KM instruments for networks of participants, i.e. active knowledge workers, in knowledge-intensive business processes along the entire knowledge life cycle. KMS can be used for a wide range of cooperative, collaborative, adhocracy and hierarchy communities, virtual organizations, societies and other virtual networks, to manage media contents; activities, interactions and work-flows purposes; projects; works, networks, departments, privileges, roles, participants and other active users in order to extract and generate new knowledge and to enhance, leverage and transfer in new outcomes of knowledge providing new services using new formats and interfaces and different communication channels.

The term KMS can be associated to Open Source Software, and Open Standards, Open Protocols and Open Knowledge licenses, initiatives and policies.

Benefits & Issues of knowledge management

Some of the advantages claimed for KM systems are:

1. Sharing of valuable organizational information throughout organizational hierarchy.
2. Can avoid re-inventing the wheel, reducing redundant work.
3. May reduce training time for new employees
4. Retention of Intellectual Property after the employee leaves if such knowledge can be codified.
5. time management

Knowledge Sharing remains a challenging issue for knowledge management, and while there is no clear agreement barriers may include time issues for knowledge works, the level of trust, lack of effective support technologies and culture (Jenner).

- Innovation Diffusion Models

  1. General vs. Domain specific

  2. Conceptual vs. Mathematical

3. Focus on innovation vs. adopters
4. Organizational vs. Individual
5. Process vs. Outcome
6. Proximity vs. Network
7. Rate-oriented vs. Threshold

- Gabriel Tarde (1903)

  - S-shaped curve for diffusion processes

- Ryan and Gross (1943): adopter categories

  1. Innovators
  2. Early adopters
  3. Early/Late Majorities
  4. Laggards

- Original Theorists

- Katz (1957) :

  - media à opinion leaders à opinion followers

- Everett M. Rogers

  Diffusion of Innovations (1962-95)

  - the process by which an innovation is communicated through certain channels over time among the members of a social system

- Rogers'(1995)

  - Diffusion of Innovation

- Hagerstrand (1965) studied diffusion of hybrid corn in farmers. Model based on proximity.

- Bass (1969) developed differential equations borrowed from physics to model diffusion of innovation

- Midgley & Dowling (1978):

  - Contingency model.

- Mahajan & Peterson (1985):

  - Extension and simplification of Bass model (has 2 parameters, internal & external influence)

- Abrahamson & Rosenkopf (1990): Bandwagons & Thresholds

Rational efficiency vs. Fad theories

- *Rational Efficiency*: The more organizations adopt an innovation, the more knowledge about the innovation's true efficiency is disseminated
- *Fad theories*: The sheer number of adopters creates "bandwagon pressures"
  - Institutional pressures: Adoption of innovation can become a social norm
  - Competitive pressures: Fear that not adopting will lead to loss of competitive advantage
  - Valence (1996)

Social network thresholds

- Personal network thresholds: number of members within personal network that must have adopted before one will adopt
  - Accounts for some variation in overall adoption time
  - "Opinion leaders" have lower thresholds and influence individuals with higher thresholds
- Factors affecting diffusion

  A. Innovation characteristics

  B. Individual characteristics

  C. Social network characteristics

A. Innovation characteristics

1. Observability
   - The degree to which the results of an innovation are visible to potential adopters
2. Relative Advantage
   - The degree to which the innovation is perceived to be superior to current practice
3. Compatibility
   - The degree to which the innovation is perceived to be consistent with socio-cultural values, previous ideas, and/or perceived needs

4. Trialability
    - The degree to which the innovation can be experienced on a limited basis
5. Complexity
    - The degree to which an innovation is difficult to use or understand.

B. Individual characteristics

- ❒ Innovativeness
    - Originally defined by Rogers: the degree to which an individual is relatively earlier in adopting an innovation than other members of his social system
    - Modified & extended by Hirschman (1980):
- ❒ Inherent / actualized novelty seeking
- ❒ Creative consumer
- ❒ Adoptive / vicarious innovativeness
    - Other individual characteristics
- ❒ Reliance on others as source of information (Midgley & Dowling)
- ❒ Adopter threshold (e.g. Valente)
- ❒ Need-for-change / Need-for-cognition (Wood & Swait, 2002)

C.Network characteristics

1. Opinion leadership: number of nominations as source of information
2. Number of contacts within each adopter category (Valente)
3. Complex structure
    - Other possible factors:
    - Lyytinen & Damsgaard (2001)
        - Social environment of diffusion of innovation
        - Marketing strategies employed
        - Institutional structures (e.g., government)
    - Cellular Automata & Diffusion of Innovation

- Boccara & Fuks (1998)

  • CA model of diffusion based on contact theory. (Not heavily based in innovation diffusion theory)

- Strang & Macy (2001)

  • Used decision rule: if current practice is unsatisfactory, evaluate "best practices". Fad-like behavior emerged

- Goldenberg, Libai, & Muller (working paper)

  • Used CA to model Bass parameters in individuals and observed aggregate-level behavior.

□□□

Chapter 3

Research Methodology

Foundations of Social Research

Sociological positivism

The origin of the survey can be traced back at least early as the Doomsday Book in 1086 whilst some scholars pinpoint the origin of demography to 1663 with the publication of John Graunt's *Natural and Political Observations upon the Bills of Mortality*. Social research began most intentionally, however, with the positivist philosophy of science in the early 19th century, which originates with founding sociologist August Compte.

Statistical sociological research, and indeed the formal academic discipline of sociology, began with the work of Emil Durkheim (1858–1917). While Durkheim rejected much of the detail of Comte's philosophy, he retained and refined its method, maintaining that the social sciences are a logical continuation of the natural ones into the realm of human activity, and insisting that they may retain the same objectivity, rationalism, and approach to causality. Durkheim set up the first European department of sociology at the University of

Bordeaux in 1895, publishing his Rules of the Sociological Method (1895). In this text he argued: "ur main goal is to extend scientific rationalism to human conduct... What has been called our positivism is but a consequence of this rationalism."

Durkheim's seminal monograph, *Suicide* (1897), a case study of suicide rates amongst Catholic and Protestant populations, distinguished sociological analysis from psychology or philosophy. By carefully examining suicide statistics in different police districts, he attempted to demonstrate that Catholic communities have a lower suicide rate than that of Protestants, something he attributed to social (as opposed to individual or psychological) causes. He developed the notion of objective *suis generis* "social facts" to delineate a unique empirical object for the science of sociology to study. Through such studies he posited that sociology would be able to determine whether any given society is 'healthy' or 'pathological', and seek social reform to negate organic breakdown or "social anomie". For Durkheim, sociology could be described as the "science of institutions, their genesis and their functioning".

Modern Methodologies

In the mid-20th century there was a general—but not universal—trend for U.S.-American sociology to be more scientific in nature, due to the prominence at that time of action theory and other system-theoretical approaches. Robert K. Merton released his *Social Theory and Social Structure* (1949). By the turn of the 1960s, sociological research was increasingly employed as a tool by governments and businesses worldwide. Sociologists developed new types of quantitative and qualitative research methods. Paul Lazarsfeld founded Columbia University's Bureau of Applied Social Research, where he exerted a tremendous influence over the techniques and the organization of social research. His many contributions to sociological method have earned him the title of the "founder of modern empirical sociology".Lazarsfeld made great strides in statistical survey analysis, panel methods, latent structure analysis, and contextual analysis. Many of his ideas have been so influential as to now be considered self-evident.

Research : According to best research is considered to be the more formal, systematic and intensive process of carrying on a scientific method of analysis. It is a more systematic activity directed towards discovery and the development of an organized body of knowledge.

Characteristics

1. Solution of problem directed
2. Generalization, principles and theories development oriented
3. Observable experience or empirical evidence based
4. Demand accuracy in observation and description
5. Involve data gathering or using data
6. Require expertise
7. Strives to objective and logical
8. Patient and unhurried activity
9. Carefully recorded and reported
10. Require strong determination to know the unknown
11. Expected be something original
12. Seek forward something new

Social Research

Social Research is useful in social sciences for the sake of knowledge and use of this knowledge in practical life.

"Social research is a scientific undertaking which by means of logical and systematized techniques, aims to

- Discover new facts or verify and test old facts
- Analyze their sequence, interrelationships, and casual explanations which were derived within an appropriate frame of references
- Develop new scientific tools, concepts and theories which would facilitate reliable and valid stage of human behavior

Social research is a systematic method of exploring, analyzing and conceptualizing social life in order to extend, correct or verify knowledge whether that knowledge aid in the construction of a theory or in the practice of an art social research seeks to find explanations to unexplained social phenomena, to clarify the doubtful and correct the misconceived ideas of social life.

Steps in conducting research

The following general steps are involved in conducting a research.

1. Formulating the research problem
2. Survey and review of literature

3. Developing hypothesis and degasifying the concepts.
4. Deciding on the research design.
5. Defining population and selecting sample.
6. Collecting data with the help of tools and techniques available.
7. Analysis and interpretation of data.
8. Testing the hypothesis
9. Conclusions, implication, recommendations.
10. Preparation and submission of report.

A research finding is considered as credible and adds to our knowledge only when a proper and scientific research methodology is adopted.

Types of Research

Research is generally classified as a fundamental or basic

1. *Fundamental/basic*: To gather knowledge for the sake of know ledge. It is not concerned with solving any practical problem, but with discovering underlying and if possible, universal laws and theories.
2. *Applied or practical research* : gathering of knowledge that could aid in the betterment of human destiny.
3. *Adoptive research* : research, to be socially useful should be conducted in situation where people live and work. e.g. on farm research.

Example

Fundamental discovery of nitrogen	Applied/practical development of the process of manufacting of nitrogenous fertilizer	Adaptive or on farm determination of the dose time, methods of applcation of nitrogenous fertilizers in different crops under varying situations.

4. *Action Research :* is a form of social research which aims at better insight into the problem by learning from the experience gained in attempt to solve the problems.

Three steps

1. Analyzing the problem with the local people and decide on the objectives.

2. Inducing appropriate action in the situation and working on it with the local people and
3. Monitoring and evaluation of the results with the local people, identify constraints, and suggest measures to improve the situation.

Types of educational research

1. *Historical research:* A historical research is one which investigates records, analyses and interprets the events of the past for the purpose of discovering sound generalizations that are helpful in understanding the past and the present end to a limited extent, in anticipating future describe what was.
2. *Descriptive research:* Describes, records, analyses and interprets the existing conditions, an attempt is made to discover relationship between existing non-maturation variables apart from some comparisons or contrast among those variables basically describes what is.
3. *Experimental research:* Is one in which we found some relationship of variables - among channeled or manipulated and other - effect of controlled variable on others. Thus it describes what will be
   a) Laboratory Experiments
   b) Field Experiments

2. Science and scientific approach

Science is a body of knowledge systematically built up based on theories and their testing for internal as well as external consistencies independent of social, cultural and other such factors author:

Views of science

- Static view
- Dynamic view
- Heuristic view

The scientific approach

Scientific approach is a special systematized form of all reflective thinking and inquiry

1. Problem - obstacle - idea stage - intellectualizing
2. Hypothesis formulation - if such and such occurs, then so and so will happen
3. Reasoning - deduction frustration leads to aggression
4. Observation - test - experiment

Conclusion : First there is doubt, a barrier an intermediate situation carrying out to be made determinate. The scientist experience vague doubts, emotional disturbances

- Studying the literature scanning of experience to formulate problem.
- Hypothesis construction based on problem.
- After this its empirical implications.

Are deducted in this process the original problem and of course the original hypothesis may be changed it may be broadened or narrowed

On the basis of research audience the hypothesis is accepted or rejected this information is then feedback to the original problem and the problem is kept altered as dictated by the audience.

Aims of science

1. Explanation
2. Understanding
3. Prediction
4. Control

Characteristics of scientific methods

1. Objectivity
2. Generality
3. Verifiability
4. Predictability

Objectivity : Objectivity means that the procedures and findings are not influenced or changed by the reveal actions feelings, values and beliefs after the research has began objective studies are those of a detached, impersonal unbiased, or unprejudiced nature.

Generality : mean the ability to generalize from a portions phenomenon under study conclusion and generalizations drawn should be of universal application. In social science, complete universality is however, rarely achieved.

Verifiability: is a characteristic which indicates that the observed phenomenon or facts can be verified by subsequent studies.

Predictability: This means that the results of the study can predict the factor events with sufficient accuracy.

Theory

"A theory is a set of interrelated constructs (concepts) definitions, and propositions that present a systematic view of the phenomena by specifying relations among variables, with the purpose of explaining and predicting the phenomena"

Example - Theory of school failure

Variables -intelligence verbal and numerical aptitudes, anxiety, social class membership and motivation

Phenomena - School failure - or school achievement

School failure is explained by specified relations between each of the six variables and school failure or by combination of the six variables and school failure.

The scientist, successfully using this set of constructs, then " understands" school failure. He is able to "explain" and to some extent at least," predict" it

Define

1. Interrelated set
2. Systematic view
3. Variable
4. Explanation and prediction

Theory as a tool of science or role of theory

1. Orientation
2. Conceptualization
3. Summarizing
4. Prediction
5. Painting gap

3. Role of theory / or theory as tool

1. *Orientation* : Theory as orientation
2. *Conceptualization/classification* : Theory as conceptualization and classification, class system, social mobility, social distance
3. *Summarizing* : It may be empirical generalization of data
4. *Predicted* : Theory predicts facts
5. *Painting gap* : Theory points to gap in our knowledge

Facts **:** Fact are the basis for generalizations which may be interpreted as casual explanations

Fact describes the who, what, when and where of an event most analyst, however, go beyond obtaining descriptive facts and look the interpretive question of the ways a historical event occurred and the reasons for it is Identifying facts to stating generalizations to interring casual explanations Facts are weighed and judged by constancy, the accumulation of evidence, and other techniques.

Interpretation of the relationship between facts is generalizations

1. Facts of science are the product of observation that is not random but meaningful i.e. theoretically relevant.
2. Thus fact and theory are not opposed but interrelated.
3. Development of science - constant interplay between theory and fact.
4. Facts are thought to be definite, certain, without question and their meaning to be self evident.
5. Facts are regarded axiom, empirically verifiable observation.

Facts productive of theory in following ways

1. Facts help to initiate theories.
2. Leads to reformulation of existing theory
3. They cause the rejection of theories which do not fit the facts
4. They change the focus and mention of theory
5. They clarify and redefine theory

Role of facts

1. Theory and facts are, in constant interactions.
2. Facts lead to rejection and reformation of existing theory.
3. Redefines and clarify theory.

Hypothesis

Explorative Research	Looks for chance of doing research
Descriptive Research	Explains the what and why in detail
Experimental Research	Independent and dependent variable
Action Research	Enriches the field of application
Pure Research	Enriched the theory

Relationship between fact and theory

1. Fact helps to initiate theory
2. Reformulation of existing theory
3. Rejection of theory which is not suitable
4. Clarify and redone the theory
5. Theory and fact are in constant interaction

4. Selection and identification of research problems

Research problem : Difficulty which a researcher experiences in the context of either a theoretical or a practical situation and wants to obtain a solution for the same

Components of research problem

1. Individual with problem
2. Some objectives to be obtained
3. Alternative means to achieve those objectives.
4. Doubt in the mind of a researcher with regard to the selection of alternative
5. Environment to which the differece pertains

Selecting the problem

1. Subject which is overdone should not be normally chosen
2. Controversial subject should not become the choice of an research
3. Too narrow problems should be avoided
4. Subject selected should be familiar and feasible
5. Take care of the importance of subject qualification of the researcher and the costs involved
6. Selection must be provided by a preliminary study
7. Should be neither vary general nor very specific
8. It should be valuable
9. It should be feasible

Identification of research problem

The first step in designing social research is to perceive a problem either theoretical or applied. It is researcher's perception or recognition of a problem that motivates research. A problem exists when we do not have enough information to answer a question. This may happen:-

1. When there is a noticeable gap in the results of investigations.
2. When the results of several inquiries disagree.
3. When one wants to test the results of earlier investigation or their relevance in different cultural or social milieu.
4. When there is desire to follow up new ideas and to refine or qualify the findings of earlier studies.
5. When a fact exists in the form of a bit of unexplained information.
6. When there is desire for innovation gap.

Identification

Problem due to deficit/incomplete information to answer a question gap in the investigation results in disagreement.

5. Methods of data collection

There are two techniques of primary data collection as mentioned below

Non Projective	Projective
1. Observation	1. WAT (Word Association Test)
2. I/Q (Interview/Que stionnaire)	2. SCT (Sentence Completion Test)
3. Case study	3. VPT (Verbal Projection Test)
4. Content analysis	4. SCT (Story Completion Test)
5. Semantic differential	5. PS (Projective Statements)
6. Sociometric	6. RT (Rorschach Test)
	7. PT (Pictorial Technique)
	8. PT (Play Technique)
	9. TAT (Thematic Appreciation Test)

- The look of data collection - offer problem defined and research design chalkout.
- Data are special types of information, generally, obtained through observation, surveys, enquires or are generated of human activity.

  Qualitative - Descriptive in nature - good after processing

  Data : Quantitative - Expressed in the form of numerals and are mountable to statistical analysis.

  Primary - Collected afresh for the first time

  Secondary - Already collected and compiled by someone.

Observation method : Scientific observation is a methodological way of recognizing and noting a fact or assurance often involving some sort of measurement. Scientific observation should be specific and recorded immediately.

Understanding of culture and interacting ability are essential.

Advantages

- Subjective bias eliminated
- Current state of affairs related information
- Independent of respondent's willingness

Units

- A time consuming and expertise method
- Limited amount of information may be available
- Echinus factor may interfere with the task of observed

Types

- Structured and unstructured
- Participant and non participant
- Controlled and non - controlled

Case study : Case study is an empirical inquiry that investigates a contemporary phenomenon within its real life context, when the boundaries between phenomenon and context are not clearly evident, and in which multiple sources of evidence are used.

Characteristics

- Unit of study may be an individual, a family, an institution, a culture group or entire communication
- Aimed at deep and detailed study of unit
- Covers a sufficiently wide cycle of time
- Number of units, not to be small
- It is mainly quantities in nature

Basic assumptions

- Totality of the being
- Underlying unity
- Complexity of the social phenomenon
- Influence of time

Sources of case data

- Personal documents - Diaries, Autobiography, memories, letters
- Life History

Steps in conducting — Case Study

- Locating case
- Collection of data
- Treatment and interpretation
- Conduction and documentation

Advantage

- Very intensive
- Flexible
- Natural setting is the source of data
- Less expensive
- Presents real record of personal expensive
- Helps in Int/Scn construction
- Enhances experience of research

Limitations

- Less reliable results
- Result may be subjective
- Very little scope of quantitative analysis

Content analysis

Content analysis is a method of studying and analyzing communications in a systematic objective and quantitative manner to measure variables.

Method of analysis of content of communication the print or electronic media is.

Example: Leaders and conceptual complexity

Successful : Revolutionaries were characterized by a conceptually simple level of functioning before revolution, but after takeover of the government they exhibited greater conceptual complexity

Unsuccessful revolutionary - who failed in governing after the revolution but who had successful achieved the revolution were conceptually simple before the revolution and remained conceptually simple after revolution.

Dependent variable

1. *Measure*: Conceptual complexity of each leader
2. *Method*: Content analysis
3. *Source*: Letters, speeches, documents
4. *Analysis*: Pre and post revolutionary complexity scored
5. *Variable*: Conceptual complexity

Some aspects of methods in content analysis what to be analyzed

1. Letter
2. Diaries
3. Ethnographic material
4. Newspaper article
5. Editorial
6. Minutes of meetings

Steps

1. Definition and categorization of universe. e.g. achievement, power
2. Units of analysis
   - Word
   - Themes
   - Characters
   - Items
   - Space and time measures

Qualification

Content can be quantified in three ways

1. Nominal measurement
2. Ranking
3. Rating

The Halo effect: This is the tendency for an irrelevant feature of a unit of study to influence the relevant feature in a favorable or unfavorable direction.

Advantages

Immense flexibility it can be applied to a wide variety of domains

Disadvantages

1. A content analyses can only be as good or the documents on which the fractioned work.
2. It is almost impossible to device coding manual.
3. Difficult to why
4. Sometimes being (occurred) of being theoretical

Semantic Differential

The S.D. is a method for measuring the meaning of concepts, research, would often like to learn the image of an entity in the mind of the public. Using the S.D. voice, the images of brand, store political candidate, company organization, and institution as idea can be measured, assessed and compared with that of similar topic.

Applications : To measure objectively the semantic properties of words and concepts in a tridimensional semantic space and more commonly and simply, As an attitude scale restricting its focus to the affective domain.

SD scale has three elements

i. The concept to be evaluated in rooms of its semantic or attitudinal properties.
ii. The polar objective pair anchoring the scale.
iii. A series of undefined scale positions.

Three principal factors accounting for most for the semantic loadings:

- Evaluation - goo - bad
- Potential - strong - weak
- Activity - fast - slow

Pairs of these paler objectives as selected avoiding the purpose of the research and then arranged at opposite ends of a series of seven point scale.

Ex: Good Bad

The subject is asked to rate the concept 'School' according to have the perceive its tools towards it at the moment by placing.

Format and Administrating of S.D

Advantages

- This method is most effective for measuring image profiles, people, things organizations, and concepts.
- Adjective most define a single dimension and each pair must be bipolar opposites labeling the extremes.
- Precisely what the respondent is to rate not be clearly stated in the introducing instructions.
- Not more than 20 times should be used and about half should being with the most positive words.
- If the some scale is used for several topics and/or rating for an ideal entity, image profiles can be compared among.
- The same scale is used for several topics and or for rating an ideal entity image profiler can be compared among real entities and between real and ideal ones.
- Select the concepts or other stimuli to be rated these should be relevant to the research problem.
- Select the polar objective pair.
- Arrange the polar adjective pairs.

Analysis and evaluation of selected data

The semantic differential yield a large amount of data with a minimum of effort. By converting the scale positions to numerical values, various statistical assessments can be made.

Good 7 6 5 4 3 2 1 Bad

In variance terms there are main sources of variance, concepts, scalar and subjects

1. Plus the scores can be analyzed for differences between concepts, between scales, between group mean or medium for individual scale or for all scalar scanned.

2. Profile analysis
3. Distance - cluster analysis of semantic space

Sociometric : is a technique for describing the social relationship among individuals in a group.

Sociograms : are chart that depicts the sociometric choices sociogram, are constructed to identify leaders and followers

Use : To know group structure - i.e. cliques, leaders, social position

Analysis of sociometric data

Three steps are indexed

1. Sociometric matrix
2. Sociogram and
3. Socio metric indices
   i. Choice status index
   ii. Group cohesiveness index
   iii. Social impartibility index
   iv. Emotional expensiveness

Socio metric matrix : It provides initial limited information about the group e.g. which member of the group is liked by follower in order of merit.

It is method of tabulation of sociometric data

Sociogram

1. It is a diagrammatic representation of the internal structure of choices of the member of a group.
2. May be made for entire population of village.
3. It classified persons as chosen by maximum persons - stars, reflected, rejected, clique, one way choice and mutual pain.
4. It only provides which person is chosen or reflected but do not provides reason for the same.

Sociometric indices

i) Sociometric status index/choice status:

The value of choice status is the direct index of an individual's popularity in the group.

E.g. if value is .40 or .80 means B is so more status enjoying person than A.

ii) Group cohesiveness

For mutual pairs it is 11%.

For clique it is 100% - mean complexly cohesive.

iii) Social compatibility index

Choices divided by one less than the group size. eg 6/10 = .60 = 60%.

iv) Emotional expansiveness

Determined by amount of civilization of available choices, expressed in percentage. = 17/22 x 100

$$\frac{\text{No of choices made}}{\text{Total choice}} \times 100 = 77\%$$

Sound and valid ratings depend on the rater willingness to rate the object or person honestly in accordance with the given instruction and rater ability to rate acceptably and consistently.

Groups in which members are expected to know each other should be selected for sociometric analysis.

Questionnaire method: In this method question are delivered to the respondents who mate down their responses on it and send it back to the researcher. This method is known as questionnaire method.

Or

"A questionnaire is a form containing a series of questions and providing space for their replies to be filled in by the respondent himself."

Types of questionnaires

Two important dimensions along with questionnaires generally vary are:

a. Degree of structure and
b. Number of response possibilities

Methods of data collection

Schedule	Questionnaire
1. It is filled in by the interviewer and is never mailed to the respondent. The presence of the interviewer enlivens the atmosphere and increases the response rate.	1. It is filled in by the respondent himself and usually mailed to him. It, therefore, lacks personal touch and lowers the response rate.
2. It is generally used where the survey is to be conducted of relatively small geographical area.	2. It is generally used where the field of inquiry is large.
3. It's involves greater expenditure of time and money	3. Much time and money are saved in it use.
4. It can be used even where the respondents are illiterate because the investigator can explain to them anyone point on the spot and ensure completion	4. It cannot be used where the respondents are illiterate. Even literate respondents leave several inquiries incomplete

Degree of Structuredness

- Structured - precisely and concisely stated questions
- Unstructured - Specify only the broad *org* anisation of the subject and not the form or sequence of the question.

Number of response possibilities:-

a. When only two response possible
   i. Dichotomous questions
b. Designing a questionnaire
   i. Beauty and accuracy

Guideline for designing good questionnaire/questions

The questions/questionnaire must be intimately related to the final objectives of the investigation.

Survey method

It is form of expost facto research in which the researcher simply collects data about certain sociological or psychological characteristics of a sample that represents a known population in natural setting. in fact the researcher is interested in knowing something about the whole population. The data are sought directly from respondents by some systematic technique such as, interviewing, questionnaire or observation at a certain point or period of time once only. This is unlike an experiment in which repeated observations are made of the effects of certain variables on other variables overtime.

Purposes

- To explore the existence of some phenomena and knowledge in rural areas or between down sizing and employee morale in a manufacturing company.
- To describe a phenomena
- To predict future conditions
- To evaluate social programmes
- To develop social indicators

Advantages

1. Use of large representative sample in a survey reduces problems of sample bias and obviously the researches to generalize his results to the parent population.
2. Data collection can take place in any setting.
3. Data are obtained directly from respondent.
4. A variety of data collection techniques.
5. If mail questionnaire is used to collect data at the cost of collection per respondent is relatively. In relation in the light of the understanding developed in the earlier cycles. It is thus an emergent process, which takes shape as understanding increases; it is an interactive process, which converges towards a better understanding of what happens. In most of its forms it is also participative and qualitative.

Action Research

"Action research is a methodology which is intended to have both action outcomes and research outcomes".

Define: Action research is a three step spiral process of

1. Planning which involve reconnaissance.
2. Taking actions and
3. Fact finding about the results of the action.

Purpose : To develop new skills new approaches and to solve problems with direct application to the classroom or working world setting.

Steps

- Define the problem or set the goal what is it that needs improvement or that might be developed as a new skill or solution.
- Review the literature to learn whether other have met similar problems or achieved related objectives.
- Formulate testable hypothesis or strategies or approaches, stating them in clean, specific, pragmatic language.
- Arrange the research setting on spell out the procedures and can delas what are the particular things you will do in an attempt to most gear objectives.
- Establish evaluation criteria, measurement techniques and other means of acquiring useful feedback.
- Analyze the data and evaluate the outcome.

Features of Action Research

1. Cyclic	A cyclic process of planning, action and evaluation
2. Continuous feedback	A continuous feedback of the research results to all parties involved, including client
3. Cooperation	Cooperation between researches, practitioner and clients from the start and throughout the entire process.
4. Application	Application of the principles that govern social life and group decision making.
5. Differences	Taking into account differences in value system and power structures of all the parties involves in the research.
6. Problem solution	Using action research compulsory to solve a problem and to generate new knowledge.

Measurement (meaning, levels, postulates)

Definition : Measurement is assigning numbers to objects or events according to rules. The purpose is to have information in a form in which variables can be related to each other.

Components of measurement

Whenever measurement is being conducted, it is always associated with components : measurement **(M) = Characteristic being measured (C) + systematic error SE + variable error (V.E.)**

Systematic error : Is the error that occurs in a consistent manner each time something is measured. It is also called bias e.g. the respondent's tendency to give favorable responses independent of his true feelings.

Variable error : Are those errors which occur randomly each time something is measured. These may be

1. Due to short term characteristic of the object e.g. fatigue, health.
2. Situational characteristics - e.g. presence or absence of some other person.
3. Characteristics of some other persons etc.
4. Characteristics of measuring instruments e.g. unclear instruction.
5. Characteristics of response process e.g. mistake caused by checking.
6. Characteristic of analysis of mistakes caused by wrong coding tabulation etc.

Accuracy of measurement

Accuracy of measurement depends upon the extent to which it is free from systematic and variable errors. Freedom from systematic error is called validity of the measurement and freedom from variable error is known as reliability of the measurement.

Steps in the tests and scale construction

Step 1 Define the universe/content - by selecting items or statements through say study, institutions, consultations, documentations and record analysis and interview the resource persons.

Step 2 Collection, selection, screening and editing of items or statements following item analysis or suitable statistical treatments.

Step 3 Selection of testable or scalable statements

Step 4 Treatment of items and or statements with technique of test scale construction

Step 5 Testing of reliability and validity of test/scale by following method reliability -

1. Test retests method
2. Split - half method
3. Parallel test or equivalent appraisal method
4. Intra-rater and interpreter reliability

Validity - Types :

Content, concurrent, criterion, face validity, construct, convergent, discriminate

Step 6 Scoring and administrating of the test and scales

1. Define DC STTS (Code to remember)
2. Collection .
3. Selection
4. Treatment
5. Testing
6. Scoring = Administering

Report Writing

The research task is not completed until the report has been written. A research report is the ultimate output of the research process. The report helps to know what has been, why it was done, how it was done, what results were obtained and what conclusions and recommendations have been arrived at.

Following factors should be considered while writing the report

1. Outline of a research report
2. Some guidelines of writing the report
3. Criteria which are commonly followed

Outline of a research report

1. Title page
2. Declaration certificate
3. Acknowledgements
4. Table of contents
5. Summary
   - Introduction
   - Principal findings
   - Conclusion and recommendations
6. Introduction
7. Methodology
8. Findings
9. Limitations
10. Conclusions and recommendations
11. Appendices
    - Copy of questionnaire
    - References and Bibliography
    - Other relevant details

Guidelines for writing the research report

- Develop your thinking
- Divide your narration into paragraphs and be informative headlines wherever necessary.
- Use present tense and active voice
- Minimize the use of technical language or jargon.
- Use visual aids in the form of tables and figures.
- Be objectives
- Treat data confidentially
- Revise and rewrite

Criteria used for judging the suitability of a report for publication

Business reports oral presentation

A. Statement of the problem
- Clarity and operationality of statement
- Significance of problem
- Documentation

B. Description of method
- Appropriateness of method
- Adequacy of sample field
- Reliability

C. Presentations of results
- Completeness and comprehensibility
- Field

D. Interpretation
- Accuracy
- Bias
- Usefulness

Concept and construct

Concept : An abstraction formed by generalization from particulars.

Weight is a concept it express numerous observation of things that are more on less "neatly" or "light" example intelligence, aggressiveness, conformity and honesty.

Construct : A constract is a concept. It has the added meaning, however of having been deliberately and consciously invented or adopted for special scientific purpose.

Intelligence is a concept, an abstraction from the observation of presumably intelligent and non intelligent behaviors but as a scientific, intelligence means both more and less than it may mean as a concept. It means that scientists consciously and systematically use it in two ways.

1. One it enters into theoretical schemes and is related in various ways to other constructs e.g. school achievement is in part function of intelligence and motivation.

2. Two, intelligence is so defined and specified that it can be observed and measured. We can make observations of intelligence of children by administering intelligence test to them or we can ask teachers 10 tell us the relative degree of intelligence of their pupils.

20 Data : "Technology speaking processing implies editing, coding classification, and tabulation of collected data so that they are amenable to analysis".

Collecting

a. Editing
b. Coding
c. Tabulation
d. Analysis
e. Interpretation

Editing : Editing of data is a process of examining the collected raw data to detect error and omission and to correct these when possible. It involves careful scrutiny of completed questionnaires and/or schedule. Done to assure that the data are accurate, consistent with other facts gathered uniformly entered. It can be done by two ways e.g. Field editing and Control editing.

Coding

Coding refers to the process of assigning numerals or other symbols to answers so that responses can be put into a limited number of categories. Such classes should be appropriate to the research problem under consideration.

- They must possess the characteristic of exhaustiveness (i.e. there must be class for every data item) and also that of mutual exclusively which means that a specific answer can be placed in one and only one call in a given category set.
- Another rule to be observed is that of unidimensionality by which it means that every class is defined in terms of only one concept.
- Coding is necessary for efficient analysis and through it the several replies may be reduced to a small number of classes which contain the critical information required for analysis.

- Coding decisions should usually be taken at the designing stage of the questionnaire.
- This make it possible to precede the questionnaire choices and which in term helpful for computer tabulations as one can straight forward key punch from the original questionnaires.
- But in case of hand coding some standard method may be used one such standard is to code in the margin with a colored pencil.
- The other method can be to transcribe the data from the questionnaire to a coding sheet whatever method is adopted. One should see that coding errors are altogether eliminated on reduce to the minimum level.
- For pretesting of codes a mock code book is constructed which shows the various codes which have been temporarily assigned to different response categories.
- To reduce error in coding, each interview schedule should be coded twice by independent coders and result compared.
- One general rule of thumb is that if an comparison at least 95% inter-coder agreement is found to exist in the case of closed.
- Ended response coding and 70% to 80% in the case of open-ended questions, there is enough reliability in coding. If the agreement is less than the score percentage, all questionnaires should be rechecked until agreement reaches at least this level.

Following are the advantages of proceedings

1. Speedy data collection.
2. Accuracy in data collection.
3. Need to interpret response after data collection is avoided.
4. Comparison with other research becomes possible.

Disadvantages

It is easier for interviewers to introduce non random bias in answer by making them 'fit' certain preceded response categories.

Following are some important guidelines which should be borne in mind while coding questionnaires:

- Codes should be mutually exclusive.
- Set of categories should be collectively exhaustive
- Separate categories should be created for recoding 'non-response' and 'no knowledge response'.
- Non-response means 'void answer' refusal and no - knowledge response means 'cannot answer' (ignorance)
- Inter-Coder and intra coder agreement tests should be conducted.
- To help ensure that responses are being coded systematically, a senior person should be appointed as head of the coding team.

Tabulation

Tabulation is a process of summarizing raw data and displaying the same in compact form (i.e. in the form of statistical tables) for further analysis. In a broader sense is an orderly arrangement of data in columns and recurs". It is essential because :

- It conserves space and reduces explanatory and descriptive statement to minimart.
- It facilitates the process of comparison.
- It facilitates the summation of items and detection of errors and omissions.
- It provides a basis for various statistical computations.

Devices

1. By hand - small inquiry
2. By mechanical or electronic devices Large size of data

Types

1. Simple - give information about one or more group of independent questions. Simple tabulation generally results in one way tables which supply answers to questions about one characteristic of data only. Complex tabulation usually results in two way tables three way tables or still higher order tables also known as magnified tables.

2. Complex shows the discussion of data in two or more categories and as such it designed to give information concerning one or more sets of inter-related questions.

Generally accepted principle of tabulation

- Every table should have clear, concise and adequate titles as to make the table intelligible without reference to the text and this title should just above the body of the table.
- Every table should be given a distinct number to facilitate easy reference.
- The column heading captions and the row heading of the table should be clear and brief.
- The units of measurement under each heading a subheading (stabs) of the table should be clear and brief.
- The units of measurement under each heading or sub heading must always be indicated
- Explanatory footnotes if any, concerning the table should be placed directly beneath the table, along with the reference symbols used in the table.
- Source or sources from where the data in the table have been obtained must be indicated just below the table.
- Usually the columns are separated from one another by lines which make the table more readable and attractive lines are always drawn at the top and bottom of the tables and below the captions.
- Those columns whose data are to be compared should be kept side by side similarly; percentages and/or averages must also be kept close to the data.
- It is generally considered better to approximate figures before tabulation as the same would reduce unnecessary details in the table itself.
- In order to emphasis the relative significance of each categories different kind of type, spacing and indentations may be used.
- It is important that all column figures be properly aligned decimal paints and (+) or (-) sign should be in perfect alignment.

- Abbreviations should be avoided to the extent possible and ditto marks should not be used in the table.
- Miscellaneous and exceptional items, if any, should be usually placed in the last row of the table.
- Table should be made as logical, clear, accurate and simple as possible if the data happens to be very large they should not be crowded in a single table for that would make the table unwieldy and inconvenient.
- Total of recess should normally be placed in the extreme right column and that of column should be placed at the bottom.
- The arrangement of the categories in a table may be chronological, geographical, and alphabetical or according to magnitude to facilitate comparison above all, the table must suit the needs and requirements of a investigation.

It should be remembered that there cannot be any rigidity about these rules tables must suit the needs and requirements of an investigation.

Analysis and interpretation of data

Analysis : means the categorizing, ordering, manipulating and summarizing the data to obtain answers to research questions. The purpose of analysis is to reduce data to intelligible and interpretable form so that the relations of research problems can be studied and tested.

Interpretation : Takes the results of analysis, makes inferences pertinent to the research relations studied, and draws conclusion about these relations. The research who interprets research results searches them for their meaning and implications.

Analysis

1. Frequencies and frequency analysis
2. Continuous measures analysis

Categorization and its rules

First step in any analysis is categorization. It was said earlier that portioning is the foundation of categorization.

Five rules of categorization

1. Categories are setup according to the research problem and purpose.
2. The categories are exhaustive.
3. The categories are mutually exclusive and independent.
4. Each category (variable) is derived from one classification principle.
5. Any categorization scheme must be on one level of discourse

Kinds of statistical analysis

1. Frequency distribution
2. Graphs and graphing
3. Measures of central tendency and variability
4. Measures of relations
5. Analysis of differences
6. Analysis of variance and related methods
7. Profile analysis
8. Multivariate analysis
   - Multiple regression analysis.
   - Canonical correlation
   - Discriminate analysis
   - Factor analysis
9. Indices - Most important in scientific analysis.

The interpretation of research data: following points should be considered

1. Adequacy of research design, methodology, measurement, and analysis
2. Negative and inconclusive results
3. Unhypothesized relations and unanticipated findings
4. Proof, probability and interpretation

Measurement

Measurement is an inseparable part of any science, natural or social. Any science aims to obtain specific and accurate measurement of the events, of the characteristics of the different units of a phenomenon, and, of the inter-relationship **Measurement and Sampling** between the units. Measurement is assigning numbers to objects or events according to rules (Stevens, 1946.) The purpose is to have information in a forming which variables can be related to each other. In social science research we have to deal with various social and psychological variables. Their measurements one of the vital stages in the research process. Measurement of social and psychological variables is a complex and demanding task. Measurement, simply speaking, is the assignment of numerals or other symbols or signs (male, female, occupational categories, for example) to objects or events according to a set of operational rules. Measurement always refers to some property of the object or event and not the object or event by itself. In this measuring process, the observer follows a scheme or procedure by which observed events can be classified into non-overlapping categories unambiguously, and the categories are given labels - numerals or symbols.

The basic assumption in measurement theory is that every event or object in a set of events or an object possesses a specific quantity of the property under observation. This quantity of the property can be compared directly against a standard scale (as when we measure the length amass of a material) or can be evaluated fairly accurately by trained observers (judges or experts assessing the level of performance of a student in a debate or in class examination). Often the measurement operations involve the use of mechanical devices such as thermometer. Barometer, measuring tape, or weighing scale. The use of such mechanical devices is the least observer dependent. and. hence, the measurements are fairly precise. Accurate, and objective. The thermometer, for example, when applied to a given object, gives a number, the temperature. This is a technical refinement of the precision of accrued judgment into categories such as very hot, hot, warm, cold, *etc.*

Importance of Measurement

"Measurement consists of identifying the values which may be assumed by some variable. and representing these values by some numerical notation. The numerical notation is systematically and consistently assigned, that is, it is assigned according to some set of rules."

Thus. the numerals assigned to the variables, indicate different condition of the variables or different values of the variables or different degrees or intensity of equality possessed by units. From the above it is evident that

i. Measurement is a purely descriptive process

ii. Measurement implies that the attributes of persons or objects are possessed in varying degrees and the degree of variation can be measured and represented

iii. Measurement, in essence, is a numerical process.

Carmon objects of measurement in sociology are individual's characteristics, interactions, interrelations, consciousness, participation, socialization, motivation.etc. Some of the properties, of the individuals are visible and easily measurable. Such as age, to measure them. These days. both in sociology and psychology. the measurement of beliefs and attitudes is common because of the greater emphasis on a democratic form of government which demands an assessment of peoples' attitude and opinion, from time to time. Besides this. Public opinion studies are carried on regularly by various public and governmental agencies. Even the commercial organizations measure peoples' opinion and attitude to know the future market of their products. Polling agencies measure people's opinion and attitude to know the people's preferences for particular political parties and candidates. Thus they want to predict, on the basis of such polling, the possibility of winning for any particular candidate. Attitudinal studies may also help in predicting an individual's future behaviour and their possible reactions toward different developmental programmers. Such studies might also help in making policies and specially imrnplementing them. For example, the study of beliefs and attitudes of Indian people towards illness and health measures or family size would help in instituting social and educational programme to mobilize the people towards vaccination or adoption of family planning. Measurement and quantification of variables, beliefs, attitudes, etc., do help in statistical manipulation of them, in experiment, and in testing specific hypothesis related to them. Thus, ultimately. It helps in the development and advancement of a theory.

Admissible Statistical Tests for Measurement

The application of various statistical tests for different categories f measurement scales are discussed below. Statistical Tests for Nominal scale: since the symbols or labels attached to any category are arbitrary and can be interchanged without altering essential information contained in the scale. the only kind of descriptive statistics that can be used are those, which would not be affected or altered by such interchange. They are crude mode, proportion and frequency.

The nominal scale data, however, can be used for testing of hypothesis relating to distribution of events among the classes. Chi-square test, Contingency Coefficient, and certain other tests based on binomial expansion can be used for the purpose. Statistical Tests for Ordinal scale: median is the most appropriate measure of central tendency of the scores that are in an ordinal scale. There are a number of on-parametric tests to test a hypothesis with scores in an ordinal scale - runs test, sign test, median test, Mann Whitney U- test, etc. These tests are often referred to as 'order statistics' or 'ranking statics'. Interrelations can be computed from rankings of two sets of observations on the individuals. Spearman's Rank Difference or Kendall Rank Correlation coefficients are appropriate for such situations. For applying tests to measurements on an ordinal scale, we make an assumption that the observations are drawn from a distribution, which is essentially continuous. Such assumptions are also made for all parametric tests.

Acontinuous limit (interval between two classes), we can have any number of values inserted in between. With an increase in the number of observations, more and more of these values are likely to be represented. It will suffice, at this point, to remind the readers that very often the crudeness of our measuring devices obscures the underlying continuity that may exist. The classification of respondents with respect to an attitude statement into categories strongly agree, agree, neutral, disagree, strongly disagree essentially presumes the presence of a continuum. If a varieties truly continuous and if the instrument for measuring the property in questions sensitive enough, then the probability of obtaining a tied observation is extremely small. Statistical Tests for Interval Scale: the interval scale preserves both the ordering of objects and the relative differences between them, even though the numbers associated with the position of the object may be changed, following a regular system. A set of observations will be scalable by interval scale if the data permits linear transformation, That is, if the equation y = a + bx, where a and b are two positive constants, satisfies a set of real numbers, the numbers are said to be inane interval scale. All the common parametric tests - arithmetic mean, median, standard deviation. Product-moment correlation, etc., are applicable to data that follow an interval scale. Parametric tests for statistical significance like Z, t, F are also applicable.

Statistical Tests for Ratio scale: since the values in a ratio scale are real numbers with a true zero (no upper limit) and only the unit of measurement is arbitrary, the ratios between two numbers and intervals preserve all theinformation contained in the scale even if these true numbers are multiplied by a true positive constant. Any statistical test, parametric or non-parametric, is usable when a ratio scale is used, such statistical tools as geometric mean and coefficient of variation, which require knowledge of true scores, can be used with observations that are in ratio scale.

Criteria for Judging the Measuring Instruments

A measurement, too, must satisfy certain criteria. The most important criteria to be used in evaluating a measurement tool are described below.

i. *Unidimensionality*: this means the scale should measure one characteristic at a time, e.g., the ruler should measure length, not temperature.

ii. *Linearity*: this means that a scale should follow the straight-line model. Some scoring system should be devised; preferably one based another-changeable units. In a ruler an inch is an inch whether it lies at one end of the ruler or at the other, but in altitude scales such interchangeability cannot be ensured. In such cases, ranking is preferable.

iii. *Validity*: this refers to the ability of a scale to measure what it is supposed to measure.

iv. *Reliability*: this is an attribute of consistency. A scale should give consistent results.

v. *Accuracy and Precision*: a tool should give an accurate and precise measure of what we want to measure.

vi. *Simplicity*: a scale should be as simple as possible; an elaborate, complicated, and over - refined scale may become unduly cumbersome, costly or even useless.

vii. *Practicability*: this is concerned with wide range of factors like cost effectiveness and interpretability. Some trade off is usually needed between an 'ideal' tool and, that which the budget can afford. The benefit to be derived should be commensurate with the cost incurred. Should be easily understandable and conveniently arranged for easy completion. In order to enable others to interpret the results of a test, there is need for such aids as a statement of its function, its construction procedure and guides for interpreting the result.

Sources of Errors in Measurement

Measurement should be precise and unambiguous in an ideal research study. This objective, however, is often not met with in entirety. As such, the researcher must be aware about the sources of error in measurement. The following are the possible sources of error in measurement.

a. *Respondent*: at times the respondent may be reluctant to express strong negative feelings or it is just possible that he may have very little knowledge but may not admit his ignorance. All this reluctance is likely to result in an interview. Transient factors like fatigue, bourdon, anxiety, etc., may limit the ability of the respondent to respond accurately and fully.

b. *Situation*: situational factors may also clone in the way of correct measurement. Anyone places a strain on interviews can have serious effects on the interviewer-respondent rapport. For instance. if someone else is present, he can distort responses by joining in or merely by being present. If the respondent feels that anonymity is not assured, he may be reluctant to expires certain feelings.

c. *Measurer*: the interviewer can distort responses by rewording or reordering questions. His behavior, style and looks may encourage or discourage certain replies from respondents. Careless mechanical processing may distort the findings because of incorrect coding, faulty tabulation and statistical calculations, particularly in the data analysis stage.

d. *Instrument*: error may arise because of the defective measuring instrument. The use of complex words, beyond the comprehension of the respondent,anbiguous meanings, poor in printing, inadequate space for replies, response choice omissions, etc., are a few things that make the measuring instrument defective, and may result in measurement error. Another type of instrument deficiency in the poor sampling of the universe of items of concern. The researcher must know that correct measurement depends on successfully meeting all of the problems listed above. He must, to the extent possible, try to eliminate, neutralize, or, otherwise deal with all the possible sources of error so that the final results may not be contaminated. In the above sections, you read about various kinds of measurement and statistical tests to be used in measurement. Now answer the questions given in criteria for judging the measuring instruments.

Purpose : To have information in a form in which variables can be relate to each other.

Nominal	Ordinal	Interval	Ratio
Numbers are assigned to objects which can be placed into mutually exclusive categories and exhaustive categories	+ and be ordered into a greater or less than scale	categorize, ordered + equal distance between scale	
Mean to name a screen name are given to the categories with no assumption about relationship between them.	No absolute zero unequal listen between scale values		
Classificatory measurement			
Eg. people may be classified as male female, rural-urban farmer-non farmer and no are assigned e.g. 1. For farmer 2. Non farmer	eg upon middle lower class	Temperature Scale	Adoption quotient

It differentiate one categoy from other but does not give meaning greater or less than			
Puts objects into classes Mode Chi-square, i.e., male1 female, marital status, occupation	Determination of Median Rank Order greater or less, i.e., Correlation, Man preference, level of Whitney U-test education achieved	Determination of Arithmetic Correlation equality of intervals or Mean Coefficient differences, i.e., temperature	Determination of Geometric Coefficient of equality of ratios, i.e., Mean, Variation weight, income, Harmonic number of visits Mean

Sampling

Sampling is a method of selecting a fraction of the population in such a way that it represents the whole population.

Probability	Non-Probability
1. Simple random sampling	1. Quota
2. Cluster sampling	2. Purposive
3. Stratified Sampling	3. Systematic
	4. Snowball
	5. Double

Probability samling : Size of population/unit is logon to instars each individual have an equal chance of being selected/being included in sample.

Non probability : That sampling procedure in which there is no way of assessing probability of the elements of population being included in the sample.

Normal distribution

Measurement of many natural phenomena and of many mental and social traits under certain conditions tends to be distributed symmetrically about their means in proportions which approximate those of the normal probability distribution. The normal distribution is not an actual distribution of test scores, but is, instead a mathematical model.

Normal curve is one which graphically represents normal distribution and looks bell shaped.

As normal curve is symmetrical about the mean i.e. the right half of the curve is equivalent to the left half. (ii) The mean, median and mode are numerically identical and fall at the centre of the distribution.

Different percentage enforced under the normal curve at various standard deviation units taken from means

B. Analysis of variance (ANOVA) : is the technique of path timing total variation among a number of individual variants belonging to two or more classes or sets into various components such as variation due to different sets or treatments or between treatments and variation due to all other distinguishing causes, including sampling. The aim is to find out whether the variation due different treatments with variation due to indistinguishing causes termed variance.

- One way ANOVA
- Two way ANOVA

One way ANOVA: - In a simple analysis of variance the total sum of square is partitioned into two groups the sum of square based upon the variation between the group means and the sum of squares based upon variations within the several groups.

Do not permit the evaluation of interaction between two or more variables

Source of variation	S.S.	D.F.	MS	F
Between groups				MS Between
Within groups (Error)				
Total				

Two way ANOVA used where one dependent and two independent variable permits the simultaneous study of two factors or variables.

Source of variation	Sum of square (SS)	Degree of freedom (DF)	Mean square	F
A	SSA	(a-1)	SSA/dfa	MSA/MSW
B	SSB	(b-1)		
A (interaction)	SSAB	(a-1)(b-1)		
Within (error)	SSW	N-ab		
Total		M-1		

A - The independent variable

a - The number of conditions in variable A

B - Second independent variable

b - The number of conditions in variable B.

AB - Interaction between A to b

N - Within groups also called errors

Assumptions underlying analysis of variance

The assumption:

- Normality of the distribution of criterion measures.
- Homogeneity of variance
- Independency of the dependent score

Origin of Participatory Technology Development (PTD)

Origins of PTD can be found in many technologies developed/ became popular during in 1970s and 1980s i.e.

- Farming system research and extension
- Indigenous technical knowledge
- Participatory rural appraisal

P.T.D. Process

- Gaining a joint understanding of the main characteristics and changes of that particular agro ecological system.
- Defining priority problems
- Experimenting locally with a variety of options derived from indigenous knowledge and from formal science.
- Enhancing farmer's experimental capacities and farmer to farmer communication
- Sustaining PTD

PTD facilitators

Researcher as well as educationalist social workers

Elements

- Curiosity experiments
- Problem solving experiments
- Adaptive experiments

Characteristics

1. Main goal to develop techniques to meet the need of small resource poor farmer.
2. Active participation of farmer.
3. Both farmer and research knowledge is crucial.
4. Research is conducted in formed field.

Participation: Participation includes peoples involvement in decision making process, in implementing process, programmers, their sharing in benefits of development programmers and their involvement in efforts to evaluate such programmed.

Steps in participatory approach

1. Rapport building
2. Do it yourself
3. Consultation with villagers and outsiders
4. Secondary data collection
5. Training of trainers
6. Diagnosis

Participating approaches

1. RRA - Rapid Rural Appraisal
2. PRA - Participatory Rural Appraisal
3. PALM - Participatory Learning Methods
4. AEA - Agro-Eco System Analysis
5. PAR - Participatory Action Research
6. PAMA - Participatory, Assessment, Monitoring, Evaluation
7. FSR - Farming System Research
8. PRAD - Participatory Rural Appraisal and Planning

	RRA	PRA
Developed in	1970s 1980s	1980s 1990s
Key Resource	Local peoples knowledge	Local peoples analytical capability
Main innovation	Methods	Change of behavior and attitude
Mode	Extractive	Facilitating
Mode of instraction	Verbal	Visual
Ideal of objectives	Insiders by outsiders	Empowerment of local people
Outsiders role	Investigator	Initiator and catalyst
Insider role	Respondent	Presenter analyst and planner
A model	For participatory research	For Interaction intervention
Who demands	Donor Organization	Insiders

Participatory Rural Appraisal

Distinguishing features of PRA

PRA are the result of response to top down approach of development, the pitfalls of the questionnaire survey method of inquiry and the problem of biases. There have thus been major shifts and reversals in PRA including :

1. Closed to open
2. Measurement to comparison
3. Individual to group
4. Verbal to visual
5. Higher to lower
6. Reserve to report

Principles

1. Listening and learning
2. Off setting biases
3. Utilization of precious community time
4. Seeking diversity
5. Triangulation/Cross checking
6. Optimal ignorance/and appropriate impression
7. Multidisciplinary team

Techniques

1. Spatial map/Social Map
2. Transact walk
3. Time line
4. Trend analysis
5. Seasonal diagram
6. Cause effect diagram/problem tree
7. Venn diagram

PLA : PLA is an approach for learning and organizing participation of local communities and groups for interacting with them, understanding them and learning from them. It helps in initiating a participatory process in sustaining.

PLA provides an alternative framework for data generation and analysis whose the people generate different kinds of qualitative and quantitative data for their own analysis. In such participatory appraisal both economic and non economic factors are important not all of which are easily quantifiable.

Main aim : Theory building

(B) Subaims

i. Exploration
ii. Understanding
iii. Prediction
iv. Control

Theory as a tool of science/role of theory

1. Orientation
2. Conceptualization
3. Summarizing
4. Predicting
5. Painting gap

Components of research problem

1. Individual
2. Problem
3. Objective
4. Alternative
5. Doubts
6. Environment

Selection of research problem

1. Original
2. Uncontroversial
3. Broad
4. Familiarity with researcher
5. Qualification of researchers
6. Preliminary study
7. Soluble
8. Feasible
9. Simple

Scientific Observation : Scientific observation is a methodological way of recognizing and noting a fact or occurrence often involving some sort of measurement scientific observation should be specific and recorded immediately.

Advantage	Limitations
1. Elimination of subjective bias	1. Time consuming and expensive method
2. Correct state of affair related information	2. Limited information
3. Independent of respondents willingness	3. Extraneous factors may interfere with the task of controlled

Types

1. Structured and unstructed
2. Participant and non-participant
3. Controlled and un-controlled

Case study

Empirical inquiry that investigates a contemporary phenomenon within its real life context when the boundaries between phenomenon and context are not clearly evident and in which multiple sources of evidence are used.

Characteristics

1. Individual/Family - unit of study institution or cultural group
2. Deep and detailed study
3. Wide time cycle
4. Small number of units
5. Quantitative nature

Assumptions

1. Totality - Totality of being
2. Unity - Under laying unity
3. Complexity - Complexity of social phenomena
4. Time - Influence of time

Steps in conducting case study

1. Locating case
2. Collection of data
3. Treatment
4. Interpretation
5. Conclusion and documentation

Advantages

1. Very intensive
2. Flexible
3. Natural setting is the source of data
4. Less expensive
5. Presents real record of personal experience
6. Helps in interview schedule construction
7. Enhances experience of research

Limitations

1. Less reliable results
2. Results may be subjective
3. Very little scope of quantitative analysis

Content analysis

"Content analysis is the method of studying and analyzing communications of the electronic and print media in a systematic, objective and quantitative manner to measure variables."

Some aspects of methods in content analysis

1. Letter diaries
2. Ethnographic material
3. News paper articles
4. Editorial
5. Minutes of meeting

Steps

1. Define universe
2. Categorization of universe
3. Units of analysis
   i. Words
   ii. Themes
   iii. Characters
   iv. Items
   v. Space and time measures

Quantification

Content can be quantified in three ways:-

1. Nominal measurement
2. Ranking
3. Rating

Advantages

1. Immense flexibility

Disadvantages

1. As good as the research document.
2. Difficult to answer why
3. Sometime accused of being too theoritical

Example

Leaders and conceptual complexity

Successful

Before revaluation conceptually simple after gaining power became conceptually complex.

Unsuccessful

Remain simple conceptuality both before and after the evaluable.

1. *Independent variable*: Conceptual complexity
2. *Dependent variable*: Success
3. *Measure*: Conceptual complexity of each leader
4. *Method* : Content analysis
5. *Source*: Letter, speeches, documents

Semantic differential : The S.D. is a method.

The semantic differential is a method of measuring, assessing and comparing the image of a brand, political candidate, company organization etc

Applications : To measure objectively the semantic properties of words and concepts in a traditional semantic space and more commonly and simply as an attitude scale restricting its focus the effective domain?

Elements : scale bear three elements

1. Concept
2. Polar objective pair
3. Scale position

Principle factors accounting for most of the semantic loading

1. Evaluative
2. Potency
3. Activity

Advantages

- Most effective for measuring image.
- Objective must define a single dimension and each pair must be bipolar opposite.
- Precisely what the respondent is to rate must be clearly stated in the introductory instructions.
- Not more than 20 times should be used and about half should begin with most positive words.
- If the same scale is used for several topics and/or rating for an ideal unity image/profile can be compared among.
- The same scale is used for several topics or for rating an ideal entity image profile can be compared among real entities and between real and ideal ones.

Sociometry

Sociometry is a technique for describing the social relationship among individuals in a group.

Sociogram

Are charts that depicts

Analysis of sociometric data

1. Sociometric matrix
2. Sociogram
3. Sociomatric indices

- Choice status index
- Group cohesiveness index
- Social compatibility index
- Emotional expensiveness

Questionnaire Method

In this method questions are delivered to the respondents who note down their response on it and send it back to the researcher.

Types of important dimensions along with questionnaire generally vary are

a. Degree of structuredness
  - Structured
  - Unstructured

b. Number of response possibilities
  - Dichotomous
  - Multiple choices

Guidelines for designing a good questionnaire

1. Objective orientation
2. Brief
3. within the sphere of respondents experience
4. Unbiased questions
5. Clarity
6. Simplicity
7. Form
8. Logical sequence
9. Cross check
10. Letter accompanied
11. Retranslation in original languages
12. Appropriate ending

Logout

1. Prominent identification
2. Front page title
3. Authority for information collection
4. Confidentialness of data
5. Date of period covered

6. Footnote should be clearly stated
7. A place for the date on which form is filled out should be provided
8. A place for signature of respondent
9. Serial number assignment
10. Numbering of pages

Pretesting of the questionnaire

Pretesting is a preview of any possible inconsistency or ambiguity in the questions such questions are then appropriately rewritten and the questionnaire is given a final shape.

Advantage

1. Reveal the misunderstood questions.
2. Helps to decide the proper and structure of each question.
3. Solution of mechanical problems of measurement.
4. Helps to improve questionnaire design.
5. Firm estimates of amount of time, money personnel and equipments

No. Response: Reasons serious problem

1. Non coverage
2. Not at home
3. Unable to answer
4. The hard core.

Techniques to increase response

1. Fellowship
2. Sponsor
3. Length
4. Introductory letter
5. Type of question
6. Inducements

7. Method of return
8. Time of arrival
9. Format
10. Selection of respondents

Advantages

1. Relatively inexpensive mode of data collection.
2. No chance for
3. It enables the respondents to answer the questions.
4. It does not give chance to investigate

   It can be answered according to the convenience of the respondent

Limitations

1. Unsuited to illiterate
2. Inability to respond/misinterpretation
3. Non check on honesty and reliability
4. Appropriate when spontaneous onslear are wanted.
5. The structuredness of a questionnaire may limit the usefulness of information.
6. Response rate being very low.

Interview schedule

It is a meeting in which the interviewer puts question.

Kinds schedule information: Three kinds of information are included in most one schedules

1. Face sheet (identification information)
2. Census type (or sociological) information
3. Problem information

Cautions: The importance of identifying each schedule accurately and completely. However need to be mentioned. The careful researcher should learn to identify with letters numbers or other symbols, every schedule and every scale. In addition identify information for each individual must be systematically recorded.

Types of schedule item

Two types of schedule items are in common use

1. Fixed - alternative (or closed)
2. Open ended (or open)

A third type of item, having fixed alternatives, is also used:- scale items.

Fixed alternative items : It offers the respondent a choice among two or more alternatives. These items are also caved closed or pall questions. The commonest kind of fixed - alternative item is dichotomous. it asks for yes/no, agree - disagree and other two alternative answers. Often a third alternative, do not know or undecided is added.

Advantage

Achieving greater uniformity of measurement and thus greater reliability, of forcing the respondent to answer in a way that fits the response categories previously setup, and of being easily coded they have certain disadvantages.

1. Superficiality
2. Non of alternative may be suitable for respondent
3. Forcing of responses
4. Choosing of an alternative to conceal ignorance
5. Choosing alternative that do not accurately represents true facts or opinions.

Open ended items

Open ended items are an extremely important development in the technique of interviewing. Open-ended questions are those that supply a frame of reference respondent's answers, but put a minimum of restraint on the answers and their expression while their content is dictated by the research problem, they impose no other restrictions on the context and manner of respondent answers.

A special type of open-ended question is the funnel. Actually, this is a set of questions directed towards getting information on a single important topic or a single set of related topics.

The tunnel starts with a broad question and narrows down progressively to the important specific point or points

Cannel and Kahn say that the funnel has the purposes of preventing early questions in a sequence of questions from affecting those that came later and determining something that of respondents frame of reference.

Second Type - Another form of formal starts within open general question and uses follow up, specific closed questions. The best way to get a feeling for good open ended questions and formals' is to study examples.

Scale items

A third type of schedule item is the scale item. A scale is a set of verbal items to each of which an individual responds by expressing degrees of agreement or some of the mode of response.

Scale items have fixed alternatives and place the responding individual at some part on the scale.

The use of scale item in the interview schedule is a development of great promise, since the benefits of scales are combined with those of interviews.

We can include, for example, a scale to measure attitude towards education in an interview schedule, for example, a scale to measure attitude towards education. is an interview schedule on the same topics.

Criteria of question writing

Criteria or percepts of question writing have been developed through experience and research some of the most important of those given below is in the form of questions

1. Is the question related to research problem and research objectives?
2. Is the type of question sight and appropriate?
3. Is the item clear and unambiguous?
4. Is the question a leading question?
5. Does the question demand knowledge and information that the respondent does not have?

6. Does the question demand personal or delicate material that the respondent may resist?
7. Is the question loaded with social and desirability?

The value of interview/Interview schedules

Since interview schedule is self administered instrument that have items of closed-order

1. Fixed
2. Alternative type.

A self-administered instrument has certain advantages with most or all of its items of the closed type, greater uniformity of stimulus and thus greater reliability can be achieved. In this respect. It has the advantages of objective type written tests and scales, if they are adequately pretested and constructed. Second advantage is that if abnormity honesty and freakiness may be encouraged. This kind of instrument can also be administered to large numbers relatively easily. A somewhat dubious advantage is that it can be mailed to respondents further, it is economical, its cost is ordinarily a fraction of that of interviews.

Disadvantages

1. The disadvantages of self ad ministered instrument (when mailed) seem to outweigh its advantages. The principal disadvantages are that it may not be as uniform as it seems.
2. A second disadvantage is that it may not be as uniform as it seems.
3. Experience has shown that some question frequently has different meanings for different people.
4. Third if only closed items are used the instrument displays the same weakness of closed items discussed earlier.
5. On the other hand if open items are used the respondent may object to writing the answers, which reduces the sample of adequate responses.
6. Because of these disadvantages interview is probably superior to the administered questionnaire.

7. The best instrument available for sounding people's behavior.
8. The self-administered questionnaire has been used too much and the structured interview too little.

Collection of secondary data

Secondary data may either be published or unpublished and are generally available from the following sources.

1. Publications of the control, state and local governments.
2. Publications of the international organization and local govts.
3. Journals magazines and books
4. Reports and publications and various associations.
5. Public records, statistics and historical documents and
6. Personal education from individuals

Secondary data, to be useful should be reliable, relevant and adequate for the purpose.

Variables: Variable is those attributes of objects events things and beings which vary and can be measured.

Construct: When a characteristic has only one value it is a constant, not available.

1. Dependent and independent variable
2. Moderator and control variable
3. Intervening variable
4. Qualitative attributes and quantitative variables.
5. Continuous and discrete variables
6. Stimulus variable and response variable
7. Extraneous variables

Confounding effect how to control

When the dependent variable is not free from the influence of extraneous variables(s) the relationship between the dependent and independent variable is said to be confounded by an extraneous variable.

There are some basic procedures for controlling the extraneous variables. These are

1. Randomization
2. Elimination
3. Matching
4. Additional independent variable
5. Statistical control

Criteria for selection of variables

Relation with objectives mutually exclusive and none overlapping According to the level of understanding of rash measured by the available or suitable teacher required could be classified into categories limited in number so as to avoid confusion.

Hypothesis

"Hypothesis is a proposition or principle which is assumed in order to draw its logical or empirical consequences and by this method to test its accord with facts which are known or may be determined"

"Hypothesis provides objectivity in research because of the freedom from beliefs and biases its ensure in testing."

Source

1. Researcher experience
2. Review of literature
3. Finding of pilot study
4. An analysis testing validity or otherwise the existing theory
5. Creative thinking and imagination of researcher.

Functions

1. To guide the researcher to collect kind of data.
2. The way in which data can be organized most efficiently in the analysis.

Characteristics of hypothesis

1. Clear
2. Precise

3. Testable
4. Should state relationship between variables
5. Limited in scope
6. Specific
7. Simple
8. Consistent with body of facts
9. Reasonable time
10. Should be amenable to testing within a reasonable time
11. Hypothesis must explain the facts that gave rise to need for explanation.

Types of hypothesis

1. Null
2. Alternative
3. Descriptive
4. Theoretical
5. Operational
6. Relational
7. Statistical

Testing of hypothesis

For statistical point of view statement, a statement made regarding the population characteristic is known as hypothesis. Assessment of the correctness of the statement/hypothesis is known as test of significance.

Important steps in testing of hypothesis

1. Stating the hypothesis
2. Stating the level of significance
3. Selecting the test criteria
4. Computation
5. Making decisions
6. State the Ho and Ha

7. Specify the level of significance
8. Describe the correct sampling distribution
9. Select a random sample and workout appropriate value from sample data.
10. Calculate the probability that sample result would diverge as widely as it has from expectations, if Ho is have.
11. Is the probability is equal to or smaller then x value in case of one tailed test and x/2 in case of two tailed test.

Yes	No
Reject Ho	Accept Ho
There by run the risk of committing type I error	There by run the risk of committing Type II nd error

Research Design

According to Selltiz and others (1976), a research design is the arrangement of conditions for collection and analysis of data in a manner that aims to combine relevance to the research purpose with economy in procedure.

Purpose of research design

Research design has two basic purposes:

1. To provide answers to research questions
2. To control variance

Principle/Function : (Maxmincon Principle) The main technical function of research design is to control variance. A research design is in a manner of speaking, a set of instructions to the investigator to gather and analyze his data in certain ways it is therefore a control mechanism.

The statistical principle behind this mechanism is:

1. Maximize systematic variance
2. Control extraneous systematic variance
3. Minimize error variance

I.O.W. We must control variance.

According to this principle, by constructing an efficient research design the investigator attempts

a. To maximize the variance of the variable or variables of his substantive research hypothesis

b. To control the variance of extraneous unwanted variables that may have an effect on his experimental outcomes, but in which he is not interested and

c. To minimize the error or random variance, including so called errors of measurement.

This is also called maximum principle or maximin con principle maximization of experimental variance:

Refers to the variance of the dependent variable influenced by the independent variable or variables of the substantive hypothesis.

Control of extraneous variables

The control of extraneous variables means that the influence of independent variables extraneous to the purposes of the study are minimized, nullified or isolated. I.O.W. the variance to such variables is in effect.

1. To eliminate the effect of a possible influential independent variable on a dependent variable, one can choose subjects so that they are as homogenous as possible on their independent variable.

2. The second method is through randomization.

Minimization of error variance

Error variance is the variability of measures to random fluctuations whose basic characteristic is that they are self compensating, varying now this way, now that way, non-positive now negative.

Minimizing how

Reduction of errors of measurement through controlled condition - an increase in the reliability of measurement.

Criteria of research design

1. Answer research questions
2. Control of extraneous independent variables
3. Generalizability
4. Internal and external validity

Validity of research design

Internal validity for research design

May be defined as the ability to conclude that the experimental independent variable did involve create changes in the dependent variable.

External validity

Is the ability to generalize the results of the research to populations, settings, treatment, variables, measurement variables and representativeness.

Threats to internal validity of research :

1. Maturation
2. Testing
3. Instrumentation
4. Selection bias
5. Statistical regression

Threats to external validity of research:-

1. Interaction effects of selection bias
2. Prior knowledge about subjects
3. Artificial situation of experiment
4. Carry over effects
5. Outside interference
6. Hawthorne effect

Research designs

1. Exploratory or formal research design
2. Descriptive and diagnostic research design
3. Hypothesis - testing/experimental research design

Principles of experimental design

According to professor R.A. Fisher:- three principles

1. Principle of replication
2. Principle of randomization
3. Principle of local control

Experimental design can be further classified

1. True experimental deigns
   1. Post test only, equivalent group design
   2. Pretest - Posttest control group design
2. Quasi - experimental design
   1. Time series design
   2. Equivalent time sample design
   3. Non-equivalent control group
3. Ex-post-facto
   1. Ex-post-facto cause to effect design.
   2. Ex-post-facto effect to cause design.

Ex post facto research

Is a systematic empirical in which the scientist does not have direct control of independent variable because their manifestations have already occurred or because they are inherently not manipulable.

	Experimental Research	Expost facto Research
1.	Investigator has manipulative control he has at least one active variable	1. Not possible
2.	He can assign subject or treatment to a group at random	2. No randomization

Self selection

Occurs when the members of the groups being studied are in the group. The limitations of post-facto interpretation:-

1. Inability to manipulate independent variable
2. The lack of power to randomize
3. The risk of improper interpretation

The value of expost facto research

Field studies : are expost facto scientific inquiries aimed at discovering the relations and interactions among sociological, psychological and educational variables in real social structures.

Types of field studies:

1. Exploratory
2. Hypothesis testing

Exploratory field study

Exploratory study has three purposes:-

- To discover significant variables in the field situation.
- To discover relations among variables
- To lay the ground work for later, more systematic and rigorous testing of hypothesis.

Strength : Realism, significance, strength of variables theory orientation and heuristic quality.

Weakness : Expost - facto character, feasibility, cost, sampling time

Survey research:- Studies large and small population (or universe) by selecting and studying samples chosen from the populations to discover the relative incidences distribution, and interpretations of sociological and psychological variables.

Types of surveys

1. Interview and schedule
2. Questionnaire
3. Panel
4. Telephone survey
5. Controlled observation

The methodology of survey

Advantages	Weakness
Wide scope	Demanding of time and money
Accurate	Requiring good knowledge

Uses of Research Design :

1. It provides answers to various questions.
2. It acts as a standard and guide post.
3. It helps in carrying out research validity, objectively, accurately and economically
4. It forestalls the possibility of fruitless inquiry and acts as an insurance agianst future failure.

Participatory Research

Need for participatory research

People participation may be defined as the process of giving priority to local people's perspective in identifying and analyzing their problems and opportunities, and improving the situation through their mobilization

Need

i. Situation oriented
ii. Problem specific
iii. Bottom up
iv. Need based

Purpose : Gain an understanding of the complexity of the topic.

Define : Participatory research is a way of learning from, and with community members to investigate analyses and evaluate constraints and opportunities, and to make informal and timely decisions regarding development projects. It is a method by which research team can quickly and systematically called information for

i. General analysis of a specific topic, question or problem
ii. Needs assessment
iii. Feasibility studies
iv. Identifying priorities for development activities
v. Implementing development activities
vi. Monitoring and evaluation of development activities

Features

1. *Triangulation:* Comparing and crosschecking data consistency
2. *Multidisciplinary team:* Different skills and back grown
3. Techniques is
4. Flexibility = Informality
5. Involvement of community

Optimal ignorance and appropriate imprecision on the spot analysis.

Tools and techniques of participatory research

1. Review of secondary sources
2. Direct observation
3. Semi structured interview
4. Ranking
5. Construction of diagrams
   a. MAP
   b. Transect
6. Seasonal calendar
7. Historical profile
8. Daily routine diagram
9. Livelihood analysis
10. Venn diagram

Logical framework analysis

Involves the situation in its totality by the application of tools and techniques of participatory research.

According to Patton (1982), creating cross classification matrices is an exercise in logic. This procedure involve creating potential categories by crossing one dimension or typology with other and then working back and forth between data and one's logical construction, filling in the resulting matrix. This logical system will create a new typology all parts of which may or may not actually be represented by data.

Example	Narrated Statement (NS)	Objectively verifiable indicators (OVI)	Means of Verification (MOV)	Assumption/ Risks
1. Objective				
2. Purpose				
3. Activity				
4. Output				

Advantages of participatory methods

According to Kumar (2002) several advantages of participatory methods are:

1. Enable outside to get internal perspective of local community.
2. It marks a shift in emphasis from measurement to comparison.
3. Trends and changes are captured easily and quickly.
4. The group interaction provides greater and more in depth information and analysis in a much shorter time.

Measurement postulates : A postulate is an assumption that is an essential prerequisite to carry out some operation or line of thinking in this case it is an assumption about the relations between the objects being measured. Although postulates are usually assumed to be true, in measurement it is necessary to test the postulates whenever possible. More than three postulates are really necessary in order to make it possible to equate objects, to rank order them. The three postulates can be written:-

1. Either (a-b) or (a ≠ b) but not both
2. If [(a=b) and (b=c),] then (a=c)
3. If [a>b] and (b>c)], then (a>c)

The first postulate says "a is either equal to b or not equal to b, but not both." This postulate is necessary for classification we must be able to assert either that one object is the same in a characteristic as another or that it is not the same. In measurement" the same" does not necessarily mean complete identity. It can mean sufficiently the same to be classed as members of the same set.' Two boys are the same in maleness.

II The second postulate says "If a equals b and b equals c than a equal This postulate enables a researcher to establish comparison the equality of set members on characteristics by comparing objects

III This is transivity postulate most measurement in Psychology and education depends on this postulates. It must be possible to assert ordinal - or rank - order statements like" a has more of a property then b; he has more of the property the C; therefore a has more of the property than c".

The preceding statement may seen obvious it is easy and even justifiable to make many statements, but not always.

Reliability and validity of tests

Probability Sampling Methods	Non Probability Sampling Methods
1. Simple random sampling	1. Quota sampling
2. Cluster sampling	2. Purposive
3. Stratified random sampling	3. Systematic
(A) Proportionate stratified random sampling	4. Snowball
(B) Disproportionate stratified random sampling	5. Double Sampling

Normality

Normal distributing tests:- Measurement of many natural, phenomena and of many mental and social traits under certain conditions tend to be distributed symmetrically about their means in proportions which approximates those of the normal probability distribution.

The normal distribution is not an actual distribution of test scores, but is, instead a mathematical model (Garrett; 1979). A normal curve is one which graphically represents normal distribution and looks bell shaped. The important characteristics of a normal curve are:-

- A normal curve is symmetrical about the mean i.e. the right half of the curve is equivalent to the left half; and
- The mean, median and moderns numerically identical and foll at the centre of the distribution.

There is a relationship between a standard deviation units and the normal curve.

Shows the different percentages of areas falling under a normal curve at different standard deviation units.

Characteristics and types of research	Experimental	Non Experimental (descriptive research)
1. Control of variables	Variables can be manipulated or controlled	Variables are inherently not manipulable
2. Research control	Reseracher had direct control over the independent variables	Researcher has not direct control over the independent variables
3. Internal validity	Sufficient degree of internal validity is found	Level of internal validity is very low very poor
4. Scope of experimentation	Can be experimentally studied	Cannot be experimentally studied
5. Nature of events	Changed when brought into laboratory or controlled situation	Events donot changes they are studied in every day realistic setting
6. Types	Two types : Laboratory experiments, Field experiments	Six types : Field studies, expost facto research, survey research, content analysis case study and ethnographic studies

❑❑❑

Chapter 4

Information Communication Technologies

Introduction

Farming Information and awareness package through print, mass media and now emails is possible. The role of impeditive farming, economic survey and evaluation of farming and women's integration need attention. Women, the half of agricultural work overcome by educating them through TV, cassettes, e-mails or other modern indication appliances.

Extension provides refined technological participatory research and educational modules to bridge the productivity gaps.

Future growth in agriculture would have necessarily come from increased productivity from a shrinking natural resources base through efficient management.

ICT : Barriers of ICT Concepts

Applications of ICT for Agriculture/Development

1. GIS
2. Remote Sensing
3. E governance
4. Cyber Extension
   - Participation
   - Problem Solving
   - TOT
   - Innovation
   - Group formation

Role of ICT in Agriculture Development

1. Economic development : Application in planning and market information for agriculture procedures.
2. Community Development Applications.
3. Research/Education Applications
4. Small and medium enterprise development
5. New media networks
6. Providing decision support to public administration for improving planning and monitoring of development programmes.
7. Improve services to citizen, and bring transparency.
8. Empower citizens through access to information and knowledge
9. Help in training to improve the functioning of developmental organization and expand employment opportunities.
10. Market
11. Linkage
12. Services
13. Small and median farmer

Digital divides : Distance and difference caused among because of the haves and have not's of the ICT.

Impact : Hurdle in conformity in gains of the ICT.

Two biggest hurdles

1. Access and content
2. Linguistic and technological divide

Each medium of communication is dominated by specific linguistic traditions. This limits the access to the medium as terms of understanding and utilizing the medium.

In case of ICT, two kinds of divides are prominent

i. A divide between English and non English speaking communities where comprehensible content is available only to a selected few who have knowledge of English.
ii. *North south divide* : Between developed and developing countries.

National Level Initiative in ICT for Agricultural Development

1. Gyandoot project
2. Warna wired village project
3. Information village project of MSSRF
4. Kisan project of Nagarjana group of companies (Andhra Pradesh)
5. Milk collection centers of dairy cooperative of (AMUL) Gujarat.
6. Taranhoat.com by development alternatives (Uttar Pradesh and Punjab)
7. Land record computerization (Bhoomi)
8. Knowledge network for grass root innovations - SRISTI (Gujarat)
9. Mahiti Samudra (Karnataka)
10. Centre for alternative agriculture media (CAAM)

Extension Teaching Methods

Define : Devices use to create situations in which communication can take place between the instructor and the learner.

Functions

1. To provide communication so that the learner may see, near and do the things to be learnt.

2. To provide stimulation that causes the desired mental and/or physical action on the part of the learner.
3. In brief, to take the learner through one or more steps of the teaching - learning process, viz attention, interest, desire, conviction, action and satisfaction.

Classification

1. According to use

i. Individual contact
   - i. Farm and home visit
   - ii. Office calls
   - iii. Personal letters
   - iv. Result demonstration

ii. Group contacts
   - i. Method demonstration
   - ii. Leader training meetings
   - iii. Lecture meetings
   - iv. Conference and discussion
   - v. Meetings at result demonstrations
   - vi. Tours
   - vii. Schools
   - viii. Miscellaneous meetings

iii. Mass contact
   - i. Bulletins
   - ii. Leaflets
   - iii. News stories
   - iv. Circular letters
   - v. Radio
   - vi. Television
   - vii. Exhibits
   - viii. Posters

2. According to form

a. Written
   i. Bulletins
   ii. Leaflets
   iii. News articles
   iv. Personal letters
   v. Circular letters

b. Spoken
   i. General and special meetings of all kinds.
   ii. Farm and home visits
   iii. Office calls
   iv. Telephone calls
   v. Radio

c. Visual or objective
   i. Result demonstrations
   ii. Exhibits
   iii. Posters
   iv. Motion pictures, charts, slides, and other visual aids

d. Spoken and visual
   i. Method demonstration meetings
   ii. Meetings at result demonstration
   iii. Meetings involving motion pictures charts and others visual aids.
   iv. Television

Selection of extension teaching methods

Factors to be considered in selection and combination of extension methods :

1. The audience

   a. Individual and collective differences

   b. Size of audience

2. The teaching objectives - Thinking attitude/feeling action/skill

3. The subject matter

4. The stage of development of extension organization
5. Size of extension staff
6. The availability of certain communication media.
7. The relative cost of method.
8. An extension worker's familiarity.
9. The nature and complexity of the subject matter.
10. The extent to which people already know and are practicing the recommended practice.
11. The relative economic and social importance of the behavioral changes desired.
12. The amount of extension worker efforts likely to be devoted to a particular phase of a programme
13. The kind and extent of assistance that may be provided by local leaders and others.
14. The member in which the methods selected complement each other in attaining the behavioral changed.

Audio - Visual aids

According to **Haas and Packer (1964)** "an audio visual aid is an instructional device that can be heard and seen"

According to date

All materials used in the classroom or other teaching situations to facilitate the understanding of written or spoken words.

Classification

Evaluation

Senses involved

According to evolution

1. *First generation media* : Handmade charts, graphs, exhibits.
2. Models, hand written material etc
3. *Second - Generation media* : Printed/Illustrated tects, printed graphics, work books

4. *Third generation* : Photographs, slides film strips, films recordings, radio telecommunication.
5. *Fourth generation* : Television, programmed instruction, language laboratories, electronic digital computers.

According to senses involved

1. Audio Aids
    1. Radio talk
    2. Tape recording
    3. Disc recording
2. Audio - Visual Aids
    1. Motion pictures
    2. Telecasts
    3. Video recording
    4. Sound synchronized slides
    5. Radio vision
3. Visual

    1. Projected

    Films, slide, OHP, Model, specimen puppet, real objects, 3D films

    2. Non - Projected aids :

i. Presentation	1. Flash cards 2. Flannel graphs 3. Charts 4. Flip books 5. Maps
ii. Display	1. Bulletin boards 2. Photographs 3. Posters 4. Wall charts 5. Exhibits
iii. Literature	1. Illustration 2. Leaflets 3. Comics

According to contribution to learning

1. According to Edger date

- Selection of audio - visual aids :
- Factors influencing selection of audio - visual aids :
- Audio - visual aids are used singly or in combination, taking the following factors into consideration :

    1. The teaching objective - i.e. the type of behaviour change you want to bring about gaining information, or change attitudes, or learning some skill.
    2. The nature of subject matter being taught.
    3. The nature of audience age level education level, interest experience, knowledge of subject
    4. The size of audience : Flash cards can be used for a small audience only, motion picture, for a large audience.
    5. Relative cost of the various aids - effective aids needs not necessarily be expensive
    6. The teacher
        - The extension worker's familiarity with and skill in using the several aids.
        - This originality and skill in selection, preparation and use of aids.
    7. *The availability* : An effective extension worker makes use of indigenious materials, when the teaching aid he would like to use.

Evaluation of audio - visual materials

1. Do the materials give a true picture of the idea they presents.
2. Do they contribute meaningful content to the topic under study?
3. Is the material appropriate for the age intelligence and experience of the learners?
4. Is the physical condition of the material satisfactory?
5. Do they make learner better thinkers, critical minded?

6. Do they tend to improve human relations?
7. Is the material worth the time and effort involved?

The effective use of A.V. aids PPPE

1. Planning
2. Preparation
3. Presentation
4. Evaluation

1. Planning

- Know clearly the objective of presentation
- Plan well in advance
- Anticipate size of audience so closely as possible
- Plan for the use of a variety of colorful visual aids.
- Determine the appropriate timing for the presentation

2. Preparation

- Prepare by refreshing or previewing in order to make a smooth presentation.
- Select as convenient and as comfortable a meeting place as possible.
- Anticipate need for special lighting or for a lately darkness.
- Make sure that all equipment is in good working order.
- Arrange the audio - visual aids in sequence.
- Keep aids out of sight until actually required for use.

3. Presentation

- Motivate the audience and stress the key points they should observe during presentation.
- Present aids at the right moment and in proper sequence.
- Display only one aid at a time.
- Remove all unrelated material.

- Stand beside the aid, not in front of it.
- Speak facing the audience and not the aids.

4. Evaluation

- In the end evaluate by providing for discussion and application to dispel and discover misunderstanding if any.
- Undertake follow-up studies and observe results.

Research finding about the uses or merits of A.V. aids

Research has shown that audio - visual aids help :

1. The learner to learn faster, more thoroughly remember longer
2. The teacher to organize his teaching material in a systematic order.
3. Clarify ideas being presented
4. Impress ideas more indelibly on the mind vitalize and make teaching more real.
6. Picture experiences outside one's own experience.
7. Combat verbalism
8. Overcame the language barrier
9. Attract attention
10. Arouse and sustain interest
11. Stimulate thinking and vitiate action.
12. Change attitude or point of view.
13. Save time because they intake learning easier and faster.

Disadvantages

1. Learner may sometimes form mistakes and distracted impression.
2. Temptation for the teacher to narrow down his teaching to only a few big ideas.
3. Spectatorism risk.

Theories of communication distortion

1. Theory of communication distortion. States that larger the volumes of communication, larger would be the communication distortion.
2. Theory of communication distortion in transit. Emphasizes that message pass through many layers in an organization therefore repetition and reproduction of the same message is multiplied into many reduces the efficiency of communication.

Theories of communication (MOSS)

1. *Linguistic theory* : Deals with the description of a language system.
2. *Psycholinguistic theory* : Focuses attention the behavior of an individual in the use of language.
3. *Socio-linguistic theory* : Explains patterns of variation in language across different people, stimulate relationship ship, context and competence.

Theories of interpersonal communication

1. Kelnan paradigm
2. Bauers scheme
3. Festinger scheme
4. Attribution theory
5. Game theory
6. Social comparison theory
7. New comb's theory
8. Stimulus response theory
9. Linguistic theory
10. Psycho linguistic theory
11. Socio linguistic theory

In general the theory of interpersonal communication explains it as the face to face interaction between people who are consistently aware of each other. Here the mutual presence of the message generator and receiver is required. Each person assures the rules of sender and receiver of message who are independent and interchangeable. The behavior of participants in interpersonal communication are so intertwined in the process that it is difficult to separate sender from receiver.

Attribute theory : Gives stress to the psychological state, motives and intentions of the individuals involved in communication.

Game theory : View interpersonal communication as the process of transaction between the people to balance possible gains and losses.

Social Comparison theory : Persons communications in order to verify their perceptions

New comb's theory : Explains the thinking process of the who are reaching to one another simultaneously.

Networking System of Information

A networking is the means by which computers store and exchange information and resources access either short distance (LAN) or globally (wide area network) making best use of networking technology is about optimum use of business resources, enhancing productivity and efficiency, reducing costs and gaining competitive advantages.

Teleconferencing

The new integrated digital communication system is being used. This technique will add new dimensions of control timing and flexibility to communication. Instead of traveling miles for a meeting conference rooms will be wired and cameras will be used to carry visuals.

Advantages

1. Time and cost for traditional day to day communication will thus be reduced.
2. It way for international integration.
3. It supports to provide/exchange latest agricultural technologies available in different countries for the farming communities.

Two types

i. Audio conferencing

ii. Video conferencing

Problems of ICT

1. Infrastructure barrier
2. Illiteracy
3. Poverty
4. Policy hurdles
5. Low people participation
6. Lack of will to implementation
7. Heterogeneous adoption
8. Digital divide

Prospects of ICT

1. Marketing information
2. Remote sensing
3. DSS
4. Realizing the knowledge potentials
5. Revival of extension organizations
6. Convergence with agricultural extension
7. Meeting the challenges of globalization
8. Online services - information, education training, monitoring, evaluation.
9. E-commerce
10. R-E-F interaction
11. Question - answer service
12. E-governance
13. Up-to-date information service
14. Database creation
15. Predictive in for - diseases, pest, weather
16. Land record facilitation

Interpersonal communication methods : Basis and key for interaction with others.

Communication takes place in the following ways :

Criteria	Mass Media	Interpersonal
1. The senses they stimulate at a time	Limited to one or two	All senses
2. Opportunity for feed back	Minimum delayed	Maximum immediate
3. Control of the pace	Uncontrollable	Controllable
4. Message codes	Highly verbal	High Non verbal and verbal
5. Multiplicative power	High	Low
6. Direction of message flow	One way	Two way
7. Message accuracy	High	Low
8. Power to preserve message	High	Low
9. Ability to select receiver	Low	High
10. Ability to overcome selectivity and noise	Low	High
11. Ability to meet specialized need	Serves common needs	Serves specialized needs
12. Speed to a large audience	Fast	Slow
13. Possible effects	Knowledge gains	Attitude and action change
14. Neutrality to type of receiver	Not neutral need literate audience or homogenous group	Any type of audience or people

1. *Intrapersonal* - talking with self, as in thinking
2. *Interpersonal* - Face to Face communication between two or more people
3. *Mass Communication* - The use of print, radio, TV and other forms of media in addressing a public 'masses.

Interpersonal Communication : Refers to that situation where the source and receiver can see each other, talk back and forth, interrupt each other, and make responses which both can readily observe.

Non interpersonal : Refers to that particular situation where something (it may be space, time or physical barrier of some sort) is placed between two interacting individuals.

Difference between Interpersonal and Mass Communication

Ingredients/elements of interpersonal communication

1. Effective listening :
   - Wishful thinking
   - Attention
   - Semantic
   - Talk person to person
   - Attitude
   - Get the fact
   - Excessive talking
2. *Empathy* : Sensitivity
3. Self awareness
4. Memory - Short term long term
5. Handling people

Making an impression to others

- Posture
- Look, listen and learn
- Expression
- Appearance
- Speech
- Eagerness to help

Consequences of poor interpersonal communication skills

1. Stress
2. Lack of communication
3. Irritation

4. Close mindedness
5. No team sprit
6. Lack of credibility
7. Poor self esteem
8. Suspicion
9. Loss of productivity
10. Isolation
11. Poor health

Methods of communication interpersonal

(i) Official/formal

1. Radio
2. Doordarshan
3. Newsletter
4. Exhibition
5. Bulletins
6. Tour
7. Written material
8. Poster
9. Meetings
10. Agricultural Unit
11. Local authorities

(ii) Non official/Informal

1. Common talk
2. Members of the family
3. Neighbors and friends
4. Shopkeepers and leaders

Importance of interpersonal communicator

1. Important in individual contact method.
2. Provide homophiles stale of communication
3. Bridge the communication gap.

Communication methods : Oral, written, information technology based.

Empathy

Berlo 1960 "Empathy is the process through which we arrive at expectations, anticipations of the internal psychological states of man"

Theories of empathy

i. *Inference theory* : Each communicator develops a concept of self in a communication situation. He can observe his own physical behaviour directly and correlate his behavior symbolically to his own internal psychological status, feeling, emotions, thoughts etc on the basis of his prior interpretations of self. He makes inferences about the internal states of others.

ii. *Role - taking theory (Mead 1937)* : This theory developed by Mead (1937) argues that the concept of self does not precede communication, rather it is developed through communication. In the case of a child, for example, the infant actually plays other's roles without interpretation. he imitates the behavior of others. Later, on individual plays other people's role with understanding still later. No single theory is sufficient to explain the legality. Generally, human beings utilize both these approaches of empathy for the purpose of communicating with others.

Empathic accuracy

1. As group increases, empathic accuracy decreases
2. When prior communication is minimal empathic accuracy is less
3. When we are insensitive to the behavior of others empathic accuracy is less.
4. Empathic accuracy also decreases when we are not motivated in a communication encounter.

"So a knowledge of the composition and working of a social system is useful in making predictions about new members of that systems will behave in a given communication situation".

Interaction : Term interaction refers to mutual Berlo, "1960" and reciprocal influencing of each other's behavior. The term interaction names the process of reciprocal role taking, the mutual performance of empathic behaviors. If two individuals make inferences about their own rules and take the role of others same time, and if their communication behaviour depends on reciprocal taking of roles then they are communicating by interacting with each other.

All communication is not interactional; however, to the extent we are in an interactional situation, our effectiveness, our ability to affect and to be affected by others, increases. Thus, 'we can define interaction as the ideal of communication, the goal of human communication' "Berlo, 1960".

Heterophily/Homaphily

According to Rogers and Shoemaker (1971)

Heterophily is the degree to which pairs of individuals who interact differ in certain attributes, such as beliefs, values, education, social status and so forth.

Effective communication

Relationship in communication

1. Physical interdependence/feedback
2. Action-reaction interdependence
3. Empathy
4. Interaction, and
5. Heterophily
6. Homopily

Social Networks in communication

Configuration of network has a direct effect on communication behavior.

STAR

1. Four person network
2. Five person networks

Mass communication : Communication which involve large, heterogeneous and anonymous audience".

Nature of mass audience directed towards relatively large, heterogeneous and anonymous audience rapid, public and transient heterogeneous audience.

- Large
- Heterogeneous
- Anonymous

Nature of Mass Communication/Communication

1. It is directed towards a relatively large, heterogeneous and anonymous audience.
2. Messages are transmitted publically.
3. It is often timed to reach most audience members simultaneously.
4. It is transient in characters.
5. The communicator tends to operate within a complex organization
6. It may involve great expense.
7. It was a mechanism that is capable of reproducing a large no. of identical messages, almost simultaneously and is able to transmit within a relatively short period to an audience that may even be physically separated in terms of space.
8. These are new technically, based geographical areas which provides, new innovation and training and material related to them.

Technology Parks/Rural Tech Parks

Also known as science parks agricultural technology parks

Main objective : To promote Agricultural food processing, Industries in clusters in areas where there is predominant production of processible agriculture & horticulture products.

Specific objectives

Quick information dissemination for farmer's school children and general public A place to see and learn latest recommendations of crop cultivation and technologies generated by both public and private institutions.

1. A place to see and study heritage agriculture system and future high tech agriculture. Education and training mechanism for village level extension workers as face to face training of widely dispersed large number is practically impossible task.

2. Establishment of live agriculture crop museum
3. Information network among agriculture officers, government and agriculture institutes
4. Aesthetically beautiful landscaping.
5. A site for eco tourism/Agro tourism
6. Generation of funds within the system for its sustainability say sustainer
7. One stop shop - sale centre for plants, seeds and other products of government institution and the private sector entrepreneurs of instructional media material information centre etc.

* **Main elements of A Parks/T. Parks**

1. Venue : Should be at the site of majority of agril related institution.
2. Main sections and presentations rice, section, vegetable garden, roots our garden, fruit garden, dry zone crops garden.
3. Exhibition stalls or main crop categories
4. Nursery for demonstration and sales
5. Mushroom house
6. Food tech. demonstration
7. Floriculture demonstration and sales area for private sector
8. A site to demonstrate traditions/indigenous agriculture in India - again agriculture.
9. A site to demonstrate future high tech agriculture
10. Area of student for discovery learning
11. Meteorology
12. Animal production and health
13. Export agril crops
14. Sales centre
15. Information centre
16. Food courts
17. Resting areas

Policing initiative : Developing of world-class infrastructure for agro food processing industries through establishment of agro food tech parks.

These are the state technology furnished locations which provide info, being, R & D and other services related to new area

Agriculture and Horticulture farms which are located within the radius of 80-100 Km from the proposed food agro technology parks would be linked to these parks.

One of the major thrust given by the India for development of world infrastructure for Agro food industries is establishment of "Agro Food Technology Parks" for which financial assistance of Rs 4 Crore per park is available from GOI.

Common facilities and infrastruce : R & D, were housing, cold storage, quality assurance labs, common effluent treatment plants.

Advantages

- RPT - Proposal to establish
- RPP initiated in 1992
- Debated over foodies
- 1995 Min. of RD accepted proposal
- 10 crore budget estimates area submitted in place.
- Collorborative offer betweenCAPART

Scope for TOT through Technology demonatization Key communication - opinion leader :

"Key communicator or those are persons in any social system who are sought out for information and advice on general or specific topics"

Rogers and Kincaid (1981) define opinion leadership as the degrees to which on individual is able to informally influence others knowledge attitude, or overt behavior in a desired way with relative frequency"

Concept was first developed by **Lazars field (1944)**

Identification methods

i. Sociometric method
ii. Self designation technique
iii. Information ratings

Characteristics

i. External communication

ii. Accessibility

iii. Social status

iv. Innovativeness

v. Relationship to the social systems

Role

1. Indirect communication of innovations in a social system.
2. Legitimization of new ideas for diffusion into the system.
3. Self adoption of a relatively greater number of innovations.
4. Help in securing services and supplies to the members in a social system.

Characteristics affecting their role

i. Social participation

ii. Education

iii. Value orientation

iv. Socio-economic status

ATIC : "Agriculture Technology Information Centre"

The ATIC is a "single window' support system linking the various units of a research institution with intermediary users and end-users (farmers) in decision making and problem solving exercise".

The rationale for establishment of ATIC are :

1. To provide diagnostic service for soil and water testing, plant and livestock health.
2. To supply research products such as seeds and other planting materials.
3. Providing information through published literature and communication material.
4. Providing an opportunity to the institutes, SAUS to generate some resources through the sale of their technologies.

The important criteria for agriculture technology information are:

1. Availability of new technologies
2. Relevance of new technologies
3. Responsiveness of new technologies
4. Sustainability of such unit within the overall institutional system.

Target/users of information

1. Farmers
2. Farmers - entrepreneurs
3. Extension workers and development
4. NGO's and
5. Private sector organizations

Objectives

1. To provide a single window delivery for producers and services.
2. To facilitate/dissect the farmers access to the institutional resources.
3. To provide mechanism for feedback from the users to the institute.

The information services and supply include :

1. Soil and water sample testing facility.
2. Plant clinic and diagnostic centre.
3. Rhizobium culture.
4. IPM - organic and biopesticide/NPY
5. Seed and planting material, small implements
6. Fertilizer quality testing
7. Tissue culture and plant materials
8. Farm literature - leaflets, pamphlets, journals
9. Audio, video cassette on crops
10. Video show
11. Processed products - cereals milk, meat job

12. Cafeteria
13. Technology park (display) exhibition
14. Veterinary animal clinical services.
15. Poultry strains, livestock breeds, seed etc.

Management structures of ATIC

PMC

TDMC

PIU DDGAG Extn - TDU

Zonal coordinator

Director of Institute/Director of extension of SAU

ATIC Management committee

E-governance : E-governance is defined by M Bakes (Bakus, M. 2001), "Application of electronic means in

a. Interaction between government and citizen and between government and business as well as.

b. In internal government options to simplify and improve democratic, government and business aspects of governance."

"E-governance is about transforming the relationship between governments and their citizens through the use of information technologies".

Benefits - Direct

- Faster, easier more convenient service convenience, near to home, more services at one place.
- Faster processing shorter wait shorter queries at government offices.
- Less number of trips to government officer.
- Better interaction environment no harassment.
- Reduced transport cost.
- Avoids wage loss for daily wage earners
- Better quality service.
- More accurate and legible documents, easy recovery from errors.

- Lesser corruption increased transparency.
- Improved access to offices.

Indirect/Derived advantages of E-governance

1. Improved complaint handling, complaint tracking
2. Introduces transparency in data, decision/actions, rules procedures and performance of government agencies.
3. Automated processes to take away discretion.
4. Reduced effectiveness in decision making.
5. Entry point for simplification rules and engineering process.
6. Makes decision traceable takes actions.
7. Builds accountability
8. Documentation to citizens to follow-up.
9. Tracks procedural, human delay
10. Easier to fix responsibility accountability
11. Modularize tasks making outsourcing possible
12. Introduces completion among delivery channels and departments.
13. Standardized documentation of common of objects leads to effective supervision through comparative indicators.
14. Centralizes data for better audit and analysis integration of data across application.

Critical success factors

1. Political will
2. Overall vision and national strategy
3. Full ownership and support of cto.
4. Effective and appropriate project.
5. Management
6. Access
7. Education of clients

8. Participation of clients
9. Culture
10. Cooperation and collaboration with private sector.
11. Technology
12. Human Resources
13. Legal framework
14. Finances

Components of E-governance

1. Political
2. Technological
3. Social
4. Cultural
5. Psychological
6. Service

Interaction among/or ecology of E-governance various components.

Stages of E-governance

Degree of progress made by government towards the ultimate e-governance goal.

a. Presence
b. Interaction
c. Transaction
d. Transformation

According to Layne and Lee (2001)

1. Cataloguing
2. Transaction
3. Vertical integration
4. Horizontal integration

Cyber Extension

Initiatives in India and other developing countries cyber extension can be defined as the extension over cyber space'. As the word extension is subjected neutral, so is cyber extension. But in the applied context of agriculture, cyber extension means". Using the power of online computer networks with the help of communication channels to deliver the content in the form of text, graphic, audio and video either passively or interactively to facilitate dissemination of agricultural technology".

Cyber extension tools

1. E-mail
2. Interactive expert systems on crop pest and disease.
3. Internet processing for extension information.
4. Video conferencing
5. Call centers, SATCOM Networks
6. Discussion groups and news groups

The broad classification of ICT initiatives currently in place :

1. *Central government initiatives :* to provide connectivity to extension system NATP / ITD component, ICAR-extentsion component under NATP, MIC initiative in Kolhapur, Sangli district of Maharashtra.
2. *State initiative :* MH government to promote info kiosk, Kerala government Akshaya project to connect village panchayats.
3. NGOs project : information village project in Pondicherry by MSSRF Chennai. Rural connectivity and allied services providers Drishtee, N Logue.
4. Private and cooperative sector transaction related initiatives such as there of ITC, Mahindra, EDD.
5. Pure information services providers such as kisan, community India, Agriwatch, India agro net etc.
6. Multiple rural services which include agriculutral extension such as WARNA, Gyandort, Taranhaat etc.
7. Knowledge networks and knowledge banks, Honey Bee Network, Harit Gyan, Indian society of AB professional detailing

CE current benefits

- CE is now an established complementary and supplementary mechanism of agril extension. All Indian State Agriculture universities included cyber extension in agriculture extension as an integral part of their undergraduate and post graduate curriculum.
- The GOI indicated to promote the role cyber extension in the reform of extension system.
- The launch of KCC (Kisan Call Centre) has added a new dimension in to the Indian extension system.
- It can be safely concluded that the Indian agriculture is on threshold of a new communication mechanism.
- Cyber extension will further improve the international competitiveness and penetration of Indian agricultural products, by providing timely information across to farming community of supply side and the prospective consumers on the demand side.

Cyber extension

The process

- Develop state-of-the-art ICT infrastructure of connect key stack holders.
- Creating ICT awareness in all development departments.
- Create information packaging mechanism at key participating agencies.
- Network with e.g. initiatives of concerned state/district.
- Create a model cell in each state to monitor the progress of CE.
- Identify a nodal connectivity agency for "cyber extension" in India.

Services provided/benefits of CE

- Providing interaction among S-E-F and other rural people.
- Providing upto date news and information service.
- A question and answer service.

- Creation and maintenance of statistical database.
- Providing the details of poverty alleviation schemes on the internet.
- Providing states of various govt. programmes and details about their implementation mechanism.
- Hosting website by major institutions participating in the agricultural extension.
- Launching online rural development and extension journals.
- Providing internet access at district and block level agricultural and rural development offices.
- Opening of cyber cafes to enable educated rural people and extension workers.
- Providing maps that display different features such as population density crops planted.
- Providing video clips to demonstrate complex procedures.
- Providing mechanism of user/beneficiary feedback for the public sector scheme.

E-learning

Electronic learning is a type of technology supported education/learning (TSL) where the medium of instruction is through computer technology, particularly involving digital technologies. E-learning has been defined as pedagogy empowered by digital technology. In some instances no face-to-face interaction takes place.

E-learning services have evolved computers first used in education there is a trend to move toward learning services, where computer based activities are integrated with practical or classroom situation.

Pedagogical Approaches or Perspectives

It is possible to use various pedagogical approaches for e-learning which includes instructional design. The traditional pedagogy of instruction which is curriculum focused and is developed by a centralized educating group or a single teacher.

1. Social Constructivist
2. Cognitive perspective
3. Emotional perspective
4. Behavioral perspective
5. Contextual perspective

Benefits of e-learning versus traditional classroom setting

1. Reducing environmental impact
2. Quality education, made affordable
3. Convenience and flexibility to learners

Communication technologies used in e-learning

Communication technologies are generally categorized as synchronized or a synchronous.

A synchronized activities use technology such as blogs, wikis and discussion bonds.

Synchronous : Involve the exchange of ideas and information with one or more participants during the same period of time. A face to face discussion is an example of synchronous communication.

Virtual classrooms and meetings can often use a mix of communication technologies.

Goals of e-learning : E-learning lessions are generally designed to guide students perform in specific tasks. Information based learning content communicates info to the students. Examples include content that distributes the history or facts related to a service, company or product. Information based content there is no specific skill to be learned. In performance based content to lesions build off a procedural skill in which the student is expected to increase proficiency.

Theories of mass communication (Message flow models)

1. Hypodermic needle model
2. One step flow model
3. Two step flow model
4. Multistep flow model
5. Theory of mass media effects

6. Theory of individual differences
7. Psychodynamic model of persuasion process
8. Theory of social categories
9. Theory of social relationship
10. Socio-cultural model of persution process
11. Free-press theory
12. Social-responsibility theory
13. Development media theory
14. Democratic participant media theory

Psychodynamic model : Explains about the relationship between internal processes and manifest over behavior of persuader.

Theory of social category : Emphasis that people who have a number of similar characteristics are likely to have similar orientation and behavior resulting in fairly uniform response to mass media.

"These are the Information Technology enabled deupces established to provide selevant information to the formers about various aspects of agriculture, Dairing Marketing".

Information Kiosk

Features

1. Touch screen
2. Smarter buying and selling dicisions
3. Increases farmer awareness
4. Printing on demand options for information kiosks
5. Self service kiosks reduce head count by lowering employee head count
6. Reduces waiting in line by farmers
7. Virtual sale assistance increases sale

Phases/steps

1. Designing
2. Installing
3. Evaluating

Location : KVK/Veterinary centre/

Purpose : Easy access to necessary information

Goal : To strengthen the agriculture and irrigation sector for increase in agriculture production.

Facilities

1. IT
2. Training

Information available at information Kiosk/Centre

1. Project related activity
2. Printed material
3. Irrigation networks
4. Rule, regulation and agreement
5. Internet connection
6. Data about the availability of water at barrage.
7. Information regarding laser land leveling
8. A website of their own

"It is a system / device which assists in decion making in an organisation for efficient management according to the inputs provided".

MIS

Evolution 1990

1. Electronic data processing
2. Computer based information system
3. Decision support system
4. Artificial intelligence

Uses of MIS on different levels of management

1. Top management
2. Middle Management
3. Lower Management

Design of MIS

1. *Assessing* : Information needs for planning and monitoring.
2. *Deciding* : Deciding the levels of information.
3. *Ensuring* : System flexibility and adaptability.

Guidelines for effective design of MIS

1. User participation
2. Cost evaluation
3. Alternative for in-house software development
4. Relevance and selectively of information
5. Pretesting of the system
6. Training the operators and users carefully

Classification of MIS

Mason and Swanson 1981 describe four categories of management information system :

1. Database information system
2. Predictive information system
3. Decision Making information system
4. Decision Taking information system

Roles of MIS

1. To plan the most effective allocation of resources.
2. To choose between alterative course of action.
3. To control day to day operations.

A typical functional structure of ATIC

(A) Products	**(B) Services**	**(C) Information**
seeds plant materials livestock species tools equipments value added products other agro products	soil and water testing plant clinics vety clinic testing and caliberation of small equipments	exhibition audio, video, aids, farm literature

Farmers and other end users

MAKS/AKIS

Rolling and Engel 1991 : AKIS is a set of organization and or person and the link and interaction between them that are engaged in or managed such process on the anticipation, generation transformation, transmission, storage, retrieval, integration, diffusion, utilization of agricultural knowledge and information which potentially work systematically to support decision making, problem solving an innovation in agriculture.

Elements

1. Agriculture research
2. Agriculture education
3. Organization
4. Rural farming community
5. Mass media

Models of AKIS

1. The INTERPAKS MODEL SWANSON 1985

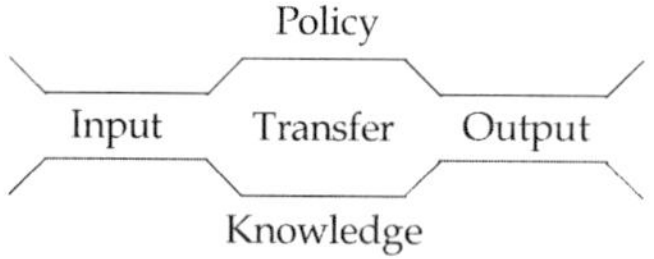

2. Elliot's ATMS model : 1987

Functions/importance of MAKS/AKIS : By Nigel 1980

1. Identification of the problem/Knowledge needs of the procedures level.
2. Generation of innovation
3. Validation under farmer's condition
4. Operationalisation for utilization
5. Dissemination
6. Utilization
7. Evaluation of experience

Expert System

"Donald 1999 Expert system is sophisticated computer programmes that manipulate knowledge to solve problems efficiently and effectively in a narrow problem area".

Steps in desiqring MIS

1. Planning
2. Designing
3. Selection
4. Diagnosis
5. Interpretation
6. Prediction
7. Decision technologies

Components

1. Knowledge base
2. Inference Engine
   i. Terminal graphic representation
   ii. Multiple character windows
   iii. Multiple graphic windows

Procedure : For development

1. Preparation of knowledge base.
2. Preparation of flow chart
3. Data entry into the programme module and computing of knowledge base where expert system are most effective.
4. DSS
5. Problem solving/ diagnosis
6. Data analysis
7. Customer/product support
8. Background monitoring

Application in agricultural extension

1. Decision
2. Advice
3. Technology transfer
4. Information holding and maintaining

Demerits

1. Lack of common sense
2. Do not make creative responses

3. Domain expert do not always able to explain their logic and reasoning.
4. Cannot adapt to changing environment unless knowledge base is changed.

Important examples of different expert systems

- AGWATER
- CROPWAT
- CIIMWAT

Development communication

The art and science of human communication linked to societies planned transformation from state of poverty to one of dynamic socio economic growth that maker for greater equity and larger unlading of individual potential.

Approaches

1. Information dissemination and education.
2. Behavioral change
3. Social marketing
4. Social mobilization
5. Media advocacy
6. Communication for social change
7. Participating development communication.

e-Choupal

e-Choupal is an initiative of ITC Limited, a large multi business conglomerate in India, to link directly with rural farmers via the Internet for procurement of agricultural and aquaculture products like soybeans, wheat, coffee, and prawns. e-Choupal was conceived to tackle the challenges posed by the unique features of Indian agriculture, characterized by fragmented farms, weak infrastructure and the involvement of numerous intermediaries. The programme involves the installation of computers with Internet access in rural areas of India to offer farmers up-to-date marketing and agricultural information.

Problems addressed

Traditionally, commodities were procured in *mandis* (major agricultural marketing centres in rural areas of India), where the middleman used to make most of the profit. These middlemen used unscientific and sometimes outright unfair means to judge the quality of the product to set the price. The difference in price between good quality and inferior quality was little, and therefore there was no incentive for the farmers to invest and produce good quality output. With e-Choupal, the farmers have a choice and the exploitative power of the middleman is neutralised.

Statement of a farmer who is a member of e-choupal : "Before ITC introduced us to e-Choupal, we were restricted to selling our produce in the local *mandi*. We had to go through middlemen and prices were low. ITC trained me to manage the Internet kiosk and I became the e-Choupal *Sanchalak* in my village. Today we are a community of e-farmers with access to daily prices of a variety of crops in India and abroad - this helps us to get the best price. We can also find out about many other important things - weather forecasts, the latest farming techniques, crop insurance, etc. e-Choupal has not only changed the quality of our lives, but our entire outlook.

Effects of e-Choupal

ITC Limited has now provided computers and Internet access in rural areas across several agricultural regions of the country, where the farmers can directly negotiate the sale of their produce with ITC Limited. This online access enables farmers to obtain information on mandi prices, and good farming practices, and to place orders for agricultural inputs like seeds and fertilizers. This helps farmers improve the quality of their products, and helps in obtaining a better price. Each ITC Limited kiosk having Internet access is run by a sanchalak — a trained farmer. The computer is housed in the sanchalak's house and is linked to the Internet via phone lines or by a VSAT connection. Each installation serves an average of 600 farmers in the surrounding ten villages within about a 5 km radius. The sanchalak bears some operating cost but in return earns a service fee for the e-transactions done through his e-Choupal. The warehouse hub is managed by the same traditional middle-men, now called samyojaks, but with no exploitative power due to the reorganisation. Indeed these middlemen make up for the lack of infrastructure and fulfill critical jobs like cash disbursement, quantity aggregation and transportantion.

Since the introduction of e-Choupal services, farmers have seen a rise in their income levels because of a rise in yields, improvement in quality of output, and a fall in transaction costs. Even small farmers have gained from the initiative. Customized and relevant knowledge is offered to the farmers despite heterogeneous cultures, climates and scales of production. Farmers can get real-time information despite their physical distance from the *mandis*. The system saves procurement costs for ITC Limited. The farmers do not pay for the information and knowledge they get from e-Choupals; the principle is to inform, empower and compete. At the same time ITC Limited has obtained benefits from the programme :

1. Elimination of non value added activities
2. Differentiated product through identity preserved supply chains
3. Value added products traceable to farm practices
4. E-market place for spot transactions and support services to futures exchange

There are presently 6,500 e-Choupals in operation. ITC Limited plans to scale up to 20,000 e-Choupals by 2012 covering 100,000 villages in 15 states, servicing 15 million farmers.

Success that followed

Today this initiative of ITC has empowered the lives of people living in 10 states where 40000 villages have 6500 e choupals and around 4 million farmers have been empowered. The initiative taken up by ITC to help the degrading reputation of a farmer has surely made a huge to difference to not only large but also medium and small scale farmers.

Milestones

Commencement of initiative : 2000

- States covered : 9
- Villages covered : 40,000
- e-Choupal installations : 6500
- Empowered e-farmers : 4 million

Through the e-Choupal initiative, ITC aims to confer the power of expert knowledge on even the smallest individual farmer. Thus enhancing his competitiveness in the global market.

The immense potential of Indian agriculture is waiting to be unleashed. The endemic constraints that shackle this sector are well known–fragmented farms, weak infrastructure, numerous intermediaries, excessive dependence on the monsoon, variations between different agro-climatic zones, among many others. These pose their own challenges to improving productivity of land and quality of crops. The unfortunate result is inconsistent quality and uncompetitive prices, making it difficult for the farmer to sell his produce in the world market.

ITC's trail-blazing answer to these problems is the e-Choupal initiative; the single-largest information technology-based intervention by a corporate entity in rural India. Transforming the Indian farmer into a progressive knowledge-seeking netizen. Enriching the farmer with knowledge; elevating him to a new order of empowerment.

Effects of e-choupal

e-Choupal delivers real-time information and customized knowledge to improve the farmer's decision-making ability, thereby better aligning farm output to market demands; securing better quality, productivity and improved price discovery. The model helps aggregate demand in the nature of a virtual producers' co-operative, in the process facilitating access to higher quality farm inputs at lower costs for the farmer. The e-Choupal initiative also creates a direct marketing channel, eliminating wasteful intermediation and multiple handling, thus reducing transaction costs and making logistics efficient. The e-Choupal project is already benefiting over 3.5 million farmers. By 2012, the e-Choupal network will cover over 100,000 villages, representing 1/6th of rural India, and create more than 10 million e-farmers.

A digital transformation

ITC began the silent e-revolution of rural India with soya growers in the villages of Madhya Pradesh. For the first time, the stereotype image of the farmer on his bullock cart made way for the e-farmer, browsing the e-Choupal website. Farmers now log on to the site through Internet kiosks in their villages to order high quality agri-inputs, get information on best farming practices, prevailing market prices for their crops at home and abroad and the weather forecast – all in the local language. In the very first full season of e-Choupal operations in Madhya Pradesh, soya farmers sold nearly 50,000 tons of their produce through the e-Choupal Internet platform, which has more than doubled since then. The result marks the beginning of a transparent and cost-effective marketing channel. Bringing prosperity to the farmers' doorstep.

Linking farmers to remunerative markets

Farmers grow wheat across several agro-climatic zones, producing grains of varying grades. Though these grades had the potential to meet diverse consumer preferences, the benefit never trickled down to the farmers, because all varieties were aggregated as one average quality in the mandis. Enter ITC's e-Choupal intervention. The e-Choupal site is now helping the farmers discover the best price for their quality at the village itself. The site also provides farmers with specialized knowledge for customizing their produce to the right consumer segments. The new storage and handling system preserves the identity of different varieties right through the 'farm-gate to dinner-plate' supply chain. Encouraging the farmers to raise their quality standards and attract higher prices.

Managing risks through technology

The and ifs in the aqua farmers' life posed daunting odds. They were haunted by the nightmare of contaminated soil, wrong levels of salinity in the water or the killer White Spot virus, any of which could wipe out an entire shrimp crop, until the e-Choupal site provided them the support and the know-how to cope with and manage such risks. Information equips farmers with comprehensive know-how to keep abreast of food safety norms to compete in the international market. Information includes parameters for antibiotic usage, hygienic washing, sanitized dressing and air-tight packing. All these factors help to neutralize the risks involved in aqua farming. Making it economically much more attractive, benefiting hundreds of aqua farmers.

A dependable knowledge partner

Coffee planters in India have for years been tossed between the highs and lows of the international coffee market. The information needed to manage risks in the volatile global coffee market, price updates and prevalent trends in coffee trading were just not available to them. Launch of e-Choupal.com has equipped India's coffee planters with appropriate knowledge base and risk management tools. The site arms them with the latest prices posted on commodity exchanges like CSCE in New York and LIFFE in London. Planters have access to technical analysis by experts to help them comprehend trends, trading ranges and chart patterns in simple language. 'Parity Chart' and the 'Calculator' on the site convert the coffee prices quoted in international auctions into raw coffee equivalent for the benefit of the small growers in India. Traders net, a special link on the site, brings together a large number of coffee planters, traders and roasters, creating a virtual market for transparent price discovery. ITC empowers Indian coffee growers with expert knowledge in logistics and risk management, thereby enabling them to face global competition.

A quiet digital revolution is reshaping the lives of farmers in remote Indian villages.

In these villages, farmers grow soybeans, wheat and coffee in small plots of land, as they have for thousands of years. A typical village has no reliable electricity and has antiquated telephone lines. The farmers are largely illiterate and have never seen a computer. But farmers in these villages are conducting e-business through an initiative called e-Choupal, created by ITC, one of India's largest consumer product and agribusiness companies".

❑❑❑

Chapter 5

Management

Introduction

One of the most important activities in business is the management of the 4M's – men, machines, material and money. The term 'management' can be interpreted differently in different contexts. Hence, it is difficult to define. In one context, it may comprise the activities of executives and administrative personnel in an organization, while in another, it may refer to a system of getting things done. In a broad perspective, management can be considered as the proper utilization of people and other resources in an organization to accomplish desired objectives. With increasing global competition, changes in the world of technology, changing business practices and increasing social responsibility of organizations, the role of managers has become all the more significant.

In this chapter, we will first examine the definitions of management given by some eminent management thinkers to understand the essence of management. Secondly, we will discuss the five basic functions of management i.e., planning, organizing, staffing, leading and controlling. The chapter also focuses on the managerial skills required at various levels of the organizational hierarchy and briefly explains the various approaches to management.

Definitions of Management

The term 'management' can be interpreted in a variety of ways. To gain a better insight into the nature of management, let us look at some of the definitions of management:

According to Koontz and Weihrich (1988)

Management is "the process of designing and maintaining an environment in which individuals, working together in groups, accomplish efficiently selected aims."

Boone and Kurtz

Defined management as "the use of people and other resources to accomplish objectives."

McFarland

Defined management as "a process, by which managers create, direct, maintain, and operate purposive organizations through systematic, coordinated, cooperative human effort."

According to Mary Parker Follet

Management is "the act of getting things done through people."

Definitions by Follet and Louis E. Boone and Kurtz call attention to the fact that managers achieve organizational goals by getting others to do the necessary tasks.

The other two definitions suggest that management is much more than "just getting the work done" and suggest the following aspects of management:

1. Managers carry out the functions of planning, organizing, staffing, leading and controlling: Henry Fayol was the first management thinker to outline the five basic functions carried out by managers. Every manager performs these basic functions.
2. Management is essential to any kind of organization: Wherever there are groups of people working together to achieve some common objectives, it becomes essential to guide, organize and control them. The term 'management' applies to any organization irrespective of the size or nature of operations.

3. Management is essential at all hierarchical levels: Management is necessary at all levels. However, the type of skills and the degree to which various skills are required at different levels of the hierarchy may vary. In order to perform their duties satisfactorily, managers need technical, human, conceptual and design skills.
4. The goal of all managers is to generate surplus: The aim of all business managers is to create a surplus. To accomplish this objective, the manager has to create an environment which encourages people to accomplish as much as possible with the least amount of resources and personal dissatisfaction. Even in non-profit organizations, the aim of managers is to accomplish their goals with the minimum amount of resources or to make as much surplus as possible with available resources.

The aim of all managers is to improve productivity, efficiency and effectiveness: Productivity is defined as "the output-input ratio within a time period with due consideration for quality." It can be expressed as:

Productivity = Outputs / Inputs (within a time period, quality considered)

Productivity can be improved in the following ways:

- By producing more output with the same inputs.
- By reducing inputs, but maintaining the same level of outputs.
- By increasing outputs and reducing inputs, thereby, making the ratio more favorable.

Productivity can be improved by ensuring efficiency and effectiveness in the operations of the firm. Effectiveness refers to achievement of stated organizational objectives while efficiency denotes the judicious use of resources to achieve organizational objectives. In the words of Peter Drucker, efficiency means "doing things right", while effectiveness means "doing the right things." In his book, "Management - Tasks, Responsibilities, Practices", Drucker states that effectiveness is the foundation of success whereas efficiency is a minimum condition for survival after success has been achieved.

The Role of Management

As mentioned earlier, managers perform five functions - planning, organizing, staffing, leading and controlling. Since these functions are very essential for effective management, they have been used as the basic framework for this book. They have been briefly explained in the next section of the chapter.

In order to understand the role of management, in the late 1960s, Henry Mint berg devised a new approach - the managerial roles approach - by observing what managers actually do. He did a careful study of five chief executives at work and found that they were involved in a number of varied, unpatented activities of short duration. Using a method called structured observation, Mintzberg isolated ten roles which he believed were common to all managers. As shown in Table 1.1, these ten roles were grouped into three categories - interpersonal roles, informational roles and decisional roles.

A manager is required to interact with many people, both within and outside the organization and hence, the need to perform interpersonal roles. The three interpersonal roles of a manager are figurehead, leader and liaison. In his role as a figurehead, a manager performs all the ceremonial or symbolic duties. Example, it would be the duty of a college dean to award diplomas at the convocation ceremony. In the leadership role, a manager is required to motivate the employees to perform at their best to achieve the company's objectives. In the liaison role, a manager is required to interact with people both within and outside the organization.

A manager acts as a channel of information within the organization. The three informational roles of a manager are that of a recipient, disseminator and spokesperson. In the role of a recipient, a manager receives information pertaining to changes, opportunities and problems that the organization may face. As a disseminator, a manager provides information to subordinates that would influence their performance at work. And finally, a manager performs the role of a spokesperson when he or she represents the organization in public.

Table 1 : Mintzberg's 10 Managerial Roles

1. Interpersonal	Figurehead	Performs ceremonial and symbolic duties such as greeting visitors, signing legal documents.
	Leader	Direct and motivate subordinates, training, counselling, and communicating with subordinates.
	Liaison	Maintain information links both inside and outside organization, use mail, phone calls, meetings

2. Informational	Monitor	Seek and receive information, scan periodicals and reports, maintain personal contacts
	Disseminator	Forward information to other organization members, send memos and reports, make phone calls
	Spokesperson	Transmit information to outsiders through speeches, reports, memos
3. Decisional	Entrepreneur	Initiate improvement projects, identify new ideas, delegate idea responsibility to others
	Disturbance Handler	Take corrective action during disputes or crises; resolve conflicts among subordinates; adapt to environmental crises
	Resource Allocator	Decide who gets resources, scheduling, budgeting, setting priorities
	Negotiator	Represent department during negotiation of union contracts, sales, purchases, budgets, represent departmental interests

Principles of management Fayol outlined fourteen

1. *Division of work*: Work specialization results in improving efficiency of operations. The concept of division of work can be applied to both managerial and technical functions.

2. *Authority and responsibility*: Authority is defined as "the right to give orders and the power to exact obedience." Authority can be formal or personal. Formal authority is derived from one's official position and personal authority is derived from factors like intelligence and experience. Authority and responsibility go hand-in-hand. When a manager exercises authority, he should be held responsible for getting the work done in the desired manner.

3. *Discipline*: Discipline is vital for running an organization smoothly. It involves obedience to authority, adherence to rules, respect for superiors and dedication to one's job.

4. *Unity of command*: Each employee should receive orders or instructions from one superior only.

5. *Unity of direction*: Activities should be organized in such a way that they all come under one plan and are supervised by only one person.

6. *Subordination of the individual interest to the general interest*: Individual interests should not take precedence over the goals of the organization.
7. *Remuneration*: The compensation paid to employees should be fair and based on factors like business conditions, cost of living, productivity of employees and the ability of the firm to pay.
8. *Centralization*: Depending on the situation, an organization should adopt a centralized or decentralized approach to make optimum use of its personnel.
9. *Scalar chain*: This refers to the chain of authority that extends from the top to the bottom of an organization. The scalar chain defines the communication path in an organization.
10. *Order*: This refers to both material and social order in organizations. Material order indicates that everything is kept in the right place to facilitate the smooth coordination of work activities. Similarly, social order implies that the right person is placed in the right job (this is achieved by having a proper selection procedure in the organization).
11. *Equity*: All employees should be treated fairly. A manager should treat all employees in the same manner without prejudice.
12. *Stability of tenure of personnel*: A high labour turnover should be prevented and managers should motivate their employees to do a better job.
13. *Initiative*: Employees should be encouraged to give suggestions and develop new and better work practices.
14. *Espirit de corps*: This means "a sense of union." Management must inculcate a team spirit in its employees.
15. *Test of Validity & Predictability* - Validity of scientific principles can be tested at any time or any number of times i.e. they stand the test of time. Each time these tests will give same result. Moreover future events can be predicted with reasonable accuracy by using scientific principles. E.g. H_2 & O_2 will always give H_2O.

    Principles of management can also be tested for validity. E.g. principle of unity of command can be tested by comparing two persons - one having single boss and one having 2 bosses. The performance of 1st person will be better than 2nd.

It cannot be denied that management has a systematic body of knowledge but it is not as exact as that of other physical sciences like biology, physics, and chemistry etc. The main reason for the inexactness of science of management is that it deals with human beings and it is very difficult to predict their behaviour accurately. Since it is a social process, therefore it falls in the area of social sciences. It is a flexible science & that is why its theories and principles may produce different results at different times and therefore it is a behaviour science. Ernest Dale has called it as a Soft Science.

Management as a Process

As a process, management refers to a series of inter - related functions. It is the process by which management creates, operates and directs purposive organization through systematic, coordinated and co-operated human efforts, according to George R. Terry, "Management is a distinct process consisting of planning, organizing, actuating and controlling, performed to determine and accomplish stated objective by the use of human beings and other resources". As a process, management consists of three aspects:

1. *Management is a social process* - Since human factor is most important among the other factors, therefore management is concerned with developing relationship among people. It is the duty of management to make interaction between people - productive and useful for obtaining organizational goals.
2. *Management is an integrating process* - Management undertakes the job of bringing together human physical and financial resources so as to achieve organizational purpose. Therefore, is an important function to bring harmony between various factors?
3. *Management is a continuous process* - It is a never ending process. It is concerned with constantly identifying the problem and solving them by taking adequate steps. It is an on-going process.

Functions of management

Management consists of the functions given below. It is based on Henri Fayol's thinking on the functions of management.

1. *Planning* : generating plans of action for immediate, short term, medium term and long term periods.

2. *Organizing*: organizing the resources, particularly human resources, in the best possible manner.
3. *Staffing*: positioning right people right jobs at right time.
4. *Directing (includes leading, motivating, communicating and coordinating)*: Communicate and coordinate with people to lead and enthuse them to work effectively together to achieve the plans of the organization.
5. *Controlling (includes review and monitoring):* evaluating the progress against the plans and making corrections either in plans or in execution.

Each of these functions is explained in some detail below.

1. Planning

- Planning is decision making process.
- It is making decisions on future course of actions.
- Planning involves taking decisions on vision, mission, values, objectives, strategies and policies of an organization.
- Planning is done for immediate, short term, medium term and long term periods.
- It is a guideline for execution/implementation.
- It is a measure to check the effectiveness and efficiency of an organization.

2. Organizing

- Organizing involves determination and grouping of the activities.
- Designing organization structures and departmentation based on this grouping.
- Defining the roles and responsibilities of the departments and of the job positions within these departments.
- Defining relationships between departments and job positions.
- Defining authorities for departments and job positions.

3. Staffing

- It includes manpower or human resource planning.
- Staffing involves recruitment, selection, induction and positioning the people in the organization.
- Decisions on remuneration packages are part of staffing.
- Training, retraining, development, montoring and counselling are important aspects of staffing.
- It also includes performance appraisals and designing and administering the motivational packages.

4. Directing

- It is one of the most important functions of management to translate company's plans into execution.
- It includes providing leadership to people so that they work willingly and enthusiastically.
- Directing people involves motivating them all the time to enthuse them to give their best.
- Communicating company's plans throughout the organization is an important directing activity.
- It also means co-ordinating various people and their activities.
- Directing aims at achieving the best not just out of an individual but achieving the best through the groups or teams of people through team building efforts.

5. Controlling

- It includes verifying the actual execution against the plans to ensure that execution is being done in accordance with the plans.
- It measures actual performance against the plans.
- It sets standards or norms of performance.
- It measures the effective and efficiency of execution against these standards and the plans.
- It periodically reviews, evaluates and monitors the performance.
- If the gaps are found between execution levels and the plans, controlling function involves suitable corrective actions to

expedite the execution to match up with the plans or in certain circumstances deciding to make modifications in the plans.

Management for any organization is a function of organizing and channelling the people together to accomplish desired goals.

In common understanding, management is also referred to the body or group of persons who perform the various functions of management.

Mary Parker Follett was the first to define management as the art of getting done through people.

Hierarchical Levels of Management

- Top management (including the members of the board of directors of the company).
- Senior management (or upper management; normally, the heads of various departments).
- Middle management (responsible for operational management).
- Junior or lower management (they include supervisors, team leaders or, foreman etc.).

Levels of Management

1. Top Level of management

It consists of board of directors, chief executive or managing director. The top management is the ultimate source of authority and it manages goals and policies for an enterprise. It devotes more time on planning and coordinating functions.

The role of the top management can be summarized as follows

a. Top management lays down the objectives and broad policies of the enterprise.

b. It issues necessary instructions for preparation of department budgets, procedures, schedules etc.

c. It prepares strategic plans & policies for the enterprise.

d. It appoints the executive for middle level i.e. departmental managers.

e. It controls & coordinates the activities of all the departments.

f. It is also responsible for maintaining a contact with the outside world.

g. It provides guidance and direction.

h. The top management is also responsible towards the shareholders for the performance of the enterprise.

2. Middle Level of management

The branch managers and departmental managers constitute middle level. They are responsible to the top management for the functioning of their department. They devote more time to organizational and directional functions. In small organization, there is only one layer of middle level of management but in big enterprises, there may be senior and junior middle level management. Their role can be emphasized as -

a. They execute the plans of the organization in accordance with the policies and directives of the top management.

b. They make plans for the sub-units of the organization.

c. They participate in employment & training of lower level management.

d. They interpret and explain policies from top level management to lower level.

e. They are responsible for coordinating the activities within the division or department.

f. It also sends important reports and other important data to top level management.

g. They evaluate performance of junior managers.

h. They are also responsible for inspiring lower level managers towards better performance.

3. Lower Level of management

Lower level is also known as supervisory / operative level of management. It consists of supervisors, foreman, section officers, superintendent etc. According to R.C. Davis, "Supervisory management refers to those executives whose work has to be largely with personal oversight and direction of operative employees". In other words, they are concerned with direction and controlling function of management. Their activities include -

a. Assigning of jobs and tasks to various workers.

b. They guide and instruct workers for day to day activities.

c. They are responsible for the quality as well as quantity of production.

d. They are also entrusted with the responsibility of maintaining good relation in the organization.

e. They communicate workers problems, suggestions, and recommendatory appeals etc to the higher level and higher level goals and objectives to the workers.

f. They help to solve the grievances of the workers.

g. They supervise & guide the sub-ordinates.

h. They are responsible for providing training to the workers.

i. They arrange necessary materials, machines, tools etc for getting the things done.

j. They prepare periodical reports about the performance of the workers.

k. They ensure discipline in the enterprise.

l. They motivate workers.

m. They are the image builders of the enterprise because they are in direct contact with the workers.

Therefore, existence of co-operation may prove to be effective condition or requisite for co-ordination. But it does not mean that co-

Human Resource Management (HRM)

The first definition of HRM is that it is the process of managing people in organizations in a structured and thorough manner. This covers the fields of staffing (hiring people), retention of people, pay and perks setting and Management, performance management, change management and taking care of exits from the company to round off the activities. This is the traditional definition of HRM which leads some experts to define it as a modern version of the Personnel Management function that was used earlier.

The second definition of HRM encompasses the management of people in organizations from a macro perspective i.e. managing people in the form of a collective relationship between management and employees. This approach focuses on the objectives and outcomes of the HRM function. What this means is that the HR function in contemporary organizations is concerned with the notions of people enabling, people development and a focus on making the "employment relationship" fulfilling for both the management and employees.

The most important resource of an organization is its human resources - the people who work in the organization. People are vital for the effective operation of a company. To meet the challenges and competitive atmosphere of today's business environment, managers must recognize the potential of human resources, and then acquire, develop and retain these resources. This forms the basis of human resource management (HRM). HRM is the management of various activities that are designed to enhance the effectiveness of the manpower in an organization in the achievement of organizational goals. Acquiring skilled, talented, and motivated employees is an important part of HRM.

Human resource management forms a crucial function in organizations of all sizes. Larger firms usually have a separate HRM department. Small organizations, however, cannot always afford to have a separate HRM department that can continually follow the performance of individuals in the organization and review their accomplishment of goals. Instead, in such organizations, each manager is responsible for utilizing the skills and talents of the employees under him, effectively.

Traditionally, HRM departments had a relatively small role to play in the organization's overall mission and plans. They developed staffing plans, handled complaints, determined benefits and compensation, and conducted performance appraisal programs. These activities were, and still are, very important in managing an organization. However, today HRM departments are playing a more strategic role in charting the course of their firms. Changes in the environment, such as increasing costs, changing demographics and limited skilled labor supply, rapid technological changes and the need for new skills, have created a strategic need for HRM expertise. These changes have led to the acknowledgment that human resources need careful attention and are vital to the success of any business.

In this chapter, we will first discuss HR planning. The other steps in the HRM process – staffing, training and development, performance appraisal, and compensation will also be discussed. The later part of the chapter will discuss the two important elements of staffing – recruitment and selection. The chapter concludes with a description of the socialization process of new employees.

Human resource management: an overview

Human Resource Management (HRM) may be defined as the organized function of planning for human resource needs, and recruitment, selection, development, compensation and evaluation of performance to fill those needs. The HRM process is an ongoing function that aims to keep the organization supplied with the right people in the right positions, when they are needed.

The HRM process includes five basic activities:

1. Human resource planning, P, S, T & D, DA, & C
2. Staffing,
3. Training and development,
4. Performance appraisal, and
5. Compensation.

Human Resource Planning

Human resource planning is the process of determining future human resource needs relative to an organization's strategic plan and devising the steps necessary to meet those needs. It involves estimating the size and composition of the future work force and helping the organization acquire the right number and the right kind of people when they are needed.

Staffing

Though the term "human resource management" is frequently used for the managerial function of "staffing," staffing is just a part of the HRM process and plays an important role. Staffing involves a set of activities aimed at attracting and selecting individuals for positions in a way that will facilitate the achievement of organizational goals. The two basic steps of staffing are recruitment and selection.

The staffing process is a systematic attempt to implement the human resource plan by recruiting, evaluating and selecting qualified candidates for job positions in the organization. Recruitment involves finding and attempting to attract job candidates who are suitable for filling job vacancies. Job analysis, job description, and job specification are important tools in the recruitment process. Once suitable candidates are attracted to the job position, the management needs to find qualified people to fill the positions through the selection process. Several methods are used in selecting prospective candidates. These include preliminary screening, application blank, selection test, comprehensive interviews, etc.

Training and development

Although organizations often recruit fully qualified individuals who require little or no training, training is usually undertaken for new recruits as well as for existing manpower, who require improved skills in order to advance in the organization. Employees at all levels – managerial, technical and operative – will require some training at some point of time in their careers. Although the objectives, methods, and course or program contents often differ, the basic principles of teaching/learning are the same. Training is formally defined as a planned effort to improve the performance of the employee in his area of work. In other words, training denotes efforts to increase employee skills in their jobs. For instance, employees might be instructed in new decision-making techniques or the capabilities of data processing systems.

Development programs are designed to educate employees beyond the requirements of their present positions in order to prepare them for promotions. They also help them get accustomed to the organizational climate. Development is long-term in nature. It helps the employee fit into the organization.

Thus, the processes of training and development aim at increasing the ability of individuals and groups to contribute to organizational effectiveness.

Performance appraisal

Performance appraisal compares an individual's job performance against standards or objectives developed for the individual's position.

The process of performance appraisal involves defining the expectations for employee performance, measuring, evaluating and recording employee performance against these expectations, and providing the employee with feedback regarding his performance. The major purpose of performance appraisal is to influence employee performance and development in a positive way. When the performance is high, the individual is likely to be rewarded (by a hike in pay or a promotion). If performance is low, some corrective action (such as additional training and development) might be arranged to make the performance meet the desired standards.

Thus, effective performance appraisal as a control technique requires standards, information and corrective action. Standards in performance evaluation are prior specifications of acceptable levels of job performance. Information must be available in order to measure the actual job performance against the standard job performance. Corrective action must be taken by managers to restore any imbalance between actual and standard job performance.

Compensation

Compensation consists of the wages paid directly to the employees for the amount of time worked or the number of units produced. It also includes the monetary and non-monetary benefits that an employee receives as part of his employment relationship with the organization. Wages paid for time worked (or number of units produced) are typically payments made in the form of cash and reflect direct work-related remuneration such as basic pay, merit increases, or bonuses. Benefits, on the other hand, are forms of supplementary non-monetary payments over and above the wages paid. They include various protection plans (such as employee insurance), services (such as company cafeteria), pay for time not worked (such as during vacations or sick leave), and income supplements (such as stock ownership plans).

A sound compensation program enhances the organization's ability to attract and retain employees. The compensation program affects every member of the organization, and it is one of the most important and time-consuming tasks of the human resources department.

Steps in Planning Function

Planning function of management involves following steps

1. Establishment of objectives
    - Planning requires a systematic approach.
    - Planning starts with the setting of goals and objectives to be achieved.
    - Objectives provide a rationale for undertaking various activities as well as indicate direction of efforts.
    - Moreover objectives focus the attention of managers on the end results to be achieved.
    - As a matter of fact, objectives provide nucleus to the planning process. Therefore, objectives should be stated in a clear, precise and unambiguous language. Otherwise the activities undertaken are bound to be ineffective.
    - As far as possible, objectives should be stated in quantitative terms. For example, Number of men is working, wages given, units produced, etc. But such an objective cannot be stated in quantitative terms like performance of quality control manager, effectiveness of personnel manager.
    - Such goals should be specified in qualitative terms.
    - Hence objectives should be practical, acceptable, workable and achievable.
2. Establishment of planning premises
    - Planning premises are the assumptions about the lively shape of events in future.
    - They serve as a basis of planning.
    - Establishment of planning premises is concerned with determining where one tends to deviate from the actual plans and causes of such deviations.
    - It is to find out what obstacles are there in the way of business during the course of operations.
    - Establishment of planning premises is concerned to take such steps that avoid these obstacles to a great extent.

- Planning premises may be internal or external. Internal includes capital investment policy, management labour relations, philosophy of management, etc. Whereas external includes socio- economic, political and economical changes.
- Internal premises are controllable whereas external are non-controllable.

3. Choice of alternative course of action
    - When forecast are available and premises are established, a number of alternative course of actions have to be considered.
    - For this purpose, each and every alternative will be evaluated by weighing its pros and cons in the light of resources available and requirements of the organization.
    - The merits, demerits as well as the consequences of each alternative must be examined before the choice is being made.
    - After objective and scientific evaluation, the best alternative is chosen.
    - The planners should take help of various quantitative techniques to judge the stability of an alternative.
4. Formulation of derivative plans
    - Derivative plans are the sub plans or secondary plans which help in the achievement of main plan.
    - Secondary plans will flow from the basic plan. These are meant to support and expedite the achievement of basic plans.
    - These detail plans include policies, procedures, rules, programmes, budgets, schedules, etc. For example, if profit maximization is the main aim of the enterprise, derivative plans will include sales maximization, production maximization, and cost minimization.
    - Derivative plans indicate time schedule and sequence of accomplishing various tasks.
5. Securing Co-operation
    - After the plans have been determined, it is necessary rather advisable to take subordinates or those who have to implement these plans into confidence.

The purposes behind taking them into confidence are :

- Subordinates may feel motivated since they are involved in decision making process.
- The organization may be able to get valuable suggestions and improvement in formulation as well as implementation of plans.
- Also the employees will be more interested in the execution of these plans.

6. Follow up/Appraisal of plans
   - After choosing a particular course of action, it is put into action.
   - After the selected plan is implemented, it is important to appraise its effectiveness.
   - This is done on the basis of feedback or information received from departments or persons concerned.
   - This enables the management to correct deviations or modify the plan.
   - This step establishes a link between planning and controlling function.
   - The follow up must go side by side the implementation of plans so that in the light of observations made, future plans can be made more realistic.

Motivation

In any type of organization, a manager must know what motivates his workers in order to make each individual employee perform to the best of his ability. It is not an easy task to motivate employees because they respond in different ways to their jobs and to organizational practices. Motivation is a human psychological characteristic that affects a person's degree of commitment. It is the set of forces that move a person towards a goal. It deals with how behaviour is energized, how it is directed and how it is sustained. The manager's challenge, then, is to channel this energy and direct this behaviour toward the organization's ends.

Factors that affect work motivation include individual differences and organizational practices. Individuals differ in their personal needs, values and attitudes, interests and abilities. Organizational practices that affect motivation include the rules, policies, managerial practices and reward systems. In order to motivate employees, managers must consider how these factors influence and affect their job performance.

Definitions and Meaning of Motivation

According to Stephen P. Robbins, motivation is the willingness to exert high levels of effort toward organizational goals, conditioned by the effort's ability to satisfy some individual need.

Fred Luthans views motivation as "a process that starts with a physiological or psychological deficiency or need that activates behaviour or a drive that is aimed at a goal or incentive."

Table 2 : Approaches to Motivation

Type	Characteristics	Theories	Managerial Examples
Content	Concerned with factors that arouse, start or initiate motivated behaviour	1. Needs hierarchy theory 2.Two-factor theory 3.ERG theory	Motivation by satisfying individual needs for money, status, and achievement.
Process	Concerned not only with factors that arouse behaviour, but also with the process, direction, or choice of behavioural patterns	1.Expectancy theory 2.Equity theory	Motivation by clarifying the individual's perception of work inputs, performance requirements and rewards.

The three key elements in the above definitions are needs, drives and goals. Needs set up drive aimed at goals; this is the basic process of motivation. Need is the origin of any motivated behaviour. Need is a felt deprivation of physiological or psychological well-being. Needs exist in each individual in varying degrees. When an individual recognizes a need, he is driven by a desire to fulfil the need. Drives are directed at fulfilment of needs. Drives are action-oriented and provide an energizing thrust toward reaching a goal. Incentives or goals are the instruments used to induce people to follow a desired course of action. Once the goal is attained, the physiological or psychological balance is restored and the drive is cut off.

Concept

One of the most important functions of management is to create willingness amongst the employees to perform in the best of their abilities. Therefore the role of a leader is to arouse interest in performance of employees in their jobs. The process of motivation consists of three stages :

1. A felt need or drive
2. A stimulus in which needs have to be aroused
3. When needs are satisfied, the satisfaction or accomplishment of goals.

Therefore, we can say that motivation is a psychological phenomenon which means needs and wants of the individuals have to be tackled by framing an incentive plan.

Motivation Theories

1. Hierarchy of needs theory

Abraham Maslow is well renowned for proposing the Hierarchy of Needs Theory in 1943. This theory is a classical depiction of human motivation. This theory is based on the assumption that there is a hierarchy of five needs within each individual. The urgency of these needs varies.

These five needs are as follows :

1. *Physiological needs* : These are the basic needs of air, water, food, clothing and shelter. In other words, physiological needs are the needs for basic amenities of life.
2. *Safety needs* : Safety needs include physical, environmental and emotional safety and protection. For instance- Job security, financial security, protection from animals, family security, health security, etc.
3. *Social needs :* Social needs include the need for love, affection, care, belongingness, and friendship.

4. *Esteem needs :* Esteem needs are of two types: internal esteem needs (self- respect, confidence, competence, achievement and freedom) and external esteem needs (recognition, power, status, attention and admiration).
5. Self-actualization need- This include the urge to become what you are capable of becoming / what you have the potential to become. It includes the need for growth and self-contentment. It also includes desire for gaining more knowledge, social- service, creativity and being aesthetic. The self- actualization needs are never fully satiable. As an individual grows psychologically, opportunities keep cropping up to continue growing.

According to Maslow, individuals are motivated by unsatisfied needs. As each of these needs is significantly satisfied, it drives and forces the next need to emerge. Maslow grouped the five needs into two categories - Higher-order needs and Lower-order needs. The physiological and the safety needs constituted the lower-order needs. These lower-order needs are mainly satisfied externally. The social, esteem, and self-actualization needs constituted the higher-order needs. These higher-order needs are generally satisfied internally, i.e., within an individual. Thus, we can conclude that during boom period, the employees lower-order needs are significantly met.

Maslow's Hierarchy of Needs

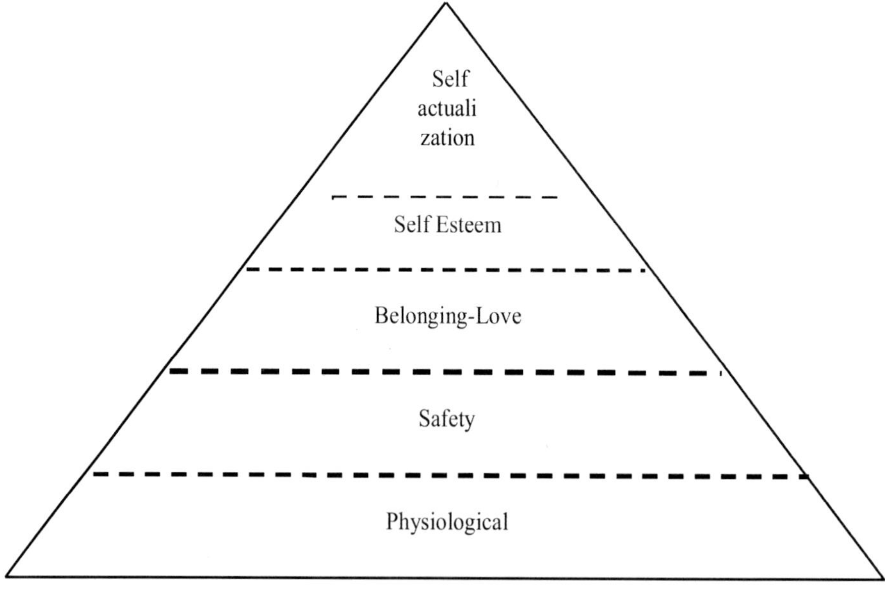

Limitations of Maslow's Theory

1. It is essential to note that not all employees are governed by same set of needs. Different individuals may be driven by different needs at same point of time. It is always the **most powerful unsatisfied need that motivates an individual**.
2. The theory is not empirically supported.
3. The theory is not applicable in case of starving artist as even if the artist's basic needs are not satisfied, he will still strive for recognition and achievement.

2. Herzberg's Two-Factor theory of motivation

In 1959, Frederick Herzberg, a behavioral scientist proposed a two-factor theory or the motivator-hygiene theory. According to Herzberg, there are some job factors that result in satisfaction while there are other job factors that prevent dissatisfaction. According to Herzberg, the opposite of "Satisfaction" is "No satisfaction" and the opposite of "Dissatisfaction" is "No Dissatisfaction".

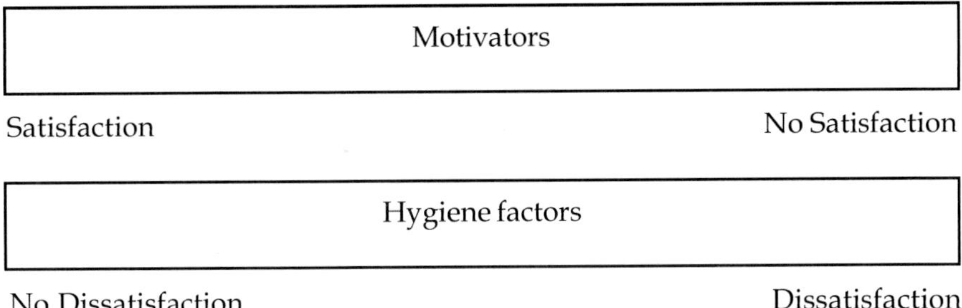

Figure : Herzberg's view of satisfaction and dissatisfaction

a. *Hygiene factors* : Hygiene factors are those job factors which are essential for existence of motivation at workplace. These do not lead to positive satisfaction for long-term. But if these factors are absent / if these factors are non-existents at workplace, then they lead to dissatisfaction. In other words, hygiene factors are those factors which when adequate / reasonable in a job, pacify the employees and do not make them dissatisfied. These factors are extrinsic to work. Hygiene factors are also called as **dissatisfies or maintenance factors** as they are required to avoid dissatisfaction. These factors describe the job environment /

scenario. The hygiene factors symbolized the physiological needs which the individuals wanted and expected to be fulfilled. Hygiene factors include:

- Pay- The pay or salary structure should be appropriate and reasonable. It must be equal and competitive to those in the same industry in the same domain.
- Company policies and administrative policies- The company policies should not be too rigid. They should be fair and clear. It should include flexible working hours, dress code, breaks, vacation, etc.
- Fringe benefits- The employees should be offered health care plans (medical), benefits for the family members, employee help programmes, etc.
- Physical working conditions- The working conditions should be safe, clean and hygienic. The work equipments should be updated and well-maintained.
- Status- The employees' status within the organization should be familiar and retained.
- Interpersonal relations-The relationship of the employees with his peers, superiors and subordinates should be appropriate and acceptable. There should be no conflict or humiliation element present.
- Job security- The organization must provide job security to the employees.

b. *Motivational factors*- According to Herzberg, the hygiene factors cannot be regarded as motivators. The motivational factors yield positive satisfaction. These factors are inherent to work. These factors motivate the employees for a superior performance. These factors are called satisfiers. These are factors involved in performing the job. Employees find these factors intrinsically rewarding. The motivators symbolized the psychological needs that were perceived as an additional benefit. Motivational factors include:

- Recognition- The employees should be praised and recognized for their accomplishments by the managers.

- Sense of achievement- The employees must have a sense of achievement. This depends on the job. There must be a fruit of some sort in the job.
- Growth and promotional opportunities- There must be growth and advancement opportunities in an organization to motivate the employees to perform well.
- Responsibility- The employees must hold themselves responsible for the work. The managers should give them ownership of the work. They should minimize control but retain accountability.

Meaningfulness of the work- The work itself should be meaningful, interesting and challenging for the employee to perform and to get motivated.

Herzberg classified these job factors into two categories

Limitations of Two-Factor Theory

The Two-Factor theory implies that the managers must stress upon guaranteeing the adequacy of the hygiene factors to avoid employee dissatisfaction. Also, the managers must make sure that the work is stimulating and rewarding so that the employees are motivated to work and perform harder and better. This theory emphasize upon job-enrichment so as to motivate the employees.

3. Douglas McGregor's theory X and theory Y

In 1960, Douglas McGregor formulated Theory X and Theory Y suggesting two aspects of human behavior at work, or in other words, two different views of individuals (employees): one of which is negative, called as Theory X and the other is positive, so called as Theory Y. According to McGregor, the perception of managers on the nature of individuals is based on various assumptions.

Assumptions of theory X

1. An average employee intrinsically does not like work and tries to escape it whenever possible.
2. Since the employee does not want to work, he must be persuaded, compelled, or warned with punishment so as to achieve organizational goals. A close supervision is required on part of managers. The managers adopt a more dictatorial style.

3. Many employees rank job security on top, and they have little or no aspiration/ ambition.
4. Employees generally dislike responsibilities.
5. Employees resist change.
6. An average employee needs formal direction.

The two factor theory is not free from limitations:

1. Herzberg assumed a correlation between satisfaction and productivity. But the research conducted by Herzberg stressed upon satisfaction and ignored productivity.
2. The theory's reliability is uncertain. Analysis has to be made by the raters. The raters may spoil the findings by analyzing same response in different manner.
3. No comprehensive measure of satisfaction was used. An employee may find his job acceptable despite the fact that he may hate/object part of his job.
4. The two factor theory is not free from bias as it is based on the natural reaction of employees when they are enquired the sources of satisfaction and dissatisfaction at work. They will blame dissatisfaction on the external factors such as salary structure, company policies and peer relationship. Also, the employees will give credit to themselves for the satisfaction factor at work.
5. The theory ignores blue-collar workers. Despite these limitations, Herzberg's Two-Factor theory is acceptable broadly.

Implications of Two-Factor theory

Assumptions of theory Y

1. Employees can perceive their job as relaxing and normal. They exercise their physical and mental efforts in an inherent manner in their jobs.
2. Employees may not require only threat, external control and coercion to work, but they can use self-direction and self-control if they are dedicated and sincere to achieve the organizational objectives.
3. If the job is rewarding and satisfying, then it will result in employees' loyalty and commitment to organization.

4. An average employee can learn to admit and recognize the responsibility. In fact, he can even learn to obtain responsibility.
5. The employees have skills and capabilities. Their logical capabilities should be fully utilized. In other words, the creativity, resourcefulness and innovative potentiality of the employees can be utilized to solve organizational problems.

Thus, we can say that Theory X presents a pessimistic view of employees' nature and behavior at work, while Theory Y presents an optimistic view of the employees' nature and behavior at work. If correlate it with Maslow's theory, we can say that Theory X is based on the assumption that the employees emphasize on the physiological needs and the safety needs; while Theory X is based on the assumption that the social needs, esteem needs and the self-actualization needs dominate the employees.

McGregor views Theory Y to be more valid and reasonable than Theory X. Thus, he encouraged cordial team relations, responsible and stimulating jobs, and participation of all in decision-making process.

Implications of theory X and theory Y

Quite a few organizations use Theory X today. Theory X encourages use of tight control and supervision. It implies that employees are reluctant to organizational changes. Thus, it does not encourage innovation. Many organizations are using Theory Y techniques. Theory Y implies that the managers should create and encourage a work environment which provides opportunities to employees to take initiative and self-direction. Employees should be given opportunities to contribute to organizational well-being. Theory Y encourages decentralization of authority, teamwork and participative decision making in an organization. Theory Y searches and discovers the ways in which an employee can make significant contributions in an organization. It harmonizes and matches employees' needs and aspirations with organizational needs and aspirations.

4. Alderfer ERG theory of motivation

To bring Maslow's need hierarchy theory of motivation in synchronization with empirical research, Clayton Alderfer redefined it in his own terms. His rework is called as ERG theory of motivation.

He recategorized Maslow's hierarchy of needs into three simpler and broader classes of needs:

1. *Existence needs*- These include need for basic material necessities. In short, it includes an individual's physiological and physical safety needs.
2. *Relatedness needs*- These include the aspiration individual's have for maintaining significant interpersonal relationships (be it with family, peers or superiors), getting public fame and recognition. Maslow's social needs and external component of esteem needs fall under this class of need.
3. *Growth needs*- These include need for self-development and personal growth and advancement. Maslow's self-actualization needs and intrinsic component of esteem needs fall under this category of need.

The significance of the three classes of needs may vary for each individual.

Difference between Maslow Need Hierarchy Theory and Alderfer's ERG Theory

ERG Theory states that at a given point of time, more than one need may be operational.ERG Theory also shows that if the fulfilment of a higher-level need is subdued, there is an increase in desire for satisfying a lower-level need. According to Maslow, an individual remains at a particular need level until that need is satisfied. While according to ERG theory, if a higher- level need aggravates, an individual may revert to increase the satisfaction of a lower- level need. This is called frustration- regression aspect of ERG theory. For instance- when growth need aggravates, then an individual might be motivated to accomplish the relatedness need and if there are issues in accomplishing relatedness needs, then he might be motivated by the existence needs. Thus, frustration/aggravation can result in regression to a lower-level need. While Maslow's need hierarchy theory is rigid as it assumes that the needs follow a specific and orderly hierarchy and unless a lower-level need is satisfied, an individual cannot proceed to the higher-level need; ERG Theory of motivation is very flexible as he perceived the needs as a range/variety rather than perceiving them as a hierarchy. According to Alderfer, an individual can work on growth needs even if his existence or relatedness needs remain unsatisfied. Thus, he gives explanation to the issue of "starving artist" who can struggle for growth even if he is hungry.

Implications of the ERF Theory

McClelland's theory of Needs / Achievement Motivation Theory

David McClelland and his associates proposed McClelland's theory of Needs / Achievement Motivation Theory. This theory states that human behaviour is affected by three needs - Need for Power, Achievement and Affiliation. Need for achievement is the urge to excel, to accomplish in relation to a set of standards, to struggle to achieve success. Need for power is the desire to influence other individual's behaviour as per your wish. In other words, it is the desire to have control over others and to be influential. Need for affiliations a need for open and sociable interpersonal relationships. In other words, it is a desire for relationship based on co-operation and mutual understanding.

The individuals with high achievement needs are highly motivated by competing and challenging work. They look for promotional opportunities in job. They have a strong urge for feedback on their achievement. Such individuals try to get satisfaction in performing things better. High achievement is directly related to high performance. Individuals who are better and above average performers are highly motivated. They assume responsibility for solving the problems at work. McClelland called such individuals as gamblers as they set challenging targets for themselves and they take deliberate risk to achieve those set targets. Such individuals look for innovative ways of performing job. They perceive achievement of goals as a reward, and value it more than a financial reward.

Managers must understand that an employee has various needs that must be satisfied at the same time. According to the ERG theory, if the manager concentrates solely on one need at a time, this will not effectively motivate the employee. Also, the frustration- regression aspect of ERG Theory has an added effect on workplace motivation. For instance- if an employee is not provided with growth and advancement opportunities in an organization, he might revert to the relatedness need such as socializing needs and to meet those socializing needs, if the environment or circumstances do not permit, he might revert to the need for money to fulfill those socializing needs. The sooner the manager realizes and discovers this, the more immediate steps they will take to fulfill those needs which are frustrated until such time that the employee can again pursue growth.

The individuals who are motivated by power have a strong urge to be influential and controlling. They want that their views and ideas should dominate and thus, they want to lead. Such individuals are motivated by the need for reputation and self-esteem. Individuals with greater power and authority will perform better than those possessing less power. Generally, managers with high need for power turn out to be more efficient and successful managers. They are more determined and loyal to the organization they work for. Need for power should not always be taken negatively. It can be viewed as the need to have a positive effect on the organization and to support the organization in achieving it's goals.

The individuals who are motivated by affiliation have an urge for a friendly and supportive environment. Such individuals are effective performers in a team. These people want to be liked by others. The manager's ability to make decisions is hampered if they have a high affiliation need as they prefer to be accepted and liked by others, and this weakens their objectivity. Individuals having high affiliation needs prefer working in an environment providing greater personal interaction. Such people have a need to be on the good books of all. They generally cannot be good leaders.

Goal Setting Theory of Motivation

In 1960's, Edwin Locke put forward the Goal-setting theory of motivation. This theory states that goal setting is essentially linked to task performance. It states that specific and challenging goals along with appropriate feedback contribute to higher and better task performance. In simple words, goals indicate and give direction to an employee about what needs to be done and how much efforts are required to be put in. The important features of goal-setting theory are as follows:

The willingness to work towards attainment of goal is main source of job motivation. Clear, particular and difficult goals are greater motivating factors than easy, general and vague goals.

Goals should be realistic and challenging. This gives an individual a feeling of pride and triumph when he attains them, and sets him up for attainment of next goal. The more challenging the goal, the greater is the reward generally and the more is the passion for achieving it.

Better and appropriate feedback of results directs the employee behaviour and contributes to higher performance than absence of feedback. Feedback is a means of gaining reputation, making clarifications and regulating goal difficulties. It helps employees to work with more involvement and leads to greater job satisfaction.

Employees' participation in goal is not always desirable.

Goal setting theory has certain eventualities such as:

a. *Self-efficiency-* Self-efficiency is the individual's self-confidence and faith that he has potential of performing the task. Higher the level of self-efficiency, greater will be the efforts put in by the individual when they face challenging tasks. While, lower the level of self-efficiency, less will be the efforts put in by the individual or he might even quit while meeting challenges.

b. *Goal commitment-* Goal setting theory assumes that the individual is committed to the goal and will not leave the goal. The goal commitment is dependent on the following factors:

c. Goals are made open, known and broadcasted.

d. Goals should be set-self by individual rather than designated.

e. Individual's set goals should be consistent with the organizational goals and vision.

Advantages of goal setting theory

1. Goal setting theory is a technique used to raise incentives for employees to complete work quickly and effectively.

2. Goal setting leads to better performance by increasing motivation and efforts, but also through increasing and improving the feedback quality.

Limitations of goal setting theory

1. At times, the organizational goals are in conflict with the managerial goals. Goal conflict has a detrimental effect on the performance if it motivates incompatible action drift.

2. Very difficult and complex goals stimulate riskier behaviour.

3. If the employee lacks skills and competencies to perform actions essential for goal, then the goal-setting can fail and lead to undermining of performance.

4. There is no evidence to prove that goal-setting improves job satisfaction.

Reinforcement Theory of Motivation

Reinforcement theory of motivation was proposed by BF Skinner and his associates. It states that individual's behavior is a function of its consequences. It is based on "law of effect", i.e, individual's behavior with positive consequences tends to be repeated, but individual's behavior with negative consequences tends not to be repeated.

Reinforcement theory of motivation overlooks the internal state of individual, i.e., the inner feelings and drives of individuals are ignored by Skinner. This theory focuses totally on what happens to an individual when he takes some action. Thus, according to Skinner, the external environment of the organization must be designed effectively and positively so as to motivate the employee. This theory is a strong tool for analyzing controlling mechanism for individual's behavior. However, it does not focus on the causes of individual's behavior.

The managers use the following methods for controlling the behaviour of the

Positive Reinforcement : This implies giving a positive response when an individual shows positive and required behaviour. For example - Immediately praising an employee for coming early for job. This will increase probability of outstanding behaviour occurring again. Reward is a positive reinforce, but not necessarily. If and only if the employees' behaviour improves, reward can said to be a positive reinforce. Positive reinforcement stimulates occurrence of behaviour. It must be noted that more spontaneous is the giving of reward, the greater reinforcement value it has.

Negative Reinforcement : This implies rewarding an employee by removing negative / undesirable consequences. Both positive and negative reinforcement can be used for increasing desirable / required behaviour.

Punishment : It implies removing positive consequences so as to lower the probability of repeating undesirable behaviour in future. In other words, punishment means applying undesirable consequence for showing undesirable behaviour. For instance - Suspending an employee for breaking the organizational rules. Punishment can be equalized by positive reinforcement from alternative source.

Extinction : It implies absence of reinforcements. In other words, extinction implies lowering the probability of undesired behaviour by removing reward for that kind of behaviour. For instance - if an employee no longer receives praise and admiration for his good work, he may feel that his behaviour is generating no fruitful consequence. Extinction may unintentionally lower desirable behaviour.

Implications of Reinforcement Theory

Reinforcement theory explains in detail how an individual learns behavior. Managers who are making attempt to motivate the employees must ensure that they do not reward all employees simultaneously. They must tell the employees what they are not doing correct. They must tell the employees how they can achieve positive reinforcement.

Equity Theory of Motivation

The core of the equity theory is the principle of balance or equity. As per this motivation theory, an individual's motivation level is correlated to his perception of equity, fairness and justice practiced by the management. Higher is individual's perception of fairness, greater is the motivation level and vice versa. While evaluating fairness, employee compares the job input (in terms of contribution) to outcome (in terms of compensation) and also compares the same with that of another peer of equal cadre/category. D/I ratio (output-input ratio) is used to make such a comparison.

Equity Theory

Ratio Comparison

Perception

O/I a < O/I b	Under-rewarded (Equity Tension)
O/I a = O/I b	Equity
O/I a > O/I b	Over-rewarded (Equity Tension)

Negative Tension state

Equity is perceived when this ratio is equal. While if this ratio is unequal, it leads to "equity tension". **J.Stacy Adams** called this a negative tension state which motivates him to do something right to relieve this tension. A comparison has been made between 2 workers A and B to understand this point.

Referents : The four comparisons an employee can make have been termed as "referents" according to Goodman. The referent chosen is a significant variable in equity theory. These referents are as follows:

An employee might compare himself with his peer within the present job in the current organization or with his friend/peer working in some other organization or with the past jobs held by him with others. An employee's choice of the referent will be influenced by the appeal of the referent and the employee's knowledge about the referent.

Moderating Variables

The gender, salary, education and the experience level are moderating variables. Individuals with greater and higher education are more informed. Thus, they are likely to compare themselves with the outsiders. Males and females prefer same sex comparison. It has been observed that females are paid typically less than males in comparable jobs and have less salary expectations than male for the same work. Thus, a women employee that uses another women employee as a referent tends to lead to a lower comparative standard. Employees with greater experience know their organization very well and compare themselves with their own colleagues, while employees with less experience rely on their personal experiences and knowledge for making comparisons.

Assumptions of the Equity Theory

1. The theory demonstrates that the individuals are concerned both with their own rewards and also with what others get in their comparison.
2. Employees expect a fair and equitable return for their contribution to their jobs.
3. Employees decide what their equitable return should be after comparing their inputs and outcomes with those of their colleagues.
4. Employees who perceive themselves as being in an inequitable scenario will attempt to reduce the inequity either by distorting inputs and/or outcomes psychologically, by directly altering inputs and/or outputs, or by quitting the organization.

The expectancy theory was proposed by Victor Vroom of Yale School of Management in 1964. Vroom stresses and focuses on outcomes, and not on needs unlike Maslow and Herzberg. The theory states that the intensity of a tendency to perform in a particular manner is dependent on the intensity of an expectation that the performance will be followed by a definite outcome and on the appeal of the outcome to the individual.

The Expectancy theory states that employee's motivation is an outcome of how much an individual wants a reward (Valence), the assessment that the likelihood that the effort will lead to expected performance (Expectancy) and the belief that the performance will lead to reward (Instrumentality). In short, Valence is the significance associated by an individual about the expected outcome. It is an expected and not the actual satisfaction that an employee expects to receive after achieving the goals. Expectancy is the faith that better efforts will result in better performance. Expectancy is influenced by factors such as possession of appropriate skills for performing the job, availability of right resources, availability of crucial information and getting the required support for completing the job.

Instrumentality is the faith that if you perform well, then a valid outcome will be there. Instrumentality is affected by factors such as believe in the people who decide who receives what outcome, the simplicity of the process deciding who gets what outcome, and clarity of relationship between performance and outcomes. Thus, the expectancy theory concentrates on the following three relationships:

Expectancy Theory of Motivation

1. Effort-performance relationship: What is the likelihood that the individual's effort be recognized in his performance appraisal?
2. Performance-reward relationship: It talks about the extent to which the employee believes that getting a good performance appraisal leads to organizational rewards.
3. Rewards-personal goals relationship: It is all about the attractiveness or appeal of the potential reward to the individual.

Vroom was of view that employees consciously decide whether to perform or not at the job. This decision solely depended on the employee's motivation level which in turn depends on three factors of expectancy, valence and instrumentality.

Advantages of the Expectancy Theory

1. It is based on self-interest individual who want to achieve maximum satisfaction and who wants to minimize dissatisfaction.
2. This theory stresses upon the expectations and perception; what is real and actual is immaterial.
3. It emphasizes on rewards or pay-offs.
4. It focuses on psychological extravagance where final objective of individual is to attain maximum pleasure and least pain.

Limitations of the Expectancy Theory

1. The expectancy theory seems to be idealistic because quite a few individuals perceive high degree correlation between performance and rewards.

   The application of this theory is limited as reward is not directly correlated with performance in many organizations. It is related to other parameters also such as position, effort, responsibility, education, etc.

   Job Enrichment

   A modern and more permanent approach to motivation is job enrichment. Here, the attempt is to build a higher sense of challenge and achievement in jobs. A job may be enriched in the following ways:

2. Allowing workers to make independent decisions on issues like work methods, sequence and pace or the acceptance or rejection of materials;
3. Encouraging involvement and participation of employees and interaction between workers;
4. Making workers feel personally responsible for their tasks;
5. Ensuring that workers get to know how their tasks contribute to the finished product and the welfare of the enterprise;
6. Giving people feedback on their job performance; and
7. Involving workers when bringing about changes in the physical aspects of their work environment, such as the layout of office or plant, temperature, lighting and cleanliness.

Monitoring and Evaluation

Monitoring is the systematic collection and analysis of information as a project progresses' is aimed at improving the efficiency and effectiveness of a project or organization. It is based on targets set and activities planned during the planning phases of work. It helps to keep the work on track, and can let management know when things are going wrong. If done properly, it is an invaluable tool for good management, and it provides a useful base for evaluation. It enables you to determine whether the resources you have available are sufficient and are being well used, whether the capacity you have is sufficient and appropriate, and whether you are doing what you planned to do.

Evaluation is the comparison of actual project impacts against the agreed strategic plans. It looks at what you set out to do, at what you have accomplished, and how you accomplished it. It can be **formative** (taking place during the life of a project or organization, with the intention of improving the strategy or way of functioning of the project or organization). It can also be **summative** (drawing learning's from a completed project or an organization that is no longer functioning). Someone once described this as the difference between a check-up and an autopsy!

Why Do monitoring and evaluation

1. Monitoring and evaluation enable you to check the "bottom line" (see Glossary of Terms) of
2. Development work: Not "are we making a profit?" but "are we making a difference?"
3. Through monitoring and evaluation, you can:
4. Review progress;
5. Identify problems in planning and/or implementation;
6. Make adjustments so that you are more likely to "make a difference".

Methods of monitoring and evaluation

1. Performance indicators
2. The logical framework approach

3. Theory-based evaluation
4. Formal surveys
5. Rapid appraisal methods
6. Participatory methods
7. Public expenditure tracking surveys
8. Cost-benefit and cost-effectiveness analysis
9. Impact evaluation

Steps in Monitoring

1. Discuss reasons for monitoring.
2. Review objectives and activities.
3. Develop monitoring questions.
4. Establish direct and indirect indicators.
5. Decide which information gathering tools are needed.
6. Decide who will do the monitoring.
7. Analyze and prevent results.

Evaluation

Steps

1. Review objective and activities.
2. Review reason for evaluaton.
3. Develop evaluation questions.
4. Decide who will do the evaluation.
5. Identify direct and indirect indicators.
6. Identify the information sources for evaluation questions.
7. Determine the skills and labour that are required to obtain information.
8. Determine when information gathering and analysis can be done.
9. Analyze the present levels.

Administrative communication

Communication, no doubt is the life and blood of any organization, and therefore the performance of any administration can be adjudged on the effectiveness and efficiency of its communication system, communication is the nervous system of an organization

Types of Administrative/Organizational communication

There are three classifications

1. On the basis of Direction of communication
   - Vertical
   - Downward
   - Upward
   - Horizontal/Lateral communication
   - Diagonal communication
2. On the basis of way of expression and body language
   - Oral communication
   - Written communication
   - Non - verbal communication
3. On the basis of organizational structure

A. Formal communication

- Horizontal
- Vertical
- Diagonal

B.Informal communication SGPC

- Single strand
- Gossip
- Probability
- Cluster

C. Barriers in communication

- External
- Emotional

- Organizational
- Personal
- Economic
- Semantic
- Political

Johari Window

	Known to self	Not Known to self
Known to self	Arena of free activity	Blind Spot
Not Known to self	Facade	Unknown

- Quadrant I: The area of free activity refers to behavior and motivation known to self and known to others.
- Quadrant II: The blind area where others can see things in ourselves of which we are unaware.
- Quadrant III: The avoided or hidden area represents things we know but do not level to other e.g. hidden agenda or matters about which we have sensitive feeling.
- Quadrant IV: Area for unknown activity neither the individual nor others are aware of certain behavior.

Communication Pattern

Every organization develops characteristic patterns over a period of time by which communication conducted within the organization pattern include

1. Direction
   - Horizontal
   - Vertical

2. Nature: Type of communication
    - Information
    - Instruction
    - Suggestions
3. Formality
    - Extent to which communication channel utilized

Principles of organizational communication

1. Communication channel to be used should have been recognized/approved by the organization.
2. Receiver must be able to learn through the channel.
3. The communication must ensure planning in the use of channels.
4. Planning for repetition of the message should also be made.
5. Use of seller methods over a period of time is necessary.
6. The communicator should consider feedback and the modification in the presentation or use of communication channels.
7. The informal network in the organization and the characteristics of individuals in the organization will have to be recognized by the communicator.

Functions of administrative communication

1. Coordination
2. Problem solving and innovating
3. Leading
4. Appraising and regulating

Decision-Making in Organization

Decision Making

Decision making is the process of selection from a set of alternative course of action which is thought to fulfill the objective of decision problem more satisfactorily then others.

Phases E.S.E.S.

1. Explorative
2. Speculative
3. Evaluative
4. Selective

Nature

1. Goal orientation
2. Implies alternative
3. Dynamic process
4. Related to environment
5. Freedom to decision maker
6. Continuous ongoing process
7. Rational process

Bases of decision making

- Experience
- Experiment
- Model (Research analysis)

Creativity in decision making PIIV

- Preparation
- Incubation
- Illumination
- Verification

Creative Techniques

1. Attribute listing
2. Brain Storming
3. Synoptic
4. Nominal group method
5. Delphi technique

Principles of decision making Marshall the facts and define goal.

Limitations of scientific decision making

1. Inadequate goal formation
2. Regularly defined problems
3. Imperfect knowledge
4. Limited time and resources
5. Human limitations
6. Power politics
7. Environmental dynamics

Types of decision: - PROI

1. Programmed and Nonprogrammed
2. Routine and strategic
3. Organizational and personal
4. Individual and group

Steps in scientific decision making

1. Identify the problem
2. Diagnose the problem
3. Discover alternative course of action
4. Evaluate the alternative
5. Choose the best alternative
6. Implement of follow-up

Theories

- Rational Economic man
- Model/Classical/optimizing
- Administrative man/Behavioral/Satisfying

Organization effectiveness

Effectiveness is the ability of an organization to mobilize its centers of power for action, production and adaption.

Effectiveness is the function of mainly three variables

1. The Manager
2. The organization
3. The environment

7s Strategy of Mc Kinsley for organizational effectiveness

The 7s models consisted of seven factors:

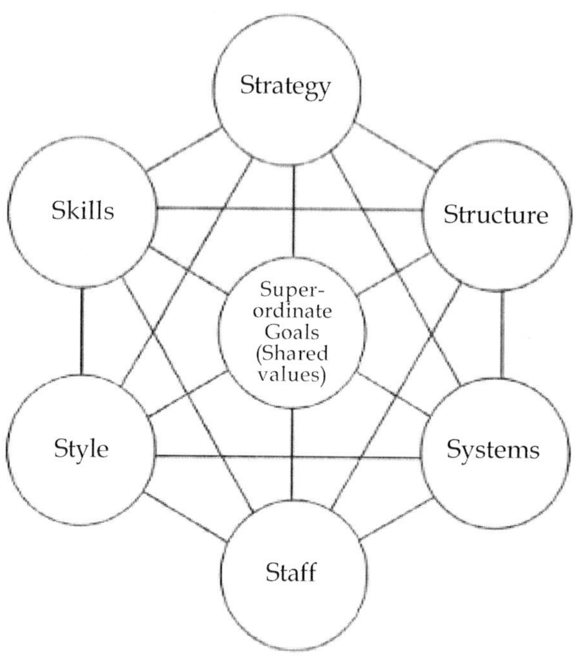

1. Strategy

The integrated vision and direction of the company as well as the manner in which it communicates and implements that vision and direction.

2. Structure

The form of the organizational chart and interconnections between positions in the organizational hierarchy.

3. Systems

The procedures and routine processes required to perform the work, including the ways information moves through the organization.

4. Staff

The personnel categories within the organization, e.g. marketers, engineers.

5. Style

The characterization of the ways key managers set priorities and behave in order to achieve the organization's goals.

6. Skills

The distinctive capabilities of the organization as a whole.

7. Shared Values

The core beliefs underlying the organization's existence and its expectations of its members. Values act as an organization's conscience and provide guidance in times of crisis.

The original intention of the model was to help guide thinking about organizational effectiveness in the broadest sense. The 7-S model turned out to be an excellent tool for judging an organization's ability to implement a given strategy. The model showed that thinking about strategy implementation was more complex than the relationship between strategy and structure as Chandler had suggested.

To be effective, an organization must have a high degree of internal alignment among all seven Ss. Each S must be consistent with the other factors for them to reinforce one another. With the exception of the skills factor, all Ss are interrelated and a change in one affects all others.

Certain key factors such as staff, strategy, structure and systems can be changed in the short term. The three remaining Ss – style, skills and shared values – are delayed factors that can only be affected long term. Skills are both hard and soft. Peters pointed out that true competitive advantage originates from these soft factors.

The model can be used as both a static checklist for analysis purposes and a tool to assess potential conflicts when a strategic program is implemented.

Organizational climate

Dimensions of organizational climate

1. Individual autonomy

2. Position structure
3. Reward orientation
4. Conflict
5. Risk taking
6. Consideration warmth, support
7. Progressiveness and development
8. Control

Determinants of the organizational climate

1. Organizational structure
2. Organic Size
3. Organic policies
4. Managerial values
5. Characteristics of members
6. Leadership style
7. External environment

Impact of organizational climate

1. Estimate for an individual
2. Puts contains on individual freedom of choice.
3. Influence behavior through evaluation of self and others
4. Organization variable influence the behavior in that they influence the individual to form the perception of the organization.

Organizational Behaviour

"Study of the behavior of the people in an organization"

Features of OB

1. It deals with the human behavior in the organization.
2. OB seeks to fulfill both employee needs and organization objectives.
3. It may individual or group behavior.

4. It has psychological foundations e.g. perception learning etc
5. OB is both art and science.
6. The basic purpose of OB is to control, understand, explain predict and control human behavior in organization.
7. OB is a dynamic concept rather then static.
8. OB is generally neither autonomous not spontaneous.

Organizational Development

Features : Organizational Development

1. Long range effort
2. Broad based
3. Systematic view
4. Dynamic process
5. Research based
6. Normative Educational strategies

Objective/ Rate of Organizational Development

1. To increase trust and mutual emotional support among members of organization.
2. To develop new organizational ways of learning and new ways of coping with problem.
3. To create a suitable environment within the organization.
4. To increase the level of say and group responsibility
5. To increase the openness of communication.
6. To find out the strategic saturation of the problem.
7. To increase the level of enthusiasm and personal satisfaction.

Process of Organizational Development

1. Problem identification
2. Diagnosis of the problem
3. Planning change strategy
4. Implementation or intervention
5. Evaluation and feedback

OD intervention technique

1. Sensitivity training
2. Process consultation
3. Transactional analysis
4. Grid training
5. Survey feedback
6. Third party peace making
7. Team building process
8. MBO

Co-ordination

Co-ordination (with respect to organization) "It is the process of integrating the objective and activities of separate work unit in order to realize goal of organization effectiveness.

"Co-ordination is the essence of managership for achieving harmony among individual effort for achieving group goals".

Problems in achieving effective coordination

1. Differentiation
2. Integration

Approaches to achieve effective coordination

i. Basic management technique
   1. Managerial hierarchy
   2. Management technique

ii. Increase coordination potential
   1. Vertical Information System
   2. Lateral Relationship

iii. Reduced need for coordination

iv. Simplified organization
   1. Departmentation which aid in coordination
   2. Clear-cut organizational procedure

v. Harmonized programmes & policies
   1. Consistency among plans
   2. Promoting

vii. Communication - It should be well designed

viii. Aids to voluntary coordination

ix. Coordination through supervision

x. Coordination through POSLC i.e. planning or ganising, staffing, leading & controlling.

Elements

1. Balancing
2. Timing
3. Integrity

Purpose

1. Reconciliation of goals
2. Total accomplishment
3. Economy and efficiency
4. Good human relations
5. Retention of managerial and other personnel

Need and Significance

1. Unity in diversity
2. Team work
3. Conflicting goals
4. Specialization
5. Synergy effect
6. Interdependence
7. Empire building
8. Human nature
9. Congruence of flows
10. Differentiation and integration

Methods

1. Automatic
2. Deliberate
   - Compulsory
   - Voluntary

Planning

Planning involves selecting missions and objectives and the actions to achieve them.

Types of plans

1. Purpose/Mission
2. Objectives/goals
3. Strategies
4. Policies
5. Procedures
6. Rules
7. Programmes
8. Budgets

Steps in Planning

1. Being aware of opportunity
2. Setting objective or goals
3. Considering planning premises
4. Identifying alternatives
5. Comparing alternative in light of goal sought
6. Choosing an alternative
7. Formulating supporting plans
8. Numbering plans by making budgets

Management by Objectives (MBO)

Management by Objectives (MBO) is a process of defining objectives within an organization so that management and employees agree to the objectives and understand what they need to do in the organization.

The term "management by objectives" was first popularized by Peter Ducker in his 1954 book 'The Practice of Management'.

The essence of MBO is participative goal setting, choosing course of actions and decision making. An important part of the MBO is the measurement and the comparison of the employee's actual performance with the standards set. Ideally, when employees themselves have been involved with the goal setting and choosing the course of action to be followed by them, they are more likely to fulfill their responsibilities.

ScanLan (1960) defined MBO system of management whereby every department and every individual has defined objectives so as to accomplish results in key areas.

Benefits

a. Improvement of management through result oriented planning
b. Clarification of organization rules
c. Delegation of authority
d. Encouragement of personal commitment to their own and organizational goals
e. Development of effective control, measuring results
f. Utilizations of individual talent and strengths to a maximized

Limitations/Failure/Recommendations

1. Philosophy not applied and thought to subordinates/mangers
2. Failure to give guide line to goal betters
3. Difficulty of setting variable goals
4. Emphasis on short term goals
5. The danger of inflexibility
6. over use of quantitative goals

Elements

Approaches in MBO system or/ Process in MBO

Robin (1971) described the following approaches in MBO system.

1. Setting organization level objectives
2. Individuals objective

3. Participation
4. Autonomy
5. Review
6. Commitment

Objectives of MBO

1. Measure judge performance
2. Individuals performance relation with organizational goals
3. Classification of jobs to be done expectations of achievement
4. To foster integrity component of subordinates
5. To enhance communication between superior and subordinates.
6. To stimulate the subordinate mediation
7. Basis for judgment about salary and promotion.

Five Step process

Step 1 : Preliminary goal setting

Step 2 : Setting subordinate objectives

Step 3 : Recycling objectives

Step 4 : Action planning

Step 5 : Periodic performance rewords

Features

1. Goal orientation
2. Participation of both top and lower management in goal setting
3. Session measureable, verifiable goals
4. Dynamic system
5. Philosophy to attain maximum result from available resources
6. Continuous process
7. An evaluative mechanism

Logical Framework Approach

It is a powerful tool for project design and management which distinguishes between higher and lower level of project objectives showing how these objectives are linked. It is an approach for any levels of planning except strategic planning.

Purpose of LFA

1. To assist in the planning management and evaluation of development activities and research projects.
2. To facilitate the systematic identification and specification of intermediate objective.
3. Plan for required input and desired output.
4. Give sound guidance for monitoring and evaluation.
5. To provide comprehensive, practice and flexible and effective means to strengthen international research project management.
6. To provide a structure which will provide project planners and evaluators to specify the components of their activities and identify the structure of LFA.

Basic

Narrative Summary	Objectively Verifiable Indicators OVIs	Means of Verification -MOVs	External Factors (Assumptions)
Development Objective			
Immediate Objective			
Outputs (Results)			
Activities			
Inputs			

Advantages

1. Provide a comprehensive, practical, flexible and effective means to strengthen international research project management.
2. Increases accountability

3. Give early warning.
4. It is an end to project management.
5. LFA specify in advance what the project is attempting to achieve by indicating the means by which this achievement may be measured.
6. It is simple to understand.
7. It is useful for impact evaluation.

Delphi Technique-Forecasting

Developed by – –OLA/Helmet and his colleagues at the RAND corporation

Process of Delphi Technique

1. A panel of experts on particular problem area is selected.
2. Experts are asked to make forecast.
3. Answer is compiled and composite results were sent back to panel.
4. With this information at hand further estimates of future are made.
5. This process may be repeated several times.
6. When the convergence of opinions beings to evolve the results are then used as an acceptable forecast.

Strategy Formulation

Tows Matrix for Strategy Formulation

Managerial Grid

Developed by Blake and Mouton to find leadership style

"It is an approach to define leadership style means which leadership style is being followed in which organization it could be defined by Managerial Grid."

Uses: Very wide applications

1. Defining/Determining leadership style
2. Training Managers
3. Identifying various combinations of leadership styles.

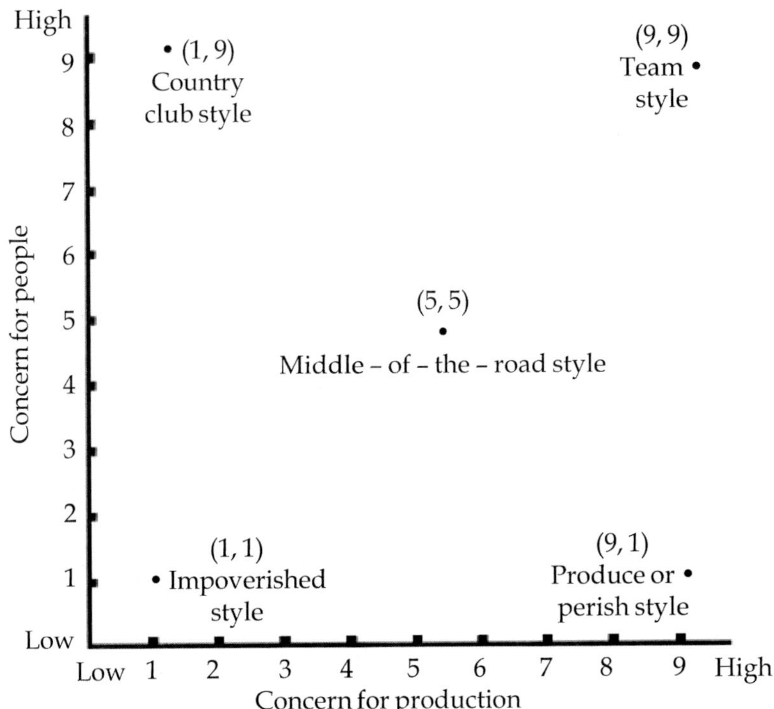

On the above four extreme by 10 are possible

Impoverished Management: Managers concern themselves very little with either people or production and have minimum involvement in their jobs.

Managers/Teamstyle : Highest possible dedication both to people and to production.

Managers/Country club management: Managers have little or no concern for production but are concerned only for people.

Managers/autocratic task manager: Who are concerned only with developing as efficient operation who have little or no concern for people and who are quite autocratic in their style of leadership.

Concern for people

Degree of personal commitment towards goal achievement, maintenance of the self esteem of workers, placement of responsibility on the basis of trust rather than obedience, provision of good working condition.

Managerial Roles

Mintzberg's 10 managerial Rules

1. Interpersonal	Figurehead Leader Liaison
2. Informational	Monitor Disseminator Spokesperson
3. Decision	Entrepreneur Disturbance handler Resource allocator Negotiator

Management Skills

Technical	Lower Level
Human/Behavioral	Middle Level
Conceptual	Top Level

Concern for production

Includes the attitudes of a supervisor towards a wide variety of things such as quality of polic decisions, procedures and processes, creativeness of research, quality for staff services, work efficiency etc.

Principle/Method

Based on the degree of the emphasis on either people or production relative place of a particular management of the grid could be divided and decisions suggestions could be taken according to that.

Project Management

A scientifically evaluated work plan decided to achieve a specific objective within a specified period for time.

Effective management of any project or task is depending on/ factors affecting project management

i. The project formulation knowledge/skills in developing a halistic perception of the project.

ii. His/her ability to precisely define the project.
iii. Appropriate sequence of these activities
iv. Identifying appropriate presence to delegate specific activities.
v. Supervising the project during the implementation stage.
vi. Mobilizing available resources for optimum utilization including appropriate choice of technology.

These are the tools in increasing the efficiency of the personnel as well as proper utilization of other resources.

- There is a wide choice among several project management techniques that helps in project planning, monitoring and controlling.
- Often modifications of several of the techniques or integration of the remove may become necessary depending upon the circumstances.

Points to be taken care of while choosing particular technique vis-a-vis project management

1. Complexity of project
2. Uncertainty involved in the project
3. Resources (Manpower, Money, Time)
4. Critical paths
5. Coordination
6. In built slack

Project Management Techniques

The project management techniques concerned with resources, cost time is as follows

1. CPM - Critical Path Method
2. PERT - Programme Eualuation & Review Technique
3. ABC (Activity Bar Chart)
4. ASBC slack
5. SBC Schedule
6. Presidency diagram

7. Manpower utilization
8. Line of balance
9. IMPACT - Implementation, plan and control technique.
10. COMET - Computer Operated Management evaluation tool.
11. Cost planning - appraisal
12. Management planning and control system
13. Programme evaluation procedure
14. PAR Project Audit Report
15. PLANNET : Planning Network
16. PRESM: Programme reliability information system for management.
17. TRACE: Task reporting and current evaluation
18. GERT: Graphical Evaluation and Radio Technique
19. Personal activity monitoring system

Project Evaluation and Review Technique (PERT)

Complex projects require a series of activities, some of which must be performed sequentially and others that can be performed in parallel with other activities. This collection of series and parallel tasks can be modeled as a network.

In 1957 the Critical Path Method (CPM) was developed as a network model for project management. CPM is a deterministic method that uses a fixed time estimate for each activity. While CPM is easy to understand and use, it does not consider the time variations that can have a great impact on the completion time of a complex project.

The Program Evaluation and Review Technique (PERT) is a network model that allows for randomness in activity completion times. PERT was developed in the late 1950's for the U.S. Navy's Polaris project having thousands of contractors. It has the potential to reduce both the time and cost required to complete a project.

The Network Diagram

In a project, an activity is a task that must be performed and an event is a milestone marking the completion of one or more activities.

Before an activity can begin, all of its predecessor activities must be completed. Project network models represent activities and milestones by arcs and nodes. PERT originally was an activity on arc network, in which the activities are represented on the lines and milestones on the nodes. Over time, some people began to use PERT as an activity on node network. For this discussion, we will use the original form of activity on arc.

The PERT chart may have multiple pages with many sub-tasks. The following is a very simple example of a PERT diagram:

PERT Chart

The milestone generally are numbered so that the ending node of an activity has a higher number than the beginning node. Incrementing the numbers by 10 allows for new ones to be inserted without modifying the numbering of the entire diagram. The activities in the above diagram are labeled with letters along with the expected time required to complete the activity.

Steps in the PERT Planning Process

PERT planning involves the following steps:

1. Identify the specific activities and milestones.
2. Determine the proper sequence of the activities.
3. Construct a network diagram.
4. Estimate the time required for each activity.
5. Determine the critical path.
6. Update the PERT chart as the project progresses.

1. Identify activities and milestones

The activities are the tasks required to complete the project. The milestones are the events marking the beginning and end of one or more activities. It is helpful to list the tasks in a table that in later steps can be expanded to include information on sequence and duration.

2. Determine activity sequence

These steps may be combined with the activity identification step since the activity sequence is evident for some tasks. Other tasks may require more analysis to determine the exact order in which they must be performed.

3. Construct the network diagram

Using the activity sequence information, a network diagram can be drawn showing the sequence of the serial and parallel activities. For the original activity-on-arc model, the activities are depicted by arrowed lines and milestones are depicted by circles or "bubbles".

If done manually, several drafts may be required to correctly portray the relationships among activities. Software packages simplify this step by automatically converting tabular activity information into a network diagram.

4. Estimate activity times

- Weeks are a commonly used unit of time for activity completion, but any consistent unit of time can be used.
- A distinguishing feature of PERT is its ability to deal with uncertainty in activity completion times. For each activity, the model usually includes three time estimates:
- Optimistic time - generally the shortest time in which the activity can be completed. It is common practice to specify optimistic times to be three standard deviations from the mean so that three is approximately a 1% chance that the activity will be completed within the optimistic time.
- Most likely time - the completion time having the highest probability. Note that this time is different from the expected time.
- Pessimistic time - the longest time that an activity might require. Three standard deviations from the mean is commonly used for the pessimistic time.
- PERT assumes a beta probability distribution for the time estimates. For a beat distribution, the expected time for each activity can be approximated using the following weighted average:
- **Expected time = (Optimistic + 4x Most likely + Pessimistic)/6**
- This expected time may be displayed on the network diagram.
- To calculate the variance for each activity completion time, if three standard deviation times were selected for the optimistic and pessimistic times, then there are six standard deviations between them, so the variance is given by:
- **[(Pessimistic - Optimistic)/6]^2**

5. Determine the critical path

The critical path is determined by adding the times for the activities in each sequence and determining the longest path in the project. The critical path determines the total calendar time required for the project. If activities outside the critical path speed up or slow down (within limits), the total project time does not change. The amount of time that a non-critical path activity can be delayed without delaying the project is referred to as slack time.

If the critical path is not immediately obvious, it may be helpful to determine the following four quantities for each activity:

1. ES - Earliest Start time
2. EF - Earliest Finish time
3. LS - Latest Start time
4. LF - Latest Finish time

These times are calculated using the expected time for the relevant activities. The earliest start and finish times of each activity are determined by working forward through the network and determining the earliest time at which an activity can start and finish considering its predecessor activities. The latest start and finish times are the latest times that an activity can start and finish without delaying the project. LS and LF are found by working backward through the network. The difference in the latest and earliest finish of each activity is that activity's slack. The critical path then is the path through the network in which none of the activities have slack.

The variance in the project completion time can be calculated by summing the variances in the completion times of the activities in the critical path. Given this variance, one can calculate the probability that the project will be completed by a certain date assuming a normal probability distribution for the critical path. The normal distribution assumption holds if the number of activities in the path is large enough for the central limit theorem to be applied.

Since the critical path determines the completion date of the project, the project can be accelerated by adding the resources required to decrease the time for the activities in the critical path. Such a shortening of the project sometimes is referred to as project crashing.

6. Update as project progresses

Make adjustments in the PERT chart as the project progresses. As the project unfolds, the estimated times can be replaced with actual times. In cases where there are delays, additional resources may be needed to stay on schedule and the PERT chart may be modified to reflect the new situation.

Benefits of PERT

PERT is useful because it provides the following information

1. Expected project completion time.
2. Probability of completion before a specified date.
3. The critical path activities that directly impact the completion time.
4. The activities that have slack time and that can lend resources to critical path activities.
5. Activity starts and end dates.

Limitations

The followings are some of PERT's weaknesses:

1. The activity time estimates are somewhat subjective and depend on judgment. In cases where there is little experience in performing an activity, the numbers may be only a guess. In other cases, if the person or group performing the activity estimates the time there may be bias in the estimate.
2. Even if the activity times are well-estimated, PERT assumes a beta distribution for these time estimates, but the actual distribution may be different.
3. Even if the beta distribution assumption holds, PERT assumes that the probability distribution of the project completion time is the same as the of the critical path. Because other parts can becomes the critical path if their associated activities are delayed, PERT consistently underestimates the expected project completion time.
4. The underestimation of the project completion time due to alternate paths becoming critical is perhaps the most serious of these issues. To overcome this limitation, Monte Carlo simulations can be performed on the network to eliminate this optimistic bias in the expected project completion time.

Critical Path Method

PERT chart for a project with five milestone (10 through 50) and six activities (A through F). The project has two critical paths: activities B and C, or A, D, and F - giving a minimum project time of 7 months with fast tracking. Activity E is sub-critical, and has a float of 2 months.

The Critical Path Method, abbreviated CPM, or Critical Path Analysis, is a mathematically based algorithm for scheduling a set of project activities. It is an important tool for effective project management.

It was developed in the 1950s by the DuPont Corporation at about the same time that General Dynamics and the US Navy were developing the Program Evaluation and Review Technique (PERT) Today, it is commonly used with all forms of projects, including construction, software development, research projects, product development, engineering, and plant maintenance, among others. Any project with interdependent activities can apply these methods of scheduling.

The essential technique for using CPM is to construct a model of the project that includes the following:

1. A list of all activities required to complete the project (also known as Work breakdown structure).
2. The time (duration) that each activity will take to completion, and
3. The dependencies between the activities.

Using these values, CPM calculates the longest path of planned activities to the end of the project, and the earliest and latest that each activity can start and finish without making the project longer. This process determines which activities are "critical" (i.e., on the longest path) and which have "total float" (i.e., can be delayed without making the project longer). In project management, a critical path is the sequence of project network activities which add up to the longest overall duration. This determines the shortest time possible to complete the project. Any delay of an activity on the critical path directly impacts the planned project completion date (i.e. there is no float on the critical path). A project can have several, parallel, near critical paths. An additional parallel path through the network with the total durations shorter than the critical path is called a sub-critical or non-critical path.

These results allow managers to prioritize activities for the effective management of project completion, and to shorten the planned critical path of a project by pruning critical path activities, by "fast tracking" (i.e., performing more activities in parallel), and/or by "crashing the critical path" (i.e., shortening the durations of critical path activities by adding resources).

Originally, the critical path method considered only logical dependencies between terminal elements. Since then, it has been expanded to allow for the inclusion of resources related to each activity, through processes called activity-based resource assignments and resource leveling. A resource-leveled schedule may include delays due to resource bottlenecks (i.e., unavailability of a resource at the required time), and may cause a previously shorter path to become the longest or most "resource critical" path. A related concept is called the critical chain, which attempts to protect activity and project durations from unforeseen delays due to resource constraints.

Since project schedules change on a regular basis, CPM allows continuous monitoring of the schedule, allows the project manager to track the critical activities, and alerts the project manager to the possibility that non-critical activities may be delayed beyond their total float, thus creating a new critical path and delaying project completion. In addition, the method can easily incorporate the concepts of stochastic predictions, using the Program Evaluation and Review Technique (PERT) and event chain methodology.

Currently, there are several software solutions available in industry that use the CPM method of scheduling, see list of project management software. However, the method was developed and used without the aid of computers.

The Critical Path is similar to the method of PERT, both consist about times and the ending of the production, if Critical Path is modified, It won't work like a real administration tool.

A schedule generated using critical path techniques often is not realized precisely, as estimations are used to calculate times: if one mistake is made, the results of the analysis may change. This could cause an upset in the implementation of a project if the estimates are blindly believed, and if changes are not addressed promptly. However, the structure of critical path analysis is such that the variance from the original schedule caused by any change can be measured, and its impact either ameliorated or adjusted for. Indeed, an important element of project postmortem analysis is the As Built Critical Path (ABCP), which analyzes the specific causes and impacts of changes between the planned schedule and eventual schedule as actually implemented.

Personnel Management

Personnel management can be defined as obtaining, using and maintaining a satisfied workforce. It is a significant part of management concerned with employees at work and with their relationship within the organization.

According to Flippo, "Personnel management is the planning, organizing, compensation, integration and maintenance of people for the purpose of contributing to organizational, individual and societal goals."

According to Brech, "Personnel Management is that part which is primarily concerned with human resource of organization."

Nature of Personnel Management

1. Personnel management includes the function of employment, development and compensation. These functions are performed primarily by the personnel management in consultation with other departments.

2. Personnel management is an extension to general management. It is concerned with promoting and stimulating competent work force to make their fullest contribution to the concern.

3. Personnel management exists to advice and assist the line managers in personnel matters. Therefore, personnel department is a staff department of an organization.

4. Personnel management lays emphasize on action rather than making lengthy schedules, plans, and work methods. The problems and grievances of people at work can be solved more effectively through rationale personnel policies.

5. It is based on human orientation. It tries to help the workers to develop their potential fully to the concern.

6. It also motivates the employees through it's effective incentive plans so that the employees provide fullest co-operation.

7. Personnel management deals with human resources of a concern. In context to human resources, it manages both individual as well as blue-collar workers.

Role of Personnel Manager

Personnel manager is the head of personnel department. He performs both managerial and operative functions of management. His role can be summarized as

1. Personnel manager provides assistance to top management - The top management are the people who decide and frame the primary policies of the concern. All kinds of policies related to personnel or workforce can be framed out effectively by the personnel manager.
2. His advices the line manager as a staff specialist- Personnel manager acts like a staff advisor and assists the line managers in dealing with various personnel matters.
3. As a counsellor - As a counsellor, personnel manager attends problems and grievances of employees and guides them. He tries to solve them in best of his capacity.
4. Personnel manager acts as a mediator - He is a linking pin between management and workers.
5. He acts as a spokesman - Since he is in direct contact with the employees, he is required to act as representative of organization in committees appointed by government. He represents company in training programmes.

Functions of Personnel Management

Following are the functions of Personnel Management:

1. Manpower Planning
2. Recruitment
3. Selection

Management Information System (MIS)

Although management information systems (MISs) vary considerably in scope and function, they are all software applications used to support basic business management activities, such as reporting, planning, and controlling. More broadly, the MIS label is also applied to the computer systems that run this software, but strictly speaking an MIS is considered software rather than hardware. In addition, some corporate information technology (IT) departments are named MIS, and the term may also be used to describe the overall management of IT resources. For purposes of this discussion, though, MIS will refer only to software systems.

Historical Development

MISs has existed since the 1960s, when mainframe computers began making significant inroads in automating information-based activities at large corporations. Historically, they represent roughly the second generation of business software applications, an intermediate step between the very basic, large transaction-centered systems of the 1950s and early 1960s and the more specialized software tools that started taking root in the 1970s, such as word processing and decision support systems. The first MISs, however, often weren't particularly useful because of equipment shortcomings, a lack of computer literacy among the intended users, and poor planning for which functions the MIS would best serve. As computers continued to grow cheaper and the software more user friendly, MISs gained wider acceptance in the 1970s.

The diversity of software suites and applications that followed, particularly in the 1980s and 1990s, substantially blurred many of the distinctions between MISs and other types of management support applications, particularly as MISs began delivering data nearly in real time. Today, systems that serve in traditional MIS-like roles may or may not be known as MISs; indeed, very few new software packages are called MISs.

Defining Features and Components

Nonetheless, MISs are still widely used and have a few distinguishing characteristics. First, they are generally used to generate routine reports for functional areas of a business, e.g., payroll statistics, financial reports, or sales reports. As such, they are mostly intended to be used by middle managers rather than entry-level staff or senior management. Finally, most MISs are linked with one or more additional business systems, such as transaction processing systems (TPSs) or executive information systems (EISs). Often this integration is achieved through networking, enabling the MIS to pull data from diverse systems such as the corporate accounting system, the human resources system, and so on.

MISs is no longer the exclusive domain of mainframes, although these powerful computers are still used effectively to house many corporate MISs. Often minicomputers, also known as midrange systems, run MISs instead, and even workstations and desktop PCs are capable of running small MISs.

The core of any MIS is a database management system (DBMS) and a series of data reporting, retrieval, and manipulation tools. Some of these tools are standardized for common or routine actions, such as generating monthly sales reports. Often these tools also allow output to be customized for special analyses. Examples of standard MIS output might include historical reports, short-term forecasts, and exception reports. Exception reports, related to the managerial philosophy of "management by exception," are triggered when some predefined unusual or problematic event occurs, such as an uncommonly large transaction, extremely high or low sales volume, or a statistical discrepancy that suggests a data error or an unexplained use of resources. Exception reports are intended to alert managers to circumstances that are either much better or much worse than normal. Custom reports may be completely user-defined, in which the user chooses which information to see and how it should be displayed, or they may be fairly standard reports executed in special ways, such as tallying weekly or daily sales rather than an end-of-the month summary.

Recent Trends Overshadow MIS

As information technology has grown more powerful and ubiquitous in business settings, the role of the traditional MIS has been obscured by newer software applications. For example, since the mid-1990s a suite of various applications known collectively as enterprise resource planning (ERP) software has taken a dominant position among large corporations' critical business applications. These packages, which typically come in functional modules such as an accounting module, a human resources module, and a manufacturing supply chain module, serve many of the functions that a traditional MIS would, and they tend to be more flexible, integrated, and user-friendly than legacy MISs. While early implementations of ERP suites were largely cross-functional databases with minimal high-level management tools, later upgrades have added decision support and data manipulation tools to facilitate a wide range of management-level—and even executive-level—analyses.

Another trend supplanting conventional MISs since the 1980s has been the adoption of executive information systems (EISs) and other high-level decision support systems. EISs are designed for top executives to glean important information from all major business systems (finance, inventory, payroll, and so on), process it in very sophisticated ways, and possibly even integrate it with outside data such as competitive intelligence or industry statistical norms. By providing such highly refined control over strategic information, EISs have diminished the importance of the more routine kinds of analyses available from an MIS.

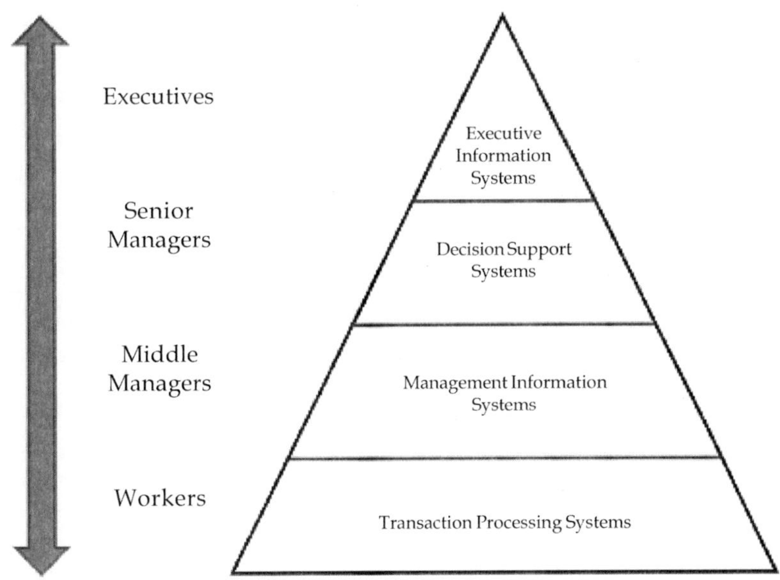

Characteristics of MIS

To function **effectively** as an interacting, interrelated, and interdependent feedback tool for management and staff, MIS must be "useable." The five elements of a useable MIS system are: timeliness, accuracy, consistency, completeness, and relevance. The usefulness of MIS is hindered whenever one or more of these elements are compromised.

1. *Timeliness :* To simplify prompt decision making, an institution's MIS should be capable of providing and distributing *current* information to appropriate users. Information systems should be designed to expedite reporting of information. The system

should be able to quickly collect and edit data, summarize results, and be able to adjust and correct errors promptly.

2. *Accuracy* : A sound system of automated and manual internal controls must exist throughout all information systems processing activities. Information should receive appropriate editing, balancing, and internal control checks. A comprehensive internal and external audit program should be employed to ensure the adequacy of internal controls.

3. *Consistency* : To be reliable, data should be processed and compiled consistently and uniformly. Variations in how data is collected and reported can distort information and trend analysis. In addition, because data collection and reporting processes will change over time, management must establish sound Procedures to allow for systems changes. These procedures should be well defined and documented, clearly communicated to appropriate employees, and should include an effective monitoring system.

4. *Completeness* : Decision makers need complete and pertinent information in a summarized form. Reports should be designed to eliminate clutter and voluminous detail, thereby avoiding "information overload."

5. *Relevance* : Information provided to management must be relevant. Information that is inappropriate, unnecessary, or too detailed for effective decision making has no value. MIS must be appropriate to support the management level using it. The relevance and level of detail provided through MIS systems directly correlate to what is needed by the board of directors, executive management, departmental or area mid-level managers, etc. in the performance of their Jobs.

An MIS provides the following advantages

1. *It Facilitates planning* : MIS improves the quality of plants by providing relevant information for sound decision – making. Due to increase in the size and complexity of organizations, managers have lost personal contact with the scene of operations.

2. *It Minimizes information overload* : MIS change the larger amount of data in to summarize form and there by avoids the confusion which may arise when managers are flooded with detailed facts.

3. *MIS Encourages Decentralization :* Decentralization of authority is possibly when there is a system for monitoring operations at lower levels. MIS is successfully used for measuring performance and making necessary change in the organizational plans and procedures.
4. *It brings Co ordination :* MIS facilities integration of specialized activities by keeping each department aware of the problem and requirements of other departments. It connects all decision centers in the organization.
5. *It makes control easier :* MIS serves as a link between managerial planning and control. It improves the ability of management to evaluate and improve performance. The used computers has increased the data processing and storage capabilities and reduced the cost.
6. MIS assembles, process, stores, Retrieves, evaluates and disseminates the information.

Disadvantages

Allocation of budgets for MIS upgrades, modifications and other revisions can be quite tricky at times. If budgets are not allocated uniformly or as per immediate requirements, key functionalities might get affected and benefits might not be realized consistently. Integration issues with legacy systems can affect the quality of output and vital business intelligence reports.

Importance of MIS

1. MIS is always management oriented and keeps in view every level of management and gets the desired information.
2. Integrated - refers to how diff components (sub systems) are actually tied up together. eg: diff departments of organization linked together.
3. Useful for planning - as every organization makes log-term and short-term plans with the help of information like sales & production, capital investments, stocks etc management can easily plan.
4. Effective MIS helps the management to know deviations of actual performance from pre-set targets and control things.
5. It's important for increasing efficiency.

6. MIS provides updated results of various departments to management.
7. MIS is highly computerized so it provides accurate results.
8. MIS adds to the intelligence, alertness, awareness of managers by providing them information in the form of progress and review reports of an ongoing activity.
9. Helps managers in decision- making.

National Academy of Agricultural Research Management (NAARM)

- Indian NARS is one of the largest systems in agricultural research in the world employing over 25,000 scientists, who work in institutes distributed throughout the entire length and breadth of the country. Judicious and proper management of this vast manpower poses a formidable challenge to the system. Keeping in view this objective, the Government of India has established the National Academy of Agricultural Research Management (NAARM), Hyderabad. Since its inception in 1976, the Academy has expanded its horizon of activities over the years. Through its innovative training and research programmes in the specialized areas of Agricultural Research and Education Management, it has helped the systems in evolving appropriate policies to improve the efficiency and effectiveness of research and education in particular and agriculture at large.

- Its scope has been further widened to encompass tailor-made training, especially for the international participants from the developing countries in Asia, Africa and Latin America. Research on management problems has formed another major activity of the Academy. This has provided an input for improvising training and the policy support. The Academy extends a facilitative role to the ICAR in developing policy guidelines on the various issues related to personnel, Organization and Management reforms, human resource development, assessment system, and others, all of which are aimed at improving the overall efficiency and effectiveness of the NARS. NAARM has been developing itself into a resource centre for management needs. On the strength of the experienced and expert faculty, the Academy has started offering specialized consultancy services in the areas of its mandated activities.

- The Academy strives to generate a sense of fraternity and inculcate a scientific work culture amongst the agricultural scientists and teachers in the country. Being a unique institution of its kind, the Academy is well poised to augment its usefulness and aims to emerge as an institution par excellence to facilitate and support a culture of dynamic management in agricultural research and education.

NIRD (National Institute of Rural Development)

It is an autonomous organization under the Ministry of Rural Development that undertakes training, research, action research and consultancy assignments in the rural development sector in India. The Institute is located in the serene rural surroundings of Rajendranagar about 15 km away from the historic city of Hyderabad in Andhra Pradesh.

The NIRD is mandated to conduct and assist in the organization of training programmes, conferences, seminars and workshops for middle and senior development functionaries; undertake, aid, promote and coordinate research on it's own and / or collaborate with state, national and international development agencies; analyze and offer solutions to problems encountered in planning and implementationof the programmes for rural development, decentralized governance, panchayati raj and related programmes; and disseminate information and transfer technology through periodicals, reports, books and other publications in furtherance of the basic objectives of the Institute.

The Institute's services are available to different ministries/ departments of the Central and State Governments, banking institutions, public and private sector organisations, civil societies, Panchayati Raj Institutions and other national and international agencies connected with rural development.

Regional Centre

The Regional Centre of NIRD, established at Guwahati (Assam) in 1983, conducts training courses, seminars, workshops and problem-oriented research studies on rural development policy and planning relevant to the North-Eastern Region covering the states of Assam, Arunachal Pradesh, Manipur, Meghalaya, Mizoram, Nagaland, Tripura and Sikkim.

Structure and Organization

The NIRD policy is determined by a forty-seven member General Council. The Union Minister of Rural Development is the President of the General Council. The management and administration of the Institute is vested in a fourteen member Executive Council with the Hon'ble Union Minister, Rural Development as its Chairman. The Director General is the CEO and is responsible for the management of the Institute. He is assisted by a Deputy Director General, Financial Advisor, Registrar and Director (Administration) and a multi disciplinary Faculty. The Academic Committee and Research Advisory Committee, under the Chairmanship of the Director General, facilitate planning of its training, research and consultancy activities.

Faculty

The faculty members are highly qualified with Doctoral Degrees, Specialised training and several years of experience in Rural Development Training, Research and Consultancy. The seventy strong multi-disciplinary faculties is assigned to eighteen thematic Centres as shown in the Annexure. For more details about the qualifications and experience of individual faculty members, please visit our website:

Training

The NIRD has a wide range of expertise and excellent infrastructure to train middle and senior development functionaries engaged in policy formulation, management, and implementation of rural development programmes. The training programmes are intended to create knowledge base, develop skills and infuse right attitudes and values. Over 300 programmes are offered every year. Besides, the Institute conducts several international training programmes sponsored by the Ministry of External Affairs and the Ministry of Rural Development. Participants from several Afro-Asian and Latin-American countries undergo training at the NIRD every year. About 8000 trainees pass through the NIRD annually.

Research and Action Research

The Institute undertakes empirical studies on specific themes like Rural Change; Socio-Economic and Human Development; Decentralization; Programme Planning, Management, Monitoring and Evaluation; Rural Technology and Environmental Concerns, among others, to assess the impact of rural development programmes, identify new areas for policy interventions and analyze the research and training needs of the Rural Development functionaries. The Action Research programmes of the institute are location–specific in nature and scope. Development models/themes are given a fair trial.

Consultancy

Consultancy assignments are taken up for Ministries/Departments of Government of India, National level organizations, state governments, public sector undertakings and premier International organisations like UNDP, FAO, UNICEF, UNESCO, WHO, ILO, World Bank, ESCAP, DFID, CIRDAP and AARDO.

Information Dissemination

The Institute disseminates the results of the research studies and recommendations of its various seminars and workshops through a number of publications. It brings out a quarterly Journal of Rural Development that is acclaimed for its quality and a bimonthly NIRD Newsletter in English and Hindi. With over 300 publications to date, NIRD is a leading publisher of Rural Development literature not only in India but also in the region. The Institute's annual publications like 'India Rural Development Report', 'Panchayati Raj Report' and 'Rural Development Statistics' have attracted the attention of academics and policy planners.

Computer Centre

The Institute has a well equipped Computer Centre for in-house and classroom Training, Management Information System (MIS), Library documentation and data processing. It has a computer lab with 'state of the art' PCs. Other facilities include INTERNET and Local Area Network (LAN) connecting all the departments and Centres of the Institute.

Rural Technology Park

The Rural Technology Park (RTP) spread over 62 acres of land aims at spreading awareness on current rural technologies and promoting sustainable rural development. The RTP showcases Rural Technologies and provides for capacity building in terms of skill development and entrepreneurship. Transfer of technology process and activities are carried out for sustainable livelihoods. The RTP has established some production-cum-training units of affordable and costeffective technologies for rural areas, and offers skill oriented training. The solar power unit is managed by illiterate women who have become barefoot solar engineers and execute turnkey solar electrification projects in the rural areas. The National Rural Building Centre and the sanitation park in RTP showcases model rural houses constructed with forty different technologies applicable to Rural India.

Documentation

The Centre for Media and Rural Documentation (CMRD) of the Institute acts as a referral and servicing agency for the rural development sector as a whole. It has a well organised library with a collection of more than lakh volumes and subscribes to over 200 journals. The Rural Development database consisting of over two lakh references is computerised. It maintains the Institute's website and operates a web-based ICT portal.

Geomatics Cell

The Centre for Geo-Informatics Application in Rural Development (C-GARD) caters to the needs of GIS, GPS and Satellite Remote Sensing Applications in rural areas. It offers custom designed solutions to Natural Resource Management and rural development oriented projects.

Indian Institute of Plantation Management Established 1990

Type of Institute-Autonomous College/Institute

Realizing the potentialities and needs of modernization of the plantation sector through management education and training, Ministry of Commerce & Industry, Government of India constituted a core group in 1990 to set up a strategic institution of management education in the plantation sector. Accordingly a blue-print for the Indian Institute of Plantation Management, (IIPM) Bangalore was prepared.

After the acceptance of the blueprint by the Government, a steering committee of the Ministry of Commerce provided the initial thrust for setting up the Institute. In November 1993, the Institute was registered as an autonomous educational institution of higher learning, under the Karnataka Society Registration Act. It is jointly sponsored by the plantation sector and the Commodity Boards viz. Coffee Board, Tea Board, Rubber Board, Spices Board and Plantation Associations, viz. UPASI and Indian Tea Association (ITA).

International Institute of Plantation of Management (IIPM)/UiTM

It is a professional body representing all sectors of the plantation industry. It was founded in May 2006 as an offshoot of the University of Technology MARA (UiTM) Malaysia. The purpose of its establishment is to promote knowledge of the science and art of planting management and to provide a source of authoritative views for communication to the government, industry and community.

The main objectives of the Institute are:

- To upgrade and uplift the profession of planters.
- To provide an organization and membership for planters and to enhance their continuing development.
- To serve as the national reference centre for the planting industry.

EEI (Extension Education Insitute)

Extension Education Institute is a regional training institute, run by Govt. of India, to impart trainig to the extension functionaries of different line departments as well as SAUs and ICAR institutes.

Differences

Table 1 : Difference in Co-ordination and Co-operation

Basis	Co-ordination	Co-operation
Meaning	It is an orderly arrangement of group efforts in pursuit of common goals.	It means mutual help willingly.
Scope	It is broader than co-operation which includes as well because it harmonizes the group efforts.	It is termed as a part of co-ordination.
Process	The function of co-ordination is performed by top management.	The functions of co-operation are prepared by persons at any level.
Requirements	Co-ordination is required by employees and departments at work irrespective of their work.	Co-operation is emotional in nature because it depends on the willingness of people working together.

Table 2 : Difference in delegation and decentralization

Basis	Delegation	Decentralization
Meaning	Managers delegate some of their function and authority to their subordinates.	Right to take decisions is shared by top management and other level of management.
Scope	Scope of delegation is limited as superior delegates the powers to the subordinates on individual bases.	Scope is wide as the decision making is shared by the subordinates also.
Responsibility	Responsibility remains of the managers and cannot be delegated	Responsibility is also delegated to subordinates.
Freedom of Work	Freedom is not given to the subordinates as they have to work as per the instructions of their superiors.	Freedom to work can be maintained by subordinates as they are free to take decision and to implement it.
Nature	It is a routine function	It is an important decision of an enterprise.
Need on purpose	Delegation is important in all concerns whether big or small. No enterprises can work without delegation.	Decentralization becomes more important in large concerns and it depends upon the decision made by the enterprise, it is not compulsory.
Grant of Authority	The authority is granted by one individual to another.	It is a systematic act which takes place at all levels and at all functions in a concern.
Grant of Responsibility	Responsibility cannot be delegated	Authority with responsibility is delegated to subordinates.
Degree	Degree of delegation varies from concern to concern and department to department.	Decentralization is total by nature. It spreads throughout the organization i.e. at all levels and all functions
Process	Delegation is a process which explains superior subordinates relationship	It is an outcome which explains relationship between top management and all other departments.
Essentiality	Delegation is essential of all kinds of concerns	Decentralization is a decisions function by nature.

Significance	Delegation is essential for creating the organization	Decentralization is an optional policy at the discretion of top management.
Withdrawal	Delegated authority can be taken back.	It is considered as a general policy of top management and is applicable to all departments.
Freedom of Action	Very little freedom to the subordinates	Considerable freedom

Table 3 : Differentiate Between PERT and CPM

PERT	CPM
PERT(Program Evaluation and Review Technique)	Critical Path Method
PERT is a probabilistic tool	CPM is a deterministic tool
This tool is basically a tool for planning	CPM also allows and explicit estimate of and control of time. Costs in addition to time, therefore CPM can control both time and cost.
PERT is more suitable for R&D related	CPM is best suited for routine and those projects where the project is performed for projects where time and cost estimates can the first time and the estimate of duration be accurately calculated. are uncertain.
The probability factor is major in PERT	The deterministic factor is more so values or so outcomes may not be exact. Outcomes are generally accurate and realistic.
PERT models have a fixed time period	whereas CPM allows for variation in time estimates/activity times
Nodes or circles to represent activities or tasks in the project for PERT	In CPM, activities are represented by (labels written on) lines with arrows between nodes or circles.

Previous Year's Questions ARS/NET

1. Define Management. Mention its function. What do you mean by management process?
2. Explain the process of Monitoring, Evaluation and impact analysis of an extension programme.

3. Write briefly about P.E.R.T. project evaluation and review technique.
4. P.R.A. /R.R.A.
5. Briefly explain need hierarchy and (motivation) self - actualization theory.
6. Write briefly about two factor theory of motivation.
7. Define the term evaluation. What are the techniques used in evaluation of extension programmes? Explain the importance of Social audit in evaluating extension programmes.
8. What are the difference between co-operation and coordination? What are the approaches envisaged in seeking coordination between State Agricultural University and state agricultural department under T&Y system?
9. What is day you mean by work motivation? How is it important for managing an extension organization?
10. What are the different types of motives? How motivation differs among people working at different levels? Illustrate with examples and theories?
11. What do you mean by decision making? What are the steps invaded in decision making? Explain how the decision making process of the farmer in different form that of a business managers.

1. MC Gregors theory of X & Y
2. Two factor theory of Herzberg
3. Equity theory
4. Self fall felling prophecy
5. Self actualization need
6. Briefly explain any four of the following
   - Commitment principle
   - Managerial principle
   - Appraisal
   - Scalar Principle
   - MBO
   - Scientific Management Theory
   - Departmentation

7. Line and Staff function
8. Managerial grid
9. Hawthorne experiment
10. Technology cluster
11. Listening
12. Stand food - Bennet Scale
13. Thematic Appreciation Test (TAT)
14. E.L.C.
15. Theory of Bureaucracy
16. Scientific Management
17. Hierarchy of needs
18. Syndicate method

❑❑❑

Chapter 6

Human Resource Development

Approaches to Training

There are three approaches

1. Discrete appraoch
2. The process appraoch
3. The system appraoch

Concept of Performance Appraisal

Concept of performance appraisal:- According to Flipps "Performance appraisal is the systematic, periodic and an impartial rating of an employess excellence in matter pertaining to his present job and his potential for a better job".

Characteristics of Performance Appraisal

1. It is systematic examination of employees strength and weakness in terms of the job.
2. PA is scientific or objective study.

3. It is an ongoing or continuous process
4. The main purpose of PA is to secure information necessary for making objective and correct decisions on employees

PA process

Consist of following steps

1. Setting the performance standards
2. Communicating standards
3. Measuring performance
4. Comparing with standards
5. Discussing results
6. Taking corrective action.

Types of PA/Kinds of PA on the basis of measuring performance against standards

1. Formal comprehensive review:- (eg annually)
2. Progress or periodic reviews:- (eg for milesner, quarterly, monthly)
3. Continuous monitoring (daily, with emphasis on self-control)

Performance Appraisal

Traditional Methods	**Modern Methods**
1. Confidential report	1. Assessment control
2. Free forum/essay	2. Human resource account
3. Straight ranking	3. Behaviourally anchored rating
4. Paired distribution	4. Appraisal through MBO
5. Forced distribution	5. 360° Appraisal
6. Graphic rating scale	
7. Checklist method	
8. Critical incident technique	
9. Group appraisal	
10. Field review	

Performance Appraisal

Systematic evaluation of the performance of the employee and to understand the abilities of a person for future growth and development PA is generally done in systematic way which are as follows:

Objectives of performance appraisal

1. To maintain record in order to determine compensation packages, wage structures salaries raises etc
2. To identify the strengths and weaknesses of employee
3. To maintain and assess potential
4. To provide feedback
5. To provide a basis to influence working habit of an employee
6. To review and retain the promotional and other training programme

Advantages of PA

1. Promotion
2. Compensation
3. Employees development
4. Selection validation
5. Communication
6. Motivation

PA biases

1. First impression (Primary effect)
2. Halo effect
3. Horn effect
4. Excessive stiffness or lenience
5. Central tendency
6. Personal biases
7. Spillover effect
8. Regency effect
9. Social differentiation

Approaches of PA Types

Three approaches are used:-

1. Casual unsystematic
2. Traditional and highly systematic
3. Behavioral approach

Management approaches (ARS 2010)

1. Classical approach	Scientific Management Bureaucratic administrative	Fredrick, Taylor, Frank and Gibirth Weber H. Fayol
2. Behavioral approach	Group influence Hawthorne studies Maslow need theory x&y model	
3. Quantitative approach	Management science, operation nanagement, management information system	
4. Modern approaches	The systems theory, contingency theory, emerging approaches, theory Z, and quality Management	William Ouchi

PI Define : Method where by a trainee teaches himself by going through a series of steps, all leading to carefully designed goals or objectives.

Elements

a. Programme - Information to be taught
b. Programmer/programme writer
c. Target population

Basic theory of programmed learning

- Dr. Skinner is the originator of what is known in programmed learning as linear programming. Every student works through all the sequences of the book straight path as it were, from page on till end. At the end of every step he is called upon to repy to questions and then correct answer is given. If he is right, he goes on to the next section. If not, he rereads the whole section/ information until he understand it.

- *Intrinsic/branching programming :* Dr. Crowder is the pioneer of this. The trainee is presented with a piece of information and called upon to give a reply. He selects a correct reply from among several possible answers, he is told that he is wrong and why. Sometimes he is asked to reread the information and try again. Sometimes he is given extra explanation and differently worded question. In this way student works through the programme by one of several paths branches. Different students would work through the programme by different paths since the students choice of reply determines which path of learning he would take through the programme this form of programming is known as intrinsic
- *Adaptive programming* : Gordon pask, a British a form of programming known as adoptive programming. This form of programming makes allowances for more variations in students ability. These programmes can only be presented in machines which adopt to the ability of trainee, hence the name.

There may be differences in the forms of programming. But they are to some extent based on a theory of:

STIMULUS	RESPONSE	FEEDBACK
Information's Question	Reply	Knowledge of Results

Basic features of programmed learning:

1. Own pace
2. Logical series of steps/or frames
3. Active responding
4. Master each step
5. Immediate feedback
6. All leading to predated goals or objectives
7. Continuous motivation
8. The programme is thoroughly tested before it is printed

Difference between programmed learning and other methods

Programmed Instruction	Other methods
1. Self instructional	No
2. Subject matter is thoroughly analyzed, and each sequence carefully designed to reach predetermined objectives	Not so
3. Trainees are called upon to take part actively at each step and they receive immediate knowledge of the results	
4. Programmes are thoroughly tested and proved to teach what they are intended to teach.	

Evaluation : Evaluation is a process of systematic appraisal by which we determine the worth value or meaning of something.

Evaluation of training may be defined as:- Determining the extent to which training programme achieved what it set out to achieve".

Levels of evaluation

1. Reaction level
2. Learning level
3. Job behavior level
4. Functional level

Evaluation methods/Tools/techniques

1. Anecdotal method
2. Briefing and debriefing
3. Checklists
4. Conversation sheets
5. Discussion with participants
6. Document analysis
7. Expert opinion
8. Ground assessment
9. Illustration technique
10. Information technique
11. Inventories

12. Job analysis
13. Known group techniques
14. Media techniques
15. Organizational diagnosis
16. Peer Rating
17. Projective technique
18. Report analysis
19. Record appraisal
20. Round table assessment
21. Self reporting surveys
22. Criterion group
23. Case studies
24. Critical incident technique
25. Historical methods
26. Observations
27. Questionnaires
28. Interviewing
29. Scaling techniques
30. Semantic tests
31. Sociometric tests
32. Test and re-testing
33. Time series analysis
34. System analysis
35. Cost benefit analysis
36. Factor analysis
37. Programme analysis
38. In-depth studies

Effective use of evaluation methods

The effective use of evaluation methods requires the following:-

1. Planning in advance and adherence to a time frame.
2. Assessment of suitability and availability of each method and desirability of combining them.

3. Selecting appropriate tools and techniques under each methods.
4. Overcoming subjective biases while using an evaluation method mix, thereby testing reliability
5. Making a dummy rehearsal and preparing blue print for overcoming practical diffciulties in facing the field realities.
6. Budgeting time, cost, resources, and logistic support for avoiding wastage and to achieve economy.
7. Testing the tools techniques and materials so prepared by asking the following questions

A. Classification of evaluation methods Types

1. Sponsor/ Institutional dominated method
2. Cooperative/Team dominated method
3. Functional/purpose dominated method
4. Project success or happiness dominated method.

1	2	3	4	5	6
Student/trainee background	Teacher/agent	Day/date	Topic	Method	Objective of Evaluation K/S/A oathers

Training for extension personnel:- May be defined as a planned and systematic effort to increase knowledge (k), improve skills(s), inculcate appropriate attitude (a), and develop other attributes (os) in extension personnel to enable them to better serve their clients the farmers.

In other words training is concerned with KSAO's and may be described as function (f) as follows:-

T=f (C,K, S, A, Os)

Training Typology

A. According to stages of career

1. Preservice training
2. Orientation training
3. Induction training

4. In-service
5. On the job
6. Refresher
7. Re-training
8. External training
9. Training for professional qualification
10. Overseas training
11. Training by overseas experts

B. According to subject matter specialization

The second criterion on the basis of which a typology of training can be developed is according to subject matter specialization. According to this training could be of 25 broad categories.

C. According to level of extension personnel

Easily understood and most popular

i. Field level extension personnel
ii. Middle level
iii. Top level

D. According to type of extension personnel

Training for

i. Field functionaries
ii. Subject matter specialists, and
iii. Extension managers

Points for making farmers training effective

1. Timeliness
2. Availability of inputs
3. A well linkage
4. Teaching illiterates
5. Use local source, conversational formats
6. Training must be followed by demonstration

The process of designing training modules:

1. Determining training need
2. Specifying training objectives
3. Organizing training content
4. Selecting training methods and techniques
5. Identifying needed training resources
6. Assembling and packaging lesson plans
7. Developing training support material
8. Developing test for measuring trainee learning
9. Trying and revising training curriculum
10. Getting feedback from trainees

Charting of training module as follows:

"Training module is a systematically designed, and well prepared document about the schedule of events and method of training to be imparted by the trainers to the trainees."

Entrepreneurship Development

Dynamics

1. Identifying/innovating ideas, product and services
2. Mobilizing resources or organizing production/services
3. Marketing

Characteristics of entrepreneurs

1. Drive energy
2. Capacity to assume risk
3. Taking initiative
4. Internal locus of control and seeking personal responsibility
5. Seek feedback of own efforts
6. Long term involvement
7. Self confidence
8. Money as a measure of performance

9. Orintach powers future goals
10. Persistent probe salary
11. Set high standard for themselves
12. Adventures
13. Tolerance to ambient

Features of HRD

- HRD is a planned and systematic approach to the development of people.
- HRD is a continuous process of developing the competencies, motivation, dynamism and effectiveness of employees.
- HRD is an interdisciplinary concept.
- Micro - HRD is concerned with improving the skills, attitudes and behavior of employees in organizational settings for the benefits of both the individual and the organization.
- Macro - HRD involves improving the authority of life of people in a country.
- HRD is a process not merely a set of mechanical and techniques.

Difference between HRM and HRD

Point of	**HRM**	**HRD**
1. Status	It is an independent function with independent sub function	It is an integrated system counting of interdependent sub system
2. Orientation	It is a sense structure function	It is a proactive function
3. Aims	It seeks to improve the efficiency of people and administration	It seeks to do the total organization and its
4. Instruction	It focuses on salary economic rewards, job simplification and job specialization for moderating people	It focuses on autonomous work groups, job challenges and creativity for motivating people

5. Responsibility of human resources manager	Responsibility of human resources manager	Responsibility of all manager
6. Morale	Improved satisfaction and morale on the cause of improved performance	It consider improved performance on the cause of improved satisfaction and morale.

Most common methods of managemnet are :

1. Observation
2. Management research
3. Group conference
4. Interviews
5. Job or activity analysis
6. Questionnaire surgeon
7. Test or examination
8. Meant or performance ratings
9. Business and production records
10. Personal records
11. Long-range organizational planning

Stress Management : Stress is a dynamic condition in which an individual is confronted with an opportunity, demand or resource related to what the individual desires and for which the outcome is perceived to be both and important:

1. Stress is not simply anxiety
2. Stress is not simply nervous tension
3. Stress is not necessity something damaging his or to be avoided.

Sources of stress

1. Environmental factor:

   a. Economic uncertainty

   b. Political anstability

   c. Technological

   d. Terrorism

2.Organizational factors

a. Lack of Coordination

b. Conflect

3. Personal factors

a. Family problem

b. Economic problem

c. Personality

4. Individual differences

a. Periphery

b. Job experience

c. Social support

d. Belief in laws of control

e. Self-efficiency

❒ Consequences

a. Physiological symptoms:-

1. Physiological

2. Behavioral

b. Psychological symptoms

1. Anxiety

2. Depression

3. Decrease in job satisfaction

c. Behavioral symptoms

1. Productivity

2. Absenteeism

3. Tumor (low)

Managing stress

There are two approaches towards managing stress:-

i. Individual approaches

ii. Organizational approaches

Individual approaches

1. Exercise
2. Relaxation
3. Behavioral self control
4. Cognitive theory
5. Networking
6. Yoga

Organizational approaches

1. Situation and placement
2. Goal setting
3. Training
4. Redesigning jobs
5. Increasing employee involvement in decision making
6. Increasing formal organizational communication
7. Organic wellness programme by the organization

TNA : Training Need Assessment

Systematic identification of the needs and problems of a specific group tend to be overlooked as part of the rural development project planning process.

Why determine training needs?

1. To increase employee productivity.
2. To make every one to perform at his optimum level.
3. To provide trainee to improve his knowledge, skill and attitude.

Formulation of training objective: They should be SMART

S = Specific

M = Measurable

A = Attainable

R = Realistic

T = Time Bound

Methods of TNA

Rational can be viewed as one that, while appearing to be logical as away of determining the training needs, donot necessary involve collection of factual data by objective means	Empirical method involve collection by objective means the relevant information on which decision about the training needs can be based
1. Observation	Job analysis
2. Informal talks	Performance evaluation
3. Complaints	Check list or questionnaire method
4. Comparisons	Tests
5. Analysis of reports	Critical incident technique
6. Opinion poll	Forced choice tech
7. The buzz group technique	
8. Analysis of new programme	

Lecture method : A lecture, thus, may now be defined as a carefully prepared oral presentation of a subject by a qualified person to a group of participants. It may be supplemented with question answer session or audio-visual aids.

- *Application :* The lecture method may be used to accomplish one or more of the following purposes. To present objective view paint, giving a fair and impartial treatment to the subject under consideration. To present a fair analysis of one side of a controversial issue. To direct the audience toward subsequent reading or inquiry. To impart information in a direct and formal manner. To provide incentive for audience participation in presentation on a particular subject using the question answer session after the speaker has finished.
- To entertain or inspire the audience
- To identify a problem.
- To expert a problem.

Physical needs facilities

A. Audience comfort

- It is important is consider temperature and ventilation of the room, hall or auditorum the venue of lecture.
- Audience may be comfortably seated.
- Audience must be able to see and hear speaker. The speaker must be able to see the entire audience.
- Audience must not face glaring light.

Other arrangements

- The room hall or auditorium selected should be appropriate to the size of group and character of seeking
- A stage or plate farm should be provided.
- There should be a speaker's stand with provision of sufficient light for speaker.
- If necessary, a public address system should be secured
- Efforts should be made to have needed audio visual aids suited to the particular be located at appropriate places.
- Serious handicap of lecture is that it is difficult to sustain interest of audience due to its passive role as mere listener.
- Are of audio-visual can help can also be organized services of lectures.

Advantages

No dependency on A.V. Aids most economical

Case methods

It involves thorough investigation of circumstances and cause of maladjustment in a person.

- In case study method, participants are required to provide solution for any one problem presented to them. This method is ideal to train participants, for tackling any problem and to find solution. The trainees study the problem and examine its different aspects. Through this method, the ability to study, analyze and take decision on any problem among trainees is developed.

- In this method every trainee studies the problem and cater on they all tally their results, and also try to find out whether there was any discrepancy in their work or not.
- In this method the trainees get the experience in solving any problem and get the advantage of natural discussion.
- Sharp analytical skill and ignoring is part of the learning objective.

Purpose of Case Study

1. Purpose of case study is to provide various experience to multiply the situations and problems participants can be exposed to during training.
2. Another is to present each sample of real life in slow action, so that through its detailed study participant can care the most from the data offered.
3. Third purpose is to help participants develop the habit of taking more factors into account than usually done to analyze them carefully and to put them together in an integrated picture for guiding understanding and action.
4. In this method instructor itself is participant.
5. Both shortage of skilled trainer and an acute shortage of adequate teaching materials severely limit the use of these training methods in developing countries.
6. Even good material is not guarantee effective incident or case sessions. Few trainers habitually use such material. Hence the danger is very real that these methods will not meet the high expectations with which they were introduced. If these methods were used without the backing of adequate material, skill and institutional support, they are likely to fail.

A case study may be distributed in the form of tape, video tape, written case or oral description. Case study provides indirect experience of a given situation. It holds interest and encourages problem solving.

	Incident	Case
Participants Role	Initiator/contributor/leader	Initiator/contributor/ leader
Data	Specific life situation else where	Specific life situation else where
Trainers role	During question time resource person; then discussion leader	Listener/Clarifies conceptualize
Time required for preparation of one session series of sessions	-	Session timex2 20x2 1/2 hours
Trainer requirement ratio	1-20-40	1-20 7
Time for preparing sessions	10x2 hours	10x2 hours
Facilities required	Circular/hares hoe seating arrangement, charge black boards	Circular/hares seating arrangements cargo black board
Special paints	High initial cost of material 2-4 weeks highly skilled work force each student.	

Group Discussion : Refers to method of learning through open discussion of topics in a small group or Refers to that form of discourse which occurs when two or more persons, recognizing a common problem, exchange and evaluate information and ideas in an effort to solve that problem. Their effort may be directed towards a better understanding of problem or towards the development of the action relating to the problem.

Discussion usually occurs in a face-to-face or co-acting situation, with the exchange being spoken and when more than two people are involved, it usually occurs under the direction of a leader.

Purposes

1. Solving a problem.
2. Exchange information each other.
3. Motivating the group members for better understanding and securing desired action.

4. Planning an action programme
5. Electing or selecting persons for different positions.
6. Entertaining the group members
7. Hearing and discussion over a report/document.
8. Developing desired attitudes
9. Releasing the tension.
10. Training the individuals and groups.

Advantages of group discussions methods

1. Individual learn more as discussion or discussion is a valuable training method.
2. Discussion stimulates attentive listening
3. Sharpens the intelligence of individuals
4. It stimulates straight thinking
5. Individual get opportunity to show his ability to have his say.
6. Provides training to speak in bigger group
7. Broader the mind - discussion makes one more broaden the mind.
8. Individual becomes more tolerant.
9. It is cooperative method knowledge is increased by hearing others.
10. Practice in discussion makes the speaker better able to present his problems to other groups.

Methods of G.D.: Lecture form, panel discussion, colloquium, case study symposium, syndicate method, sensitivity training, workshop, seminar, buzz session, round table, conference, role playing.

It is a type of small group interaction designed to encourage the free introduction of ideas on an unrestricted basis and without any limitations of feasibility.

Brain Storming

1. It is a technique of exploring possible solutions to problems.
2. In this method, there is formation of small groups of trainees.

3. In this method, there is formation of small groups of trainees. There are one or two capable and knowledgeable persons who present the problem before the group of trainees and within a definite period they have to find out the solutions, spontaneity and creativity are important.
4. Its purpose is to promote group creativity, so that all aspects of a problem are considered. It is better that it is restricted to under twenty people generally.
5. The ideas presented are recorded so that everyone can see them. When a large no. of suggestions are made the group is asked to reflect on them and evaluate their merits and demerits. A reasonably few worthwhile solutions are expected to merge out of their exercise.

Syndicate method : It is a sub group of 6-8 participants with a chairman and secretary to work on its own on an assigned problem. Its end result is a report containing recommendations for action syndicate reports are finally presented before the entire group and the trainers.

Discussion : This syndicate discussion give the solution to the problem given them. All the participants analyze the problem, collect the data from library exchange their views and derive benefits from the experience of the subject specialists.

After deliberations, a report is prepared and presented before the members of the syndicate discuses the report. In the end the chairperson of the syndicate presents his views regarding this report. Such type of syndicate method continues for a month or more than one month. In this about 10-12 meetings are organized.

In this method, especially the subject specialists participate. There is in depth discussion on the problem to find out its solution.

In this method time money and study material are required.

1. Decide the problem
2. Discuss the problem
3. Find the solution
4. Prepare report

Management Games / Business Games

It is a training method that deals with certain specific aspects of business or administration. It is a simulated exercise representing, as closely as possible, the constraints and pressures of day-to-day work environment of the participants. The participants are presented with information about operations of an organization - management of human resource, financial marketing etc. They are grouped in functional teams to consider the sequence of events and problems and manage the operations.

The task includes finding allocations to the problems, taking decisions, and dealing with people. Each team takes a decision considering the data and information available to it, within the broad parameters and objectives of the game. The consequences of this decision are to the team by the trainer. The team analyzes the outcomes of this decision in the light of this effect on the situation and other events in the operation and the team then proceeds to take further decisions addressing the new situation, events and circumstances.

Steps in a management game - an important aid in teaching production marketing and financial concepts. Treatment is the key in arriving at decisions.

In Basket Method : As the name indicates, in basket training is structured around the familiar receptacle used for collecting incoming MMFR mail, memos, forms and reports. Materials that require problem solving are put into an in basket. And the students plays the role of a manager responsible for solving the problems found in the in basket. The students are given background information on the personalities and situations involved then using their experience as a guide, students are asked to take the appropriate action with a short time period. Individual or group discussion follows the written part of the exercise. Even students of handling problems usually find that discussions broaden their knowledge of the various possible ways to approach the same problem.

So for the in basked method has been used mostly to learn about effective management and supervision. It has dealt with the concepts of organizing, delegating and planning work. The technique attempts to simulate real life situations. The problems are organized to appropriate work experience as closely or possible.

One typical in basket approach is to ask students to assume students who has just returned from a business and must leave again shortly on another trip. The student manager has twenty minutes to make decisions on materials that have accumulated in the basket. How well can the manager list priorities? What assumptions are made? Is the work distribution planned adequately? Is the work delegated appropriately?

John the approval came for a facilities foreman. Do you want to write up a formal job description for me? Do you want to have panel interview candidates for the job or do you want to handle the interview yourself.

1. Would you delegate the task of making out the job description? If so, to whom?
2. Would you use a panel interview or a personal interview? Why?

T-Groups Lab

T-Groups Lab also lawrance
Diagnostic training group
Sensitivity training group
Conceptual discussion system
Laboratory training
Encounter group

It is a method of changing behavior through instructed group interaction provides insights to one's own behavior, sensitivity to others and improvement in behavioral skills. Has not yet found any significant application in extension training in India. T-groups, or training group, differ radically from more traditional forms of training in that it has no demonstratable goal. The agenda in a sensitivity group or T-group is whatever people want to talk about. People who have been successful in getting things done "find this experience frustrating and at first idiotic with no problems to solve, participants are diverted of their worldly status and authority. They are left to deal with other people without their formal relationship to back them up. This means that relating to others must be reexamined, and new ways tried. Gradually group members are learn to be more honest in their communication. As they become more open with one another, they trust one another more and accept others point of view. They learn to give and take leadership and make group decision without using for their formal relationship to back them up. This means that relating to others must be reexamined, and new ways tried. Gradually group members learn to be more honest in their communication. As they become more open with one another they trust one another more and accept their points of view. They learn to give and take leadership and make group decisions without using formal authority as a crutch.

Sensitivity training makes certain assumption about therapy.

Interacting:- Understanding others - understanding self - sharing and disclosure of feelings - emotions acknowledged experienced and shared - they become more acceptable and more understood - by the self and by others. As are result of such disclosures, changes in behavior often result.

* The purpose of sensitivity training - to provide safe environment in which feeling are conditioned culturally not to share true feelings, T-groups help people to learn how to do this. Another assumptions underplaying sensitivity training is that it is possible.

1. Ego states

Berne defined three basic personalities or Ego states, each with characteristic attitudes, feelings, behaviors' and language. Two of the states subdivided into two further facets:

Parent	Critical Parent	makes rules and sets limits disciplines, judges and criticizes
	Nurturing Parent	advises and guides protects and nurtures
Adult		concerned with data and facts considers options and estimates probabilities makes unemotional decisions plans and makes things happen
Child	Free (Natural) Child	fun-loving and energetic creative and spontaneous
	Adapted Child	compliant and polite rebellious and manipulative

2. Life positions

The other building block of TA is the view we have of ourselves in relation to other people around us. There are four life positions, shown as a grid that became known as the "OK Corral". The quotation in each box typifies the attitude of each Life Position:

I'M NOT OK YOU'RE OK	I'M OK YOU'RE OK
"I wish I could do that as well as you do"	"Hey, we're making good progress now"
I'M NOT OK YOU'RE NOT OK	I'M OK YOU'RE NOT OK
"Oh this is terrible - we'll never make it"	"You're not doing that right - let me show you"

People will move around the grid depending on the situation, but have a preferred position that they tend to revert to. This is strongly influenced by experiences and decisions in early life.

"I'm OK, you're OK" people are in the 'get on with' position. They're confident and happy about life and work, and interact by collaboration and mutual respect, even when they disagree.

I'm OK, you're not OK" people are in the 'get rid of' position. They tend to get angry and hostile, and are smug and superior. They belittle others, who they view as incompetent and untrustworthy, and are often competitive and power-hungry.

I'm not OK, you're OK" is the 'get away from' position. These people feel sad, inadequate or even stupid in comparison to others. They undervalue their skills and contribution and withdraw from problems.

I'm not OK, you're not OK" is the 'get nowhere' position. These people feel confused or aimless. They don't see the point of doing anything, and so usually don't bother.

3. Transactions

The central concept of TA is that Transactions between people can be characterised by the Ego State of the two participants. What's more, the Ego State adopted by the person who starts the transaction will affect the way the other person responds.

For example, Mr A says "what time will they arrive?", and Mr B replies "at 2pm." This is a simple Adult to Adult transaction.

However, if Mr A adopts a Child state: "I'm worried that they might not arrive on time," that will tend to produce a Nurturing Parent response from Mr B: "Don't worry, we'll still have plenty of time to talk to them."

This model shows how the Transaction 'balances' between the two people: if one drifts into Parent, that will encourage the other to move to Child, and vice versa. The preferred state for most business transactions is Adult-Adult, but it's OK to move to another state as long as you're aware of it and are ready for the changed response from the other person.

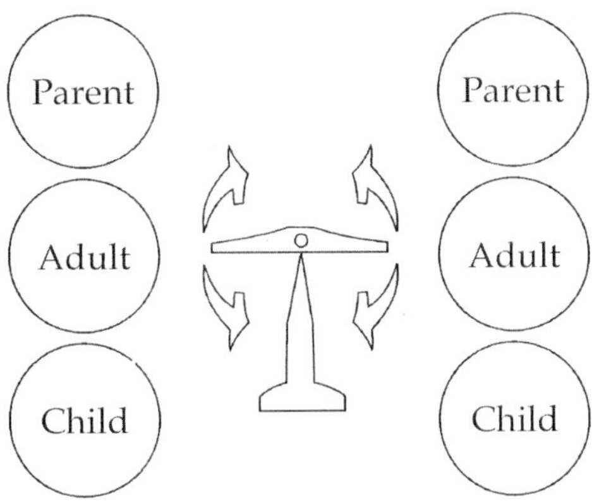

Of course it doesn't always work that way, and an Adult state can sometimes be met with, say, a Parent response: "What time will they arrive?" "How should I know, I didn't arrange it!"

4. Strokes

We all need and seek care, attention, love and recognition from others, and in TA, a stroke is defined as a unit of recognition. With children, strokes are obviously sought and given: they show off their new toy, or misbehave to get attention, and know the adults will respond right on cue. But grown-ups do the same: working hard, deliberately making mistakes, arriving late, or simply arriving home and sighing "what a day!"

Strokes can be positive or negative, and it's generally better to give a negative stroke than none at all (because that may be taken as negative anyway). But in many business organisations, strokes are subject to a set of unwritten rules:

1. Don't give positive strokes freely;
2. If you give positive strokes, make them conditional;
3. Don't ask for positive strokes - certainly not directly;
4. Most positive strokes are insincere ('plastic');
5. Never give a physical stroke - by touching someone;
6. Don't miss a chance to give a negative stroke.

The result is a cold, unfeeling environment where normal human emotions are generally suppressed. Even in 'warm' organisations where it's OK to express feelings, strokes are still subject to certain norms - such as not giving them to people above you in the hierarchy.

In the absence of a free exchange of strokes, people manipulate others in order to get the strokes they crave, and start playing games.

5. Games

The complexity of the TA model leaves it open to manipulation, or "Games". You adopt a Child state because you want someone's help, or a Parent state to make them do something for you. But often the games end up damaging the relationship, and the type of game someone plays is influenced by his or her life state.

Examples of games players are:

The Persecutor: "if it weren't for you", "see what you made me do", "yes, but".

The Rescuer: "I'm only trying to help", "what would you do without me?"

The Victim: "this always happens to me", "poor old me", "go on, kick me".

There are more themes to the TA model than we can show here, but every day you see complex transactions that fit the model. Many effective leaders - and those who are seen as "good with people" - apply the TA principles, often without being aware of it. Even though TA has fallen out of fashion, it's just as applicable today as it ever was. If you're *really* interested in how people interact - and not just wanting to hone your interpersonal skills - then TA is worth exploring.

Developed by the late Eric Bern, T.A. concentrates on styles and content of communication between people. It teaches people to send messages that are clear and responsible and to give responses that are reasonable and national. T.A. attempts to reduce destructive communication habits or games in which intent or full message is obscured.

"When two people interact with each other there it results a social transaction. Analysis of the social transactions is called Transactional Analysis (TA)." A formal definition of TA is that it is "The study of moves people make in their dealing with each other and is based on the idea that peoples interactions resemble moves in game".

It was introduced by Eric Berne and popularized by Thomas Harris in the 1960s advantage of T.A.

1. The bases of TA are simple to learn.
2. The validity of TA is demonsratable
3. It provides a means for reducing the amount of bad feeling experienced of an individual
4. It increases efficient use of time
5. Helps promote effective communication
6. It is applicable at home as it is at work
7. Reinforces and implements other management development activities, including communication, leadership, brainstorming management by objective, job enrichment, conflict resolution, and the like.

TA has certain limitations which are not to be ignored they are :

1. Few scientific studies of its outcome are available they are.
2. Ego states, basic to understand and utilize TA, are difficult to define.
3. If inappropriately applied, TA tends to encourage.
4. TA Jargon may lead to more "Cuteness" than insight into human psychologising.
5. It can be used as a put down, or a discount in inter-personal relations.

Experiential Learning

Refers to the type of learning obtained through transformation of experience into knowledge.

Experience

1. *Training as process :* Training process consists of 4 components and three phases.

Phases

1. Initiation or pertraining phase
2. Motivation or training phase
3. Development or post training phase

Components

1. Planning:
    1. Job analysis - trainee analysis
    2. Training need assessment
    3. Determining training objective
2. Preparation:
    1. Select and organize content.
    2. Select training techniques, methods and aids
    3. Prepare lesson plan
    4. Plan evaluation
3. Presentation:
    1. Conduct training
    2. Evaluate
    3. Review and revise training for future programme
4. Training evaluation
    1. Job improvement plan
    2. Review and revise tray
    3. Post training contact

Basic Training Methods : T.S.D.

1. Telling	2. Showing	3. Doing
Include both talking and listening		
Lecture	Written words	Role playing
Seminar	Pictures	Business games, skits
Symposium	Movie films/TV	Syndicate
Panel discussion	Slide/Film strips	Programmed instruction
Workshop	Charts/Diagrams	Project, committee work
Group discussion	Physical objects	Field work
Forum	Tour	
Debate		
Case study		
Incident process		
Lab/sensitivity		

Factors determining selection of training methods:

1. Objectives to be achieved
2. Number of the participants
3. Background of the participants
4. Time available for training
5. Skill level and preferences of the instructor
6. Facilities available
7. Cost involved

Kurt lewin - changes in behavior

a. Unfreezing
b. Changing
c. Refreezing

Result Demonstration

RD is a way of showing people the value or worth of improved practice whose success has already been established on the research station, followed by district trials or observation plots.

In this method the new practice is compared with the old. Such demonstration required a substantial period of time and records need to be maintained. It is in no sense an experiment or a trial except perhaps in the mind of the co-operator.

It may include single, series of practices or entire form.

RD may be:

1. Varietal
2. Manurial
3. Cultural
4. Combination of above

There are two common sense principles underlying this method.

What a farmer himself does or sees he will believe?

What is good for one person will have general application to others.

Objectives

1. To show utility and feasibility of a rase practice under village condition.
2. Chiefly to establish confidence on the part of the farmer as well as the extension teacher.

Procedure or technique

1. Analyze situation and determine need
2. Decide upon specific purpose
3. Plan the result demonstration
4. Select demonstrators
5. Select the plot
6. Structure the demonstration
7. Supervise the demonstration
8. Complete the demonstration
9. Follow up

Advantages

1. Give the extension worker extra assurance that recommendation is practical and furnishes local proof of its advantage.
2. Increases, confidence of farmers in extension worker and his recommendations.
3. Useful in introducing new practice
4. Contributes to discovery of local leaders.
5. Provides teaching material for further use by extension workers

Limitations

- Requires lot of time and preparation on the part of extension worker.
- A costly teaching method
- Difficult to find good demonstrators who will keep records
- Teaching value frequently destroyed by unfavorable weather and other factors
- Few people see the demonstration at the stage when it is most convincing.

Table : Difference between baining education

Training	Education
1. Directed towards improving individuals job proficiency	Concerned with personal development
2. Need based problem centered learning and skill oriented learning	Focused on general learning
3. For less structured and of less repetitive nature	Learning with strict schedule with tight curricula and course material
4. Course have greater variety and organization at much shorter notice	Less variety
5. Heterogeneous group	Homogenous group in terms of age, mental level and interest and experience
6. Variety of instructional methods are used	Imparted through lecture method
7. Not disciplinary problem	Disciplinary problem
8. Short duration	Long duration

- Unsuccessful demonstrations may undermine the prestige of extension and entail loss of confidence.

Method demonstration

It is relatively short time demonstration given before a group to show how to carry out an entirely new practice or an old practice in a better way. It deals with how to do something e.g. Pruning of Grapes.

Trainer shows the step by step procedure in the operation, explaining each succeeding step as he proceeds. The learners watch the process, listen to the oral explanation, and ask questions during or at the close of the demonstration.

Objectives or purposes

1. To enable the people to acquire new skills
2. To enable people to improve upon their aid skills
3. To make the learners do things more efficiently, by getting rid of defective practices.
4. To save time, labour and increase satisfaction of learner.
5. To give confidence to the people that a particular recommended practice is a practicable proposition in their own situation.

Procedure or steps to be followed :

1. Analyze the situation and determine the need.
2. Plan the demonopation in detail.
3. Rehearse the demonstration.
4. Give the demonstration.
5. Follow up

Advantages

1. Peculiarly suited in teaching skills to many people.
2. Seeing, hearing, discussion and participation increases and stimulates interest and action.
3. The costly trial and errors procedure is eliminated.

4. Acquirement of skills is speeded.
5. Builds confidence of extension worker in himself, and also confidence of people in the extension teaching.
6. Simple demonstration readily lands themselves to repeated use by local leaders.
7. Introduces changes of practice at a low cost.
8. Provides publicity material.

Limitations

1. Suitable only for practices involving skills.
2. Needs good deal of preparation, equipment and skill on the part of extension worker.
3. Considerable equipment transportation.
4. Requires certain amount of showmanship not possessed by same extension workers.

Table : Difference between method demonstration and result demonstration

Method demonepation	**Result demonepation**
It shows how to use a practice	It shows the end result/outcome
It is quite appropriate to show working of new instrument or skills in a practice	Result of a practice or product can be shown
It takes less time	It is time taking involving all the steps to get results
No comparison is required	Comparison with old practice is essential
Often paves the way for result demonstration	May involve one or more method demonstration
It is for the persons present at the demonstration	May involve one or more method demonstration
It is for the persons present at the demonstration	For the purpose of demonstration as well as other formers.
Conducted by extension worker himself or local leader specially trained for the purpose	Farmer under the guidance of extension worker

Fish Bowl Technique

Fishbowl is another special kind of discussion the title comes from structuring where at certain phases some of the discussants sit in the center (the fishbowl) to discuss the issue other member observes this fish bowl.

Fish bowl may be used as the analysis of group processes or as an Instrument to control the content of a topical discussion Technically, the method works about like this:

- Choose the subject.
- Prepare six to eight catalytic questions for the moderators
- Assign or draw numbers for all participants
- Arrange chairs into an inner ring and an outer ring
- Place participants with the odd number in the outer ring.
- Designate one inner ring moderator to use the questions in arousing and maintaining discussions.
- Tell others in the fish bowl (inner ring) that their role is to keep the discussion going bal to unit their inputs to not more than one minute per input.
- Approach the outer ring member with people specially assigned to observe. The number of times each person speaks. The number of times the discussion drifts to other topics.
- Facial and other nonverbal communications. Disagreements with and by whom? Any conclusion about the subject or the process.
- At the appointed time, stop the fish bowl and feed back the data gathered in step 9.
- Repeat steps 7,8,9 and 10 but with the even numbered people in the fish bowl and those with odd numbers assuming the observer rules.
- Used in the manner described there the fish bowl is clearly more concerned with group process than with course content. If instructors wish content to be the focus they can divide the group into small buzz groups each of which discuss all of the catalytic questions. Each group also designates the sequence in which its members will participate in the fish bowl. At the appointed time,

the fishbowl starts only fish bowlers may speak but if they misrepresent their group a caucus can call them back for short re-instructions or for a change of representative.

- When fish bowl takes their form, participants can acquire both the content of a discussion and skill in reflecting group opinions. The instructor must watch time units carefully to make sure serve as representatives in the fish bowl for a useful period of time.

Advantages

1. It is suitable for large groups
2. They do not make any difference between speaker and audience.

Laboratory Training

- Sensitivity training: "Members are brought together in a free and open environment in which participants discuss themselves and interact. The discussion is lightly directed by a behavior expert, who creates the opportunity for participants to express their ideas, beliefs and attitudes."
- The objective of this method is to provide the participants with increased awareness of their own behaviour and of how others perceive them, greater sensitivity to the behaviour of others and increased understanding of group processes. Specifically it aims at increased ability to empathize with others, improved listening skills, greater openness, increased tolerance for differences and improved conflict resolution skills.
- The goal of sensitivity training is for individuals to understand how they function in a group setting and how a group functions.
- Sensitivity training is totally dedicated to learner participation
- Participants role himself
- Trainers non-authoritarian, resource person
- Data-generated by group itself, here now feedback
- Trainer-trainee ratio - 2:12-15
- Time required 15x2 hours
- Time for preparation very little
- Facilities required - Small room tape recorder
- Special points - Require specific training skills

Model of lab training / sensitivity training

Upset expectations undefined situations open feedback trainers' intervention	Unfreezing highlighted motivation need to extra. Now norms	Moving establish new norms feedback experimentally	Refreshing development stabilize adjustment

Goal

1. For individual how they function in a group setting
2. How a group functions

Objectives

1. To increase empathy
2. To increase/develop sensitivity to other
3. To know own and other behavior

Advantages

1. Increased sensitivity empathy
2. Increased efficiency
3. Increased organization climate
4. Progressive environment
5. Harmony/cordial atmosphere
6. Reduction of stress

Disadvantage

1. Informal atmosphere

Brain Storming

"Brain Storming is a type of small group interaction designed to encourage the free introduction of ideas on an unrestricted basis and without any limitation to feasibility".

Use

It is commonly used in

i. Real problem solving situations- as a training method its most frequent use is

ii. To teach learners to suspend judgment until a maximum no of ideas have been generated

iii. A second use is to train people to listen positively to the ideas of others, refraining from negative comments which might cause the creative process to dry up.

Theory

Brain storming applies the synergistic theory that group can generate more ideas, and better ideas then all the individual members could do if they worked independently. It is therefore a useful method only when there are no of trainees five or six are minimum for a workable brainstorm.

The learning occurs because participants must discipline their inputs to the discussion. The controls occur through the instructions and through the better behaviour of leader. Those instructions usually include these points.

1. Generate, do not evaluate - There will be time for evaluation. Later in brainstorming quantity is goal quality matters only when people stop to challenge quality during the creative processes, they inhibit their creativity by moving into a judgment mode too soon.
2. Create new ideas by amending those which have already been suggested. This amendment can take such forms as increasing, decreasing, adding, deleting, consolidating and reversing. The "reversing" specifically prohibits negative statements
3. Post all suggestions on a visible list on front of the group. This serves as a reinforce for those who contribute and as an encouragement to further participates on when participants begin to analyze or question or debate the leader used "neutral reinforcement."

Phases

Total brainstorming/brainstorm include three phases generation, analysis and action planning. In the second step analysis, participants ask the contributors to explain strange storms or unfamiliar ideas. Analysis also includes evaluation criteria for selecting the best ideas, and then tests each idea against those criteria. As ideas 'fall out' because they do not meet the criteria, the group is left with a workable list of options. From these they can select the best solution. They are then ready to move to the third phase, action planning. This consists of outlining the steps needed to put the adopted solution into operation.

Advantages

1. Create atmosphere for free expression of ideas this develops the ability to develop suppressed thoughts.
2. Creative thinking ability.
3. Useful for problem requiring creative solution.

Disadvantages

1. Participants do not get exposed to a full length presentation.
2. The method has the limitations of the basic requirements.
3. It needs a dynamic trainer.

Human resource development : Process is denoted as the means by which personnel are recruited, trained, and put into use depending upon their skills knowledge and potential as per the need of the job and keeping in line with organizational objectives.

HRD Development Directions:

1. Individual
2. Role
3. Dyads
4. Interterm
5. Teams
6. Organization

HRD instruments/techniques :

1. Role analysis
2. Performance planning
3. Performance appraisal
4. Induction training
5. Job rotation

6. Potential appraisal and development
7. Stress Management
8. Quality circles
9. Task forces
10. Meetings
11. Appraisal system
12. Career system
13. Training system
14. Work system
15. Cultural system
16. Self renewal system

❑❑❑

Chapter 7

Issues Related to Extension Policies

Technology Assessment and Refinement through Institution Village Linkage Programme (IVLP)

It has been observed / recognized that before the new technologies are transferred from the research system to the extension system programme. Approach that involves different components of the new technology which are relevant to the farmers in varying agro-ecological conditions will be needed for triggering a perceptible change. The innovative technology TAR through IVLP has been launched by the ICAR in 1995 by involving certain ICAR institution and SAUs.

This IVLP concept is based on the participatory mode, which insures greater scientists and farmers linkage in bottom up approach. It insures access to new agricultural technology generated by the entire agriculture research system. It has given promising results, which has encouraged the ICAR in proposing to extend IVLP in near future to many more institutions in the national linkage.

Objectives of TAR

i. To introduce technological interventions with emphasis on stability and sustainability along with productivity of small farm systems.

ii. To introduce and integrate the appropriate technologies to sustain technological interventions and their integration to maintain productivity and profitability taking environmental issues into consideration in a comparatively well defined farm production systems.

iii. To introduce and integrate the appropriate technologies to increase the agricultural productivity with marketable surplus in commercial on and off farm production system.

iv. To facilitate adoption of appropriate post harvest technologies for conservation and on farm value addition of agriculture products and wastes for greater economic dividend and national priorities.

v. To facilitate adoption of appropriate technologies for removal of drudger increased efficiency and higher income of farm women.

vi. To monitor socio-economic impact of the technological intervention for different farm production systems.

vii. To identify extrapolate domains for new technology, technology modules based on environmental characterization at meso and mega levels.

Technical Programme

i. *Participatory institutions*: The ICAR institutes Headquarters, Regional research stations/Zonal research stations of ICAR institutes and SAUs and KVKs will implement the programme. The participating centers identified are those who have required infrastructure a team of scientists, laboratory facilities, transport etc.

ii. *Selection of villages*: Each participating centre is to identify and adopt one village or a cluster of villages to cover about 1,000 farm families. The selected village or cluster of villages should be representative of district, should not be for away from the centre and must be approachable by road for frequent visits by the scientists, like co-operative, school, panchayat etc. Shall be an added advantage for having better linkages in the programme. The states of the present level of farm production systems and farming intuitions should also be taken into account while identifying the village(s).

iii. *Agro-Eco-System analysis*: The agro-eco-system analysis survey in an important activity before the programme is launched. It will be done using the PRA tools. It provides information above the resource availability with the farmer's present production practices, interaction within and amongst various enterprise on spatial and temporal basis and an ultimate base to plan the work.

iv. *Multi disciplinary team*: The size of such team is limited up to 4-5 multidisciplinary team disciplines depends on the need of the area. In the multidisciplinary, team leaders will be identified by the head of the implementing centre that shall be accountable directly to the council for technical matters.

Training Activities

Farmers will be provided adequate training on the concept and operation of the project. To complete this, the following steps will be taken.

1. Identification of core trainers at the ICAR HQ and zonal levels.
2. Development of the core training for the core trainers.
3. Preparation of training materials or core trainers.
4. Training of core trainers at ICAR Head Quarters and zonal levels.
5. Preparation of operational guide lines at ICAR head quarters for implementation of the project.
6. Developing training plans which may include venue, duration tentative calendar, etc, for the scientists of core and optional
7. Implementation of training programmes for the multidisciplinary team of scientists at the zonal and regional level.
8. The importance of training in seed innovative project needs no emphasis. The consultant training has to be engaged where ever required to see that all in its proper perspectives understands the project.

Plan

- In small farm production system.
- In well defined production system.

Technical activities

1. Agro-ecosystem analysis.
2. Tailoring of appropriate technologies.
3. Collection and assimilation of information.
4. Testing of research information.
5. Retaining technologies.
6. Collection and classification of data base.
7. Development of appropriate expert system.

Plan Strategy for Technology Assessment Systems

A. *In small farm production system*: emphasis on training of technologies for different farming situations and environments, through on farm research. In well defined production system the emphasis well be on farm trails and testing and demonstrations. In the entire programme, the emphasis would be to develop multiple options (Basket approach) for different forget groups through participated approach. In case of small farm production systems, foremost priority will be on agro-ecosystem analysis including grain water availability situation.

B. The well defined farming conditions will be divided into fully and partially irrigated systems. In farmer situation emphasis will be on sustained higher productivity. In all these programmes, emphasis will be on participatory approaches and on development of multiple options to suits the needs and goals of farmers having diverse social, economic and cultural backgrounds. Emphasis will also be conservation of natural resources.

The detailed technical programme, for each centre, will be developed for the participation families/problems after detailed agro-eco analysis and problem identification by the multidisciplinary teams.

Conclusion

- ICAR sponsored project implemented by Agricultural universities /ICAR institutions.
- Whole village (100-1200 farm families) is included in the project.
- Assessment of new innovations is done for different resource groups of farmers in different farming situations.
- Technologies related to agriculture, horticulture, livestock and other enterprises are to be assessed as per problem perceived by the villagers and according to their priorities.
- Direct linkage between scientists and farmers is ensured. Participatory technology assessment and refinement takes place in this project.

National Agricultural Technology Project

Thrust areas

1. Shift from research orientation to programme/production
2. Teamwork
3. Mission mode approach
4. On farm development and assessment of technologies
5. Commercialization of technology
6. Promoting global competitiveness
7. Improve efficiency of agricultural research system
8. Reform in organization and management system
9. Strengthening - R.-E.-F. Linkage
10. Policy reform
11. Decentralization
12. Devaluation

Objectives

1. Research and extension should be farmer driven.
2. Reallocation of central and state government resources to finance bottom-up research extension.
3. Empowering farmers by getting them organized in SHGS, FIGs, for creating demand driven extension.

4. Empowering farmers by getting them organized in SHGs, FIGs, FAs and FAs for creating demand driven extension.
5. Making extension and research system financially sustainable.
6. HRD human resource development has to be key component.
7. ATMA (Agriculture Technology Management Agency)
8. Programme co-ordination
9. R-E (Research extension)
10. SREP (Strategic Research Extension Plan)

The diverse activities of the project have been planned under three Major Components.

1. Agro-Eco system research
2. Innovations in technology dissemination
3. Organization and management systems

1. Farmer driven
2. Bottom up
3. Organizing farmer into SHG etc
4. Empowering farmer
5. Financial sustainability of research
6. HRD
7. ATME
8. R-E-F
9. SREP

NATP was launched by the ICAR on June 30, 1998, with the support of the World Bank, to strengthen and complement the existing resources and to augment the output of the National Agricultural Research System.

The NATP implemented its objectives through strategies.

Coverage

Implemented in 28 districts covering/states, viz. Andhra Pradesh, Bihar, Jharkhand, Himachal Pradesh, Maharashtra, Orissa and Punjab over a period of five years.

Components

NATP has three components implemented by

- Research component - ICAR
- Extension components - Department of Agriculture and Cooperation
- ITD Component - 5(PIA)s project implements
- Directorate of extension (DOE)
- National Institute of Agriculture Extension Management (MANAGE)
- NATP cell at state HQ
- State Agricultural Management and Extension Training Institute (SAMETI)
- District Level Agricultural Technology Management Agencies (ATMAS)
- Organization and Management Systems Under this component, following activities will be supposed an
  - Organization and Management reform in the CIAR.
  - Strengthening of ICAR headquarters including the national agriculture science centre (NASC).
  - Directorate of Information and Publications of Agriculture (DIPA).
  - Institutionalization of priority setting mechanism
  - Information system development ISD.
- NATP was world's biggest World Bank assisted Agriculture project worth Rs. 992 crores developed and executed by NARS.
- NATP life span was seven years from June 30, 1998 to June 30, 2005.
- NATP was the first project in NARS to shift the focus from discipline oriented research to production system research.
- NATP was the first project in NARS to involve competitive funding and have pluralistic approach to involve and fund partners from outside NARS.
- NATP successfully completed a whopping total of 852 projects.

NATP Thrust Areas

1. Shift from commodity/discipline oriented research to programme/production system research.
2. Team working along institutions in mission mode, teams of excellence, production systems research.
3. On farm development and assessment of technologies.
4. Commercialization of technologies and promoting global competitiveness.
5. Utilization of national capabilities and promotion of pluralism ICAR, SAU, State Agricultural Universities, General Universities, non-government organizations, private sectors.
6. To improve the efficiency of agricultural research and improve its relevance to farmers and agribusiness.
7. Reforms in organization and Management Systems to efficiency in public sector.
8. Strengthening research-extension-farmer linkages.
9. Key policy and institutional reforms.
10. Decentralization and devolution.

Objectives of NATP

The main objective of project is to address key constraints which currently limit, and which if not addressed would in future increasingly restrict, the efficient use of the public resources which India devotes to the generation, assessment and transfer of agricultural technology. Thereby it will improve the relevance of technology generation, refinement, assessment and transfer programmes to the changing needs of farmers and processors, and thus contributions made by improved agricultural technology to key national objectives of food security, economic growth, equity, the alleviation of rural poverty and the conservation of natural resources.

To address the key constraints

Strategies of NATP

- Improving the efficiency of public research and technology assessment and dissemination institutions by strengthening management tools and procedures and rising human skills stressing improvements in priority setting and resource allocation.

Stranger liks and cooperation with external sources of expertise in agricultural sciences, the improvement of staff incentives and accountability and concurrent monitoring and impact evaluation.

- Adopting the organization and management, staff skills modes of operation and programmer of public research and technology assessment and transfer services to make them more relevant to current and upcoming needs especially for location specific inter disciplinary programmer of production system research and development focused on technologies in the public goods categories.
- Where ever appropriate, facilitating the entry of farther put actors or producers themselves in research and extension activities which give privately appropriate able benefits or for which cost can be recovered directly from users.
- The accumulation of replicable experience to guide further change.

Impact of NATP Processes on NARS

Stragetic Research and Extension Plan (SREP)

Strategic Research and Extension Plan is a demand driven, situation specific, multifactor oriented, pre-determined course of action to accelerate agriculture development in the district. ATMA of each district is required to develop a SREP by involving all stakeholders.

To address

The present mechanism of planning and implementation of agriculture and allied development programme is centralized in nature. This top down approach focuses on individual commodities/ enterprises rather than on holistic integrated approach. It is ad-hoe in nature and does not involve all actors. The farmers are considered as receivers of benefits rather than as responsible partners who can influence the production process. So address the afore mentioned issues under extension reforms scheme, ATMA of each district is required to develop SREP by involving all stacke holders.

1. Get an overview of the prevailing scenario in the district.
2. Explore and understand the problems and opportunities in different farming systems, preference and priorities of the farming community.

3. Facilitate long term visioning and strategic planning for agriculture development in the district.
4. Facilitate involvement of all actors at different levels in the development process and in the long run, share the lead on public extension system.
5. Development of annual action plans for each block in respect of the prevailing agro-ecological situation.
6. Develop farmer centered market oriented extension research management system.

Contents of SREP

A. Under diagnostic section, there are again two parts

   Information of the district includes general features, agro climatic, information, agro ecological situations, demographic data, information on land based systems, infrastructural facilities, rainfed area, irrigated area, ongoing development programmes, information on market.

B. *Analysis of data*: Each identified farming system needs to be analyzed in terms of its interaction with other options of farming systems.

C. Under strategy section/stagey for militia a strategy subsection This subsection would be developed on the basis of prioritized strategic issues, separately for extension and research. This exercise is carried out by team constituted of line departments, KVK and ZRS scientist in the district.

D. *Activity subsection*: All the formulated strategies are spell out to the programme content.

How to develop SREP

1. Orientation of district team
2. Identification of Agro-ecosystem situation (AES) within the district.
3. Training of AES team
4. Data collection through participatory approach
5. Data analysis, identification and prioritization of research and extension issue.

6. Developing strategies for research and extension issues in the district.
7. Developing activity schedule.
8. Approval of SREP.

SREP through PRA

What/Why

Earlier	In SREP
Top down approach	Bottom up approach
Commodity Oriented	Farmer participation
Centralized	Programme oriented
Farmer as a receiver of technology not partner	Decentralized

ATMA. At district level is responsible to develop SREP to address above problems by involving all stock holders.

Research-Extension-Farmer Interface

Existing pattern of interface

1. The challenge for researchers today is to develop economically viable technology that is early adaptable to the rural society much of the developed world follows "top down", approach or model of research and demonstration where farmers are the passive recipients of the research results based on perceived needs identified by scientists with little input from the end user's. Here new technologies developed are passed to the extension staff that refine the technology and disseminate it to produce in a easily understandable manner. This model assumes minimal interaction among the various groups. This has resulted in uni-directional information flow, i.e. from researchers to producers.
2. According to FAO (Anon 1984), Public agricultural research institutions often have poor relations with extension agencies. The World Bank states that, bridging the gap between research and extension is the most serious institutional problem in developing research and extension programmes. Extension workers often are researchers working in an ivory tower generating technologies not applicable to the farm where as researchers often question the ability of the extension agents to perform their jobs effectively.

3. Farmer's involvement in research has been limited by inadequate funding, institutional policies and hierarchies, specialization, and incompatible personalities. He suggest that priorities for agriculture research using public funding should be identified through a democratic process involving farmers in order to provide a better balance between basic/applied research and demonstration.
4. Better communication among the researchers, extension workers and farmers is an essential component for improving transferability of technology.
5. What should be so shift must occur away from the top down hierarchical approach, criticized by farmers as elitist, to an egalitarian an participating approach in which farmers, researchers and extensionists serve as peers.

Research Extension Farmer Market Linkage

Why linkage

1. Several measures have been taken to improve this strategic linkage but the results have not been as expected because of top-down flow of information leading to a work feedback mechanism.
2. The extension role in assessment, refinement and validation of technology particularly the ITK (Indigenous Technical Knowledge) continues to be weak because of lack of competent and qualified staff at district level. This is on account of inadequacy of institutional mechanisms and a lack of focus on location specific issues/gaps in adoption and weak participation-integration-coordination particularly across crop and livestock systems.
3. The research recommendations from the institutes are too general with a little focus on available technologies to fill the gaps.
4. The farmer participation in research priority setting is also very poor mainly due to inadequate research documentation/ information sharing and a very little, negligible monitoring of research outcomes by the farmers.
5. This is on account of inadequacy of institutional mechanisms and a lack of focus on location specific issues/gaps in adoption and weak participation-integration-coordination particularly across crop and livestock systems.

6. The research recommendations from the institutes are too general with little focus on available technologies to fill gaps.
7. The farmer participation in research priority setting is also very poor mainly due to inadequate research documentation/ information sharing and a very little, rather, negligible monitoring of research outcomes by the farmers.
8. It is no exception that farmers do not adopt technologies recommended by extension, because they realize that these are not the technologies which help them best to realize their goals in their situation.

Scientists-Extension-Farmers Interface

Any kind of technology carrying out of the research system is to be transmitted to farmers system must pass through extension system. All three stakeholders must indicate their views, interact with each other and find out common solutions to the farmers problem so as to push the technology for its easy adoption by the farmers.

Linkage between R-E-F-M Research Extension Farmer Markets

- Earlier sources of growth irrigation and intensive input was no more effective growth from these sources is now slowing down. Now latest technologies are needed to push out the yield frontier, utilize the inputs more efficiently, diversify to more sustainable and higher value cropping patterns to realize more farm income.
- These knowledge intensive technologies require strong research - extension farmer and market linkage mechanism to achieve the desired results. The links between research - extension farmers and market linkage mechanism to achieve the desired results.
- The links between research, extension and farmers often tended to weaker over the time. Lack of a close working relationship between agricultural research and extension organizations, and with different categories of farmers and farm organizations, is one of the most difficult institutional problem.
- The ineffective link between research and extension has impeded the development and TOT appropriate for small scale, resource poor farmers particularly those who work in low relatively low-potential, heterogeneous agro-ecological areas problems in technology development and transfer functions arise because these functions are treated in isolation.

- According to W.B.R. "Bridging the gap between research and extension is the most serious institutional problem in developing an effective research and extension system.
- Research and extension organizations generally compete over the same score government resources and, frequently, these institutions do not see them as part of a broader system.

Utility and Types of Linkages

Indian agriculture has made rapid strides in activity self-sufficiency in food production. To create a demand driven research and extension system, it is essential that farmers be given access to linkage mechanisms through which they can articulate their problems and needs.

1. Linkage mechanisms are used to channel information between groups and to conduct required tasks in the process of getting relevant technologies to farmers it is also used
2. To network all concerned and bring them together to store ideas, materials, information, expertise by the development chain.
3. To institutionalize the mechanisms of information flow, material flow and providing necessary services.
4. To maintain continuous flow of updated information to the client system.
5. To meet the unforeseen circumstances and contingencies like floods, cyclones and other natural disasters (drought, pest outbreaks etc.).
6. The true test of linkage is the achievement of the ultimate goal of increasing production, productivity, profitability.

Classification of linkages

1. Production linkages
2. Technology generation
3. Post production - Market agencies

Reasons for linkage failure

1. Political problems - pressure
2. Technical problems too much specialization lack of ethics and values

3. Organizational problems
4. Differences in orientation and work style
5. Heads of institutions treats the linkages with least priority
6. Motivational and incentive problem
7. Resource problems
8. Communication problems

Linkage principles and mechanism

Linkage principles

1. They should have a common purpose i.e. (domain consequences)
2. They should perceive that it is advantageous for them to participate in linkage activities
3. There should be common ground or proximity of location to facilitate collaboration.
4. Linkage activities should be compatible with other activities of each group.
5. There should be rewards for individuals participating in linkage activities.
6. Communication between members of different groups should be effective and there should be free flow of information groups.

Linkage mechanism

- The linkage mechanism is the concrete procedure, regular event, arrangement, device or channel which bridges the gap between components of the system and allows communication between them.
- These mechanisms over which managers or policy makers may have some control are categorized into
  1. Structural
  2. Organizational
  3. Managerial

The structural and organizational mechanisms are

1. Combining research and extension functions into one unit
2. De-centralizing research and extension activities into regional institutions

3. Staffing extension liaison positions in research institutions
4. Establishing communication cum information departments
5. Redefining roles and responsibilities between research and extension units
6. Creating inter-agency committees or councils
7. Developing inter-agency agreements for collaboration.
8. Farmer participation in research activities
9. Leasing with private organizations and NGOs.
10. The managerial mechanisms are
11. Redefining job description to strengthen relationships
12. Establishing joint reviews of research and extension activities
13. Improving individual incentives for collaboration
14. Exchanging personnel, e.g. posting extension staff in a research organization
15. Joint training for expanded roles in a technology system
16. Joint use of facilities and services such as soil testing laboratories
17. Joint participation in functions such as field testing and demonstrations.
18. Promoting informal linkages
19. Exchanging information using jointly developed protocols.
20. Approaches in strengthening linkage
    - Institutional mechanism
    - State level inter department
    - Working group
    - ATMA
    - FIAC
    - CIGS Commodity interest group

Linkage through strategic research and extension plan

In addition to these farmer linkage mechanisms both research and extension personnel would be expected to have regular, informal contacts with different groups of farmers while carrying out joint on farm and adaptive trails, frontline demonstrations, exposure visits to success stories and field days.

Advantages of REF Interface

1. More problems oriented and need based technology generation.
2. Area specific technology generation
3. Bridging the gap R.E.F.
4. Quick and stable technology adoption
5. Very less discontinuance
6. Empowered farmers
7. Faith in farmer community
8. Sustainable adoption and use of technology generated
9. More user oriented technology generation
10. Change of research system orientation
11. Very less wastage of resources due to gap.

Front Line Extension Programmes of ICAR

1. The frontline system of ICAR presently indeed a network of KVKs and eight Trainers Training Centers (TICs)
2. In order to provide specialized training to the
3. Teaching staff of the KVKs as well as other extension service staff in different institutions engaged in providing the vocational training programmes, the ICAR has established eight TTC in different part of the country with following different objectives
4. To impart being to the teachers of the KVKs both in agricultural technology as well as in pedagogy with a view of improve their subject matter background in agriculture, home economics and the skills necessary for the teaching learning procedures and techniques.
5. To train the teachers
6. To train trainers of the Gram Sevika/gram
7. To organize vocational training progammes in agriculture and home economics
8. To organize and conduct other short or long term vocational courses for teachers/trainers as and when required by the government and agricultural and home science institutions.

9. The TTCs will have a core staff of 5-6 specialists headed by a Chief Training Organizer. Where over necessary. To and training associates, assistants are also being provided.
10. In addition, guest/specialists/visiting specialists area invited for a specified period of line.

KVK - Krishi Vigyan Kendra

Krishi Vigyan Kendra is an innovative science-based institution that undertakes vocational training to inservice extension personnel, farmers and farm women and rural youths, conduct on farm trials for technology refinement and frontline demonstration to promptly demonstrate the latest agriculture technology to the farmers as well as extension workers.

1. K.V.K. functions on the principle of collaborative participation of scientists, SMSs, extension workers and farmers.
2. Dr. Mohan Singh Mehta (1974) committee appointed by ICAR to provide skill oriented vocational training support to farmers.
3. Approach Imparting learning through work experience to those who are engaged in farming.
4. The syllabus and programmes of each KVK is tailored to the felt needs of the farmers, resources and potential for agricultural growth in a particular area. "Teaching by doing" and 'Learning by doing' are the main methods of imparting skill trainings.
5. New approach with effect from 1st April, 1992 all the first line transfer of technology projects of the ICAR viz (i) National Demonstrations (2) Operational Research Projects and (3) Lab to land programme have been integrated into KVKs and hence there has been shift in approach and objectives of the KVK.
6. Initially emphasis was on human resources development through training the farmers and extension workers now the KVK also participate in the process of agricultural technology development and refinement.
7. Emphasis also on employment generation to bring about equality of opportunities for socio economically deprived class of farmers and traditionally backward areas in the country including dry land areas.

Aim

Speedy TOT, to reduce the time lag between generation of technology at the research institution and its transfer to the farmers fields for increasing production and productivity and income from the agricultural and allied sectors on a sustained basis, in order to achieve this goal, following four mandates have been envisaged in the design of KVK.

Mandates

1. Collaborate with the SMS of the SAVS/Scientists of the Regional Research Stations (NARP) and the State Extension Personnel in on-farm testing, refining and documenting technologies for developing region specific sustainable land use system.
2. Organize training to update the extension personnel within the areas of operation with emerging advances in agricultural research on regular basis.
3. Organize long-term vocational training courses in agriculture and allied vocations for the rural youths with emphasis on "Learning by doing" for generating self employment through institutional financing and Organize front-line demonstrations in various crops to generate production data and feed back information.

The specific objectives

1. To demonstrate the latest agricultural technologies to the farmers as well as the extension workers of the State Department of Agriculture/NGOs with a view to reduce the time log between the technology generation and its adoption.
2. To test and verify the technologies in the socio-economic condition of the farmers and identifying the production constraints.
3. To get first hand scientific feedback from the fields and passing it to the research system in order to keep the scientist abreast with the performance of the technologies and the forming problems so that they re-orient their research, education and training programmes accordingly.
4. To impart training to the farmers, farm women, rural youth and field level extension functionaries by following the principles of "Teaching by doing" and "Learning by doing".

5. To provide training and communication support to the line department of the state/NGOs and
6. To develop extension models to be adopted by general extension system for large scale multiplications.

The objectives stated here are common for all the KVKs. In addition each KVK has to have a mandate that is a specific set of responsibilities to perform. The mandate of a KVK is unique for it and is determined on the basis of most important need of the clientele, their resources and constraints and nature of the eco-system.

Implementing agencies

1. The ICAR institutes, agricultural universities, selected NGOs and some State department of agricultural implement the KVK projects.
2. The first KVK was established in 1974. The priority for the establishment of KVKs is given to hilly areas, drought prone areas, forest areas, costal areas, flood prone areas and areas as dominated with tribal farmers, weaker sections, small farmers and landless labors. The objectives are to gradually cover the entire country with one KVK in each district, priority being given to the backward areas.
3. Monitoring and evaluation of KVKs
4. The coordination, monitoring and implementation of the KVKs are done through and zonal units.

Activities of KVK

Based on mandates the following are the main activities of KVK

1. Organization of OFTs
2. Organization of in-service training programmer for extension personal
3. Organization of vocational training programme for farmer, form women and youth
4. Organization of frontline demonstrations and other extension activities.

Role of KVKs in agricultural extension for agricultural and rural development.

1. Rural development could be attained through improving agriculture, forestry, animal husbandry,
2. Dairying sericulture, fisheries, rural engineering and rural crafts.
3. Agricultural extension here has an important contribution to make. It is a tool that can be used to fight poverty too.
4. Empowering rural masses to enable them to practice these occupations to earn more.
5. Empowering rural masses to enable them to practice these occupations to earn more and live better in a more specific terms, agril extension here has an important contribution to make. It is a tool that can be used to fight poverty to faster education of rural people's, and to promote behaviour and technology that link high productivity with natural resources sustainability.
6. In promoting local organizations, extension also plays a vital role in empowering rural communities.
7. Further, sustainable agriculture is termed as the system of raising crops for human utility through utilization of resources with greater efficiency without disturbing, unbalance or polluting the environment.
8. Agriculture however must be more than economically competitive. It must also be suitable, which entails conserving natural resources such as soil water and biological diversity, and taking into account agricultures social and cultural context. This complex challenge can also be overcome if traditional and new knowledge are effectively combined in new production systems that are compatible with the cultural and social values of rural societies. Contributing to the development of such system will be one of the most important tasks of agril. extension.
9. In India the extension efforts have largely been taken up by the state departments of agriculture and other disciplines as a subject. The Indian Council of Agriculture Research (ICAR) as the apex body to provide new technologies in agriculture and allied aspects has its own transfer of technology activities too. The extension efforts of ICAR have evolved through national demonstration projects, ORPS, LLPS and integration of these approaches to KVKs since 1974. Thus KVKs are attaining the focal point for front line transfer of technologies for all developmental activities related to agriculture, community and industries in rural India.

10. Farmer's capacity building is often seen within the perspective of giving them the knowledge and skills required to practice crop animal husbandry in better way.
11. Though knowledge and skills are fundamental to efficiency in any enterprise, the Indian farmers need more than that because of the limitations and complexities under which they operate.
12. The KVKs which have been mandated to work with farmers farm workers and rural youth directly as well as through field extension functionaries have the greatest challenge to make their clients more efficient, specialized and to be economically active. The fact that the need for agricultural and rural information and advisory services is to intensify into he foreseeable future exerts more pressure on their performance and have the expectation runs higher.
13. K.V.K. for knowledge and skill building among farmers through training

Fundamental Principles of KVK

1. Agricultural production as the prime goal.
2. Work-experience as the main method of imparting training and
3. Priority to weaker sections of the society is the back bone of the KVK programmes.

Frontline Demonstrations

The field demonstrations conducted under the close supervision of the scientists of the NARS are called front line demonstrations because the technologies are demonstrated for the first time by the scientists themselves before being passed on to the main extension system of the State Department of Agriculture.

The objective is to demonstrate newly released crop production and protection technologies and their management practices on the farmer's field under different agro-climate regions and farming situations.

On-Farm Testing

These are conducted on the farmer's field on such problems where the appropriate technologies are not available for particular agro climatic situation to transfer, and the relevant research information available does not suit the situation from the point of view of the farmers. The main objective is to give overriding importance for farmer's perspectives and participation at all the steps of on-form testing viz

1. Problem diagnosis
2. Experimentation and
3. Extrapolation

The importance of location specificity in development of appropriate technologies keeping socio-economic, ecological and cultural parameters in view, is gradually being appreciated. This has plover the way for technology assessment in different micro-environments and its refinement to suit varied situations through participatory approaches.

ITK Indigenous Technical Knowledge

Are the time tested knowledge and experience of peoples accumulated experience in dealing with solutions and problems in varying concepts of life and such knowledge and practices are special to particular culture.

Indigenous knowledge are not primitive, left over from the past lost on the contrary, are systems of finally tuned and adopted both biologically and socially to counter the process of plot are often harsh

Characteristics of ITK

1. ITK is not static or dynamic
2. ITK study needs a holistic approach
3. ITK if properly trapped can provide reliable insight into resources, processed, possibilities and problem in particular area.
4. ITK is mainly realistic in nature
5. ITK is recorded and transferred through oral tradition
6. ITK is honed through observation and hands on experience
7. ITK forms an information base for a society.
8. ITK reflects local tradition.

Role of ITK

1. ITK can provide development efforts.
2. ITK can facilitate local peoples participation.

3. ITK is reliable source of appropriate technologies.
4. Process and methods of ITK analysis.

Identification and Collection of ITK

Methods and techniques

1. Documentation of oral stories
2. The Delphi methods/technique
3. Agro eco system analysis
4. Manual discrimination analysis
5. Decision free analysis
6. Use of local resource person
7. Linguistic and holistic analysis of concepts
8. Ethnography method
9. Critical incident to analysis CIT
10. Analysis of peasants journals and news papers
11. Arranging competition
12. Conducting documentation workshops
13. Continuous interactions during on farm experiments
14. Anthropological methods

Documentation

1. Documenting large variety of practices without scientific validation.
2. Documenting prevalent practices and comparing them with traditional one.
3. Documentation of the practices/details of experimentation on a specific aspect and understanding the various linkages.
4. Documenting the practices enrolled to mitigate specific problems of farming or for their survival under conditions of ecological and economic stress.
5. Documenting practices that had enrolled in response to specific external intervention.

Methods and techniques

1. Notes
2. Photos
3. Audio and Video recording

Testing and validation

1. Prepare a list of all collected ITK practices.
2. Decide the continuum for rating the rationality of ITK with specific weight ages.
3. Send the list of ITK practices to experts for their opinion and judgment on each practice.
4. Calculate the weight age mean score of individual practices.
5. Select practices above mean score as rational.

S.No	Continuum	Weightage
1.	Very rational	5
2.	Rational	4
3.	Undecided	3
4.	Irrational	2
5.	Very Irrational	1

Scope of ITK analysis

1. Hold biological and ecological innovation
2. Resources management
3. Protected areas and conservation
4. Development planning
5. Environment assessment

Privatisation of Extension

Concept : "Privatization of agril extension services refers to the services rendered in the area of agriculture and allied aspects by extension personnel working in private agencies or organizations for which farmers are expected to pay a fee (or free) or it can be reviewed as supplementary or alternative to public extension services".

This concept about privatization emphasized three aspects, they are :

1. It involves extension personnel from private agency/ organization.
2. Clients are expected to pay the service fee (private extension may not expect).
3. Act as supplementary or alternative to public extension service.

Reasons behind the privatization issue

1. Financial burden of govt/to cut down govt. expenses.
2. Disappointing performance of public extension service.
3. Commercialization of agriculture

Prospects and need for privatization of extension/reasons

1. To cut down govt. expenses.
2. To get rid an embarrassment to public extension on account growing dissatisfaction with the quality of extension services available.
3. Increase of influence and involvement of farmers in extension organization
4. Developing trends towards improved agriculture from subsistence to commercialized agril business
5. Expectation for supply of quality services.
6. Increase in efficiency and specialization among farmers.
7. Competition among different extension services providers for assured and prompt services.
8. Gradual change of technology from being largely a public good to private sector to invest in its dissemination and adoption by clients
9. Technological developments in mass media and information technology in place of traditional extension method.

Problems of Privatization of Extension

1. Increasing specialization need more updated/up to dating on the part of farmers.

2. The consequences of privatization are users free-cost from beneficiaries. The feasibility of charging fee for extension service raises a question on the paying capacity of the farmer. Normally, traditional farmers are not even willing to avail free extension service what to speak charging fee.
3. The public extension system aims at providing free information, training motivation and persuasion to attract the farmers to by the technology still the response is poor particularly in case of small and marginal farmers.
4. Under such situations, if there is any possibility of privatization of extension service then the question arises what are the different practices which should be privatized, to what extent these should be privatized and what type of privatization should be tried in India.

Positive Points

1. It would induce competition among different agencies.
2. In commercial agricultural areas when training is given free without any fee, its worth is not recognized by the farmers. On the other hand if the farmers pay some amount of fee, which they earn well effort it creates sense of risibility, recognition, realization and importance of training and feeling of belonging only those farmers who are in need of training will attend.
3. Privatization always looks out of progressive and wealthy farmers who can afford.
4. High technology again.
5. Failure of high technology may result in sudden collapse of that.
6. Privatization does not care for sustainability instead advocate exploitation of natural resources to the maximum extent.
7. The privatization extension usually concentrates on commercial crops, horticulture and floriculture which have good market potential both at domestic and global lend.
8. The commercial interacts of private agencies jeopardizes the efforts made by public research and extension towards ecofriendly and sustainable agriculture.
9. Private Extension restricts flow of information among farmers.
10. 70% of operational holding of India are small and marginal so their ability to pay for the extension service is very poor.

Merits of Privatization

1. Extension generates new income, extension become economic input.
2. Provide demand driven service.
3. Increase the voice of farmers in the extension service.
4. Extension service becomes more cost effective with efficient and quality service.
5. Privatization complements or supplements the efforts of public extension.
6. Extension personnel become more clients accountable.
7. Private extension increases staff professionalism.
8. Clients (farmers) are more committed to service.
9. Private extension contributor's big and progressive farmers and areas having favorable environment.
10. Private extension is less education oriented and more commercial in nature.
11. Private extension does not concentrate only on food grain production but will promote crop diversification.

Demerits of privatization of extension services

1. Social distance
2. Social conflict
3. Social disequilibrium
4. Higher gap between have and have not's
5. Side lining coming of farmers
6. Selectiveness of technologies for adoption
7. Commercialization with human value
8. Shortage if essential food grain required at rural village level.
9. With temptation to act as middleman rather than food producers.

Factors for privatization of agriculture extension service in India.

The following factors have to be considered for privatization of argil extension service in India.

1. Types of crops cultivated through out the country.
2. Socio-economic strata differences among the forming community.
3. Entirely drastic agro-climatic variation along and across the length and breath of the country.
4. Different approaches followed in various developing and developed countries and their utility possibilities in our country environment etc.
5. Change in rainfall pattern, vagaries of monsoon, floods, cyclone etc.

Market Led Extension

With the globalization of market farmers has to transform themselves from mere producer-seller in the domestic markets to producer cum seller in a wider market sense to best realize the return for his increments, risk and efforts. This to be achieved farmers need to know answers to like what to produce. When to produce, how much to produce when and where to sell at what price and form to sell his produce farmers received most of the production technologies from extension systems extension system new needs to be outdated with knowledge and skills related to the market.

Stages in promoting marketing extension

In the marketing system stages, the extension worker must need to understand

1. Produce distribution system.
2. Understanding how the marketing system works.
3. Marketing margins at various cards.
4. Whole sales selection or trade partners.
5. Information services.

Enhanced roles of Agril extension personnel light of market led extension

1. SWOT analysis of the market.
2. Organization of farmer's interest group on commodity basis.
3. Supporting and enhancing the capacities of locally established groups.

4. Enhancing the interaction and communication skills of the farmers to increase their bargaining power.
5. Establishing REFM linkage.
6. Advice in product planning.
7. Educating the farming community.
8. Direct Marketing.
9. Capacity building of farmers.
10. Providing market intelligence
11. Organisation of study tours of FDG
12. Production of radio films of success stories of community specific farmers.

Need of agril marketing extension

The agriculture marketing extension is an immensely important due to following reasons

1. Complex agriculture input marked.
2. Efficient use of valuable inputs in decision making
3. Supply driven output market.
4. Need for enterprise mix
5. Widening consumer preference due to illiteracy, nutritional awareness and increasing income.
6. Increasing financial market.
7. Averting risk
8. Competition in international market
9. Sanitary phytosanitary measures, codex standards, GAP norms etc.
10. Quality product requirement for agro-process industries.
11. IPR issues (branding trademarks, patents, PPV and Fact)

Which factors forcing for changing the orientation?

Market oriented agricultural extension is the need of the time, which is a real challenge for our public sector dominated agricultural extension system.

Following factors prompt us for a sound agricultural marketing extension for the sustainable agricultural development

1. Farmer-Agro-Industry-consumer linkage.
2. Shift in physical output to value realization.
3. Agro-processing and value addition of agricultural produce.
4. Grading and packing.
5. Diversification of agriculture
6. Augmented infrastructural facilities
7. Cost effectiveness
8. Expert promotion
9. Global market research and information centre.
10. Upgrading of human resource through trainings.
11. Involvement of private sector.
12. Marketing information network.
13. Marketing of agro inputs.
14. Complex clauses on AOA under WTO.
15. Declared EXIM policy (2002-07) by govt. by India.

ATMA: Agricultural Technology Management Agency

ATMA is registered society of key stackholders involved in agricultural activities for sustainable agricultural development in the district. It is a focal point for integrating research's and extension activities and decentralizing day to day management of agricultural extension system.

Objectives/why ATMA

- To strengthen research extension farmer linkage.
- To provide an effective mechanism for co-ordination and management of activities of different agencies involved in technology adaptation validation and dissemination at the district level and below.
- To increase the quality and type of technologies being disseminated.

- To move towards, shared ownership of the agricultural technology system of key stake holders.
- To develop new partnerships with the private institutions including.

The members of ATMA includes the representatives of department of Agriculture, Horticulture & Fisheries' etc and research and extension units within the district such as ZRS - Zonal Research Stations or substations KVK sets ATMA is supported by governing board (GB) and management committee (MC).

The governing board is a policy making body and provide guidance as well as review the progress and functioning of AIMA. The chairman of GB is DC.

The Organizational Structure of ATMA

The management committee is responsible for planning and executing the day to day activities of ATMA. The chairman is the project director of ATMA.

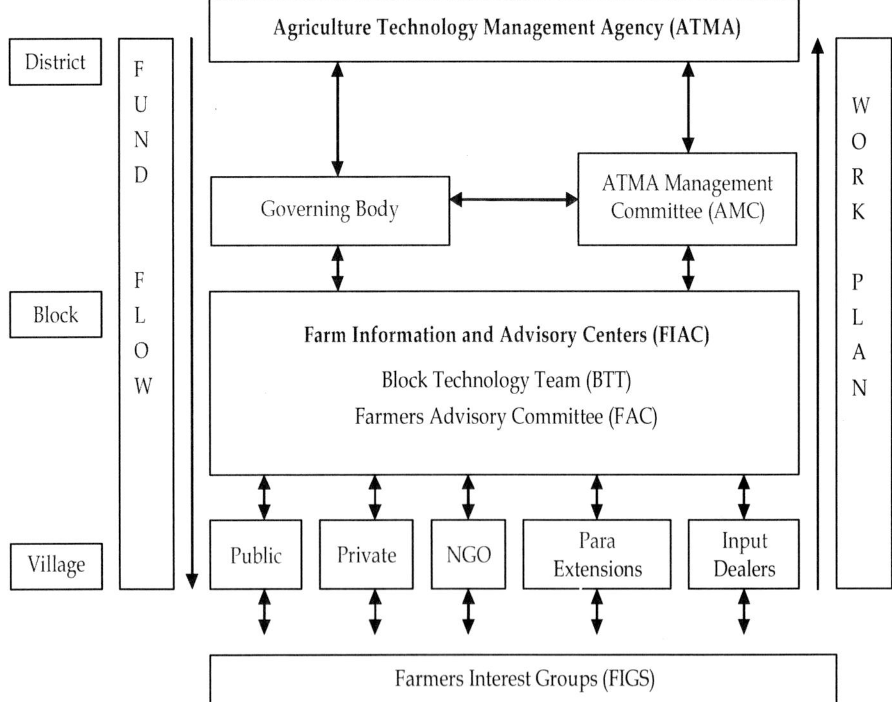

At the block level: Under each ATMA, FIACs are created at the block level. It consists of two bodies namely farmer's advisory committee (AAC) and Block Technology Team (BTT). The FAC is a body of farmer's representatives. The BTT on the other hand is a group of technical advisor operating at block level representing agriculture and allied sectors.

- FAC and BTT, taken together act as planning and operation arm of ATMA (BTT operation alive the SREP in each block) FAC provides farmer feedback mechanism.
- Commodity interest groups (CIGs) are promoted at block/village level to make the technology generation/dissemination farmer driven and farmer accountable these village level CIGs are ultimately federated at block/district level and represented in FACs and GB. In order to address the FACs and GB. In order to address the extension needs of these groups, ATMA will establish close linkage with various players at cutting edge technology viz public, private, NGOs, para extension workers input dealer etc.
- State Agricultural Management and Extension Training Institute have been established to provide needed HRD in innovative areas of extension delivery.
- The interdepartmental working group (IDWG) closely monitors the project activities of ATMA at state level.

1. Developmental Strategies : "Wastershed Development Programmes"

It is a realistic approach of development of any area where all kinds of activities are accounted for. The main actors may it be planning or development to be the human population.

Development Approaches

- Empowerment of village communities for local level planning by the panchayati raj institutions has been pledged by enacting 74th amendment to the constitution of India.
- Management of natural resources by communities is the current paradigm of research and development strategies.
- Ninth plan documents of various ministries have accepted a natural unit of water shed for harmonizing synergies of different resources to realize livelihood aspirations of agrarian community.

- In the participatory made meaning fall demand is created from within the system where accountability, responsibility, equity and sustainability automatically get passed on to the local community.

Watershed Developmental Approach

- Evolutionary Approach.
- Participatory approach of development is evolutionary in nature, social psychological factors viz. conflicts negotiations, interactions; prohibitions, persuasion and violence have been observed to be active during planning. These factors are generally guiding principles in participatory activities.
- In order to accomplish desired objectives and smooth working environment, certain rules and norms within the social system are required to be framed through common agreement.
- In every society or in a social system people have a common interest which act as a deriving force for community initiatives.
- Such common interest is fundamental of evolution in which nor mass and values play significant role promoting participation.
- It has been observed that evolutionary nature of participation does not occur without facilitators or catalysts.
- A professional service provider, who is the exogenous individual, may be a "catalyst" in a group because of his leadership ability and suggestions for fundamental problems requiring participatory solutions.
- Hence, an approach which empowers the beneficiary or the farmers for poverty alleaviation and food security through judicious natural resources management in a watershed is an important component of sustainable development.
- In the traditional mind-set of top down processes farmers are regarded as passive receptor, ignorant, conservative and unwilling to charge. Where as participatory mode accepts them as equal partners.
- In real sense, farmers have valuable knowledge and wisdom which is dynamic to catch up with new paradigms.
- They have well tested and proven traditional technologies for their own specific local situations. Therefore farmer should be allowed to play significant role in decision making process.

- It is a matter of great realization that the process of depletion of natural resources is best demonstrated by local people. It is the local perception which helps in identifying the process of natural resource regeneration and conservation through 'social fencing.'
- Blueprint to learning process
- Rapport with villager's creates conductive and democratic to express their own analysis through diagrams, maps, narration of micro situations and expected outcome.

Operational Guidelines for Watershed Development

- All development activities within a watershed should be pooled into a unified group action.
- Independent and uncoordinated sectional development does not fit into this concept.
- Simultaneously and harmonized use of natural resources producers several additional extensionalities of livelihood gathering through collective actions of community.
- Integrated development of watershed through participatory approach has been conceptualized on way of real development of rural section by in the policy and decision makers.
- During eight five year plan emphasis was given that planning for development should be undertaken on watershed basis and in view of that new guidelines were delineated to achieve success in improvement of socio-economic condition of people.

The Guidelines are :

1. Emphasis on people's participation.
2. Identifying problems and planning of activities by people.
3. Contribution from people living in watershed.
4. Implementation of programme by people.
5. Operation and maintenance by people.
6. Transparency in utilization of funds and maintenance of accounts.
7. Involvement of different agencies in development process.

8. Focus on group approach.
9. Convergence of development and welfare service to watershed area.
10. PIA should be made responsible for planning and implementation of watershed development projects.
11. Provision for entry point activities to build credibility.
12. Full time WDT exclusively for watershed development.
13. Provision for human resource development, training, gender issues, and income generating activities.
14. Focus on local institutional building through participatory approach in the watershed.

Gram Panachayat in Development of Watershed

1. Preparation of annual plans for the development of village panchayat.
2. Preparation of annual budget for village development.
3. Organizing voluntary labor and contribution for community works.
4. Maintenance of essential basic data of villages.

Agriculture

1. Promotion and development of agricultural activities and technology extension.
2. Promotion and development of horticulture.
3. Development of waste lands.
4. Development their unauthorized alienation and use.

Watershed as a developmental approach is adopted by different programmer of GOI i.e.

1. DPAP
2. DDP
3. Integrated wasteland development project scheme WDP 1989.

4. NWDPRA - National Water Shed Development Programme in Rain fed Areas (NWDPRA) (MOA).
5. Following watershed development programmes are implemented by ministry of agriculture, govt. of India.
   - NWDPRA - National Watershed Development Project for rain fed areas.
   - RVP and FPR - Soil conservation in the catchments of river valley project and Flood Prone Rivers.
   - WDPSCA - Watershed Development Projects in Shifting Cultivation Areas.
   - WDF - Watershed Development Fund.

New Framework for Next Generation Watershed Programmes

Key features of this unified approach

1. Delegating power to states to sanction and the implementation of watershed projects.
2. Dedicated institutions
3. Dedicated implementing agencies with multidisciplinary professional teams at National, State and district level.
4. Financial Assistance to dedicated institutions
5. Duration of the programme 4 to 7 years
6. Livelihood orientation
7. Cluster approach - geohydrological units normally of average size of 1000 to 5000 hect. Comprising clusters of micro watersheds.
8. Scientific planning
9. Capacity building
10. Multitier approach - Multitier ridge to valley sequenced approach.

Guiding Principles

1. Equity and gender sensitivity
2. Decentralization
3. Facilitating agencies

4. Centrality of community participation
5. Capacity building and technology inputs.
6. Monitoring, evaluation and learning
7. Organizational restructuring.

Technology inputs

To strengthen programme management and co-ordinations

1. Document best practices and case studies.
2. GIS facilities augmented with satellite imagery data.
3. Remote sensing data
4. Information connectivity would be extended to the entire district.
5. National portal creation for hosting data generated for data generated for all watershed projects.

Institutional Arrangements at National State and District Levels

1. Rate of National Rain fed Area Authority.
2. Institutional arrangements at ministry level.
3. National Level Data centre and National Portal.
4. Institutional arrangements at project level.
   i. PIA project implementing agency
   ii. Watershed development team
   iii. Rules and responsibilities of WDT
5. Institutional arrangement of village level
   i. SHGS
   ii. User groups
   iii. Watershed committee (WC)

Criteria for selection of watershed projects

1. Acuteness of drinking water security.
2. Extent of over exploitation of ground water resources.
3. Preponderance of wastelands/degraded lands.

4. Contiguity to another watershed that has already been developed/treated
5. Willingness of village community to make voluntary contribution, enforce equitable social regulations for sharing of common property resources.
6. Proportion of Scheduled Castes/Scheduled Tribes.
7. Area of the project should not be covered under assured irrigation.
8. Productivity potential for condition.

Project Management

1. Preparatory phase - 1-2 years
2. Watershed works phase - 2-3 years
3. Consolidation and withdrawal phase 1-2 years

Techonology Mission

- A powerful group of front rank experts charged with attainment of specific objectives and goals in a thrust area at the national level like - technology mission on oil seeds.
- The organizations and agencies involved in the programmes are given specific responsibilities and the progress of work is closely monitored at the highest level.
- As quickly as possible programme implemented with liberal financial assistance so taht non-availability of funds does not hinder their implementation.

Technology mission is a special assignment of task to a group of people to fulfill a technological programme within a given period of time. It is generally taken up at the national level to remove a persistent problem pervading the life of the people through their time bound result oriented. Multidisciplinary, highly intensive programmer.

Horticulture Mission

1. National horticulture mission for all the states and UTs except N.E. States, HP and J & K and Uttaranchal.

2. Second N.H. mission for the states J & K, N.E., H.P., Uttaranchal

    i. To promote growth of the horticulture sector covering fruits vegetables, roots and tuber crops, mushroom, spices, flowers, aromatic plants, cashew and cocoa programme for the development of coconut will be implemented by the coconut development board (CDB) independent of the mission.

    ii. This will be a centrally sponsored scheme in which GOI shall provide 100% assistance to the state missions during Xth plan.

Mission Objectives

The main objectives of mission are

i. To provide holistic growth of horticulture sector, through an area based regionally differentiated strategies which include research technology promotion, extension root harvest management, processing one marketing in consonance with comparative, advantage of each state/region and its diverse agro-climatic features.

ii. To enhance horticulture production improve nutritional security and income support to farm household,

iii. To establish convergence and self energy among multiple on-going and planned programme for horticulture development.

iv. To promote, develop and dissemination of technologies, through a seamless blend of traditional wisdom and modern scientific knowledge.

v. To create opportunities for employment generation for skilled and unskilled persons, specially unemployed youth.

Strategy

- To achieve the above objectives of the mission adopt the following strategies.
- Ensure an end to end holistic approach covering production, holistic end to end PHTM, processing and marketing to ensure appropriate returns to growers/producers.
- Promote R&D technologies for production post harvest management and processing.

- Enhance average, coverage and productivity through.
- Diversification - from traditional crop to plantation, orchards, vinegars, flower and urged gardens.
- Extension of appropriate technology to the farmers for high tech horticulture cultivation and precision farming.
- Assist setting up post harvest technology facilities such as pack have ripening chamber, cold storage, controlled atmosphere (CA) storage, processing units for value addition and marketing infrastructure.
- Adopt a coordinated approach and promotion of partnership convergence and synergy among RID, processing and marketing agency in public as well as private sectors, national, regional, state and substate levels.
- Where appropriate and fusible, promote MDDB model of cooperatives to ensure support and adequate returns to farmers.
- Promote capacity building HRD at all levels.

Mission Structure

a. National Level

General council

Chairman - Union Agriculture Minister

Executive Committee - Secretary agriculture department head

b. State level

State Level Executive Committee

c. District Level

District Mission Committee

d. Technical - support- group

National level

State level

District level

Technical support group

Key Elements of NHM

1. Base line survey.
2. Area based annual and perspective plans based on end to end approach with backward and forward linkages.
3. Research to be guided Research Advisory Committee or the basis of felt need by Boards such as ATMA and SHM and funded by perspective organizations and of their own budget.
4. Demand driven production based on cluster approach for potential crops having comparative advantage in different areas.
5. Best quality feed and planning material to be produced and made available.
6. Technology driven programmer to improve productivity and quality
7. Introduction of improved varieties.
8. Rejuvenation with improved cultivars.
9. Use of plastics.
10. High density plantation
11. Capacity of farmers and personnel.
12. Marketing infrastructure development linked with reforms
13. Meticulous reporting and monitoring

Report of National Farmer's Commission

Policies for sustainable livelihood of farmers

A national land use advisory service should be immediately established and linked to state and block level land use advisory services on a hub and spoke mode.

1. The scope of MSP programme should be expanded to cover all crops of importance to food and income security for small farmers. Arrangements should be made to ensure MSP at the right time and at the right place particularly in the areas coming within the scope of National Rain fed Area Authority.
2. A market price stabilization fund should be established jointly by the central and state govts and financial institutions to protect farmers during the period of violent fluctuations in prices.

3. An Agriculture Risk Fund should be setup to insulate farmers from risk arising from recurrent droughts and other weather aberrations.
4. The scope of agricultural insurance policies should become wider and should also cover health insurance, as envisaged under private Bima policy of seed companies should provide insurance in the case of G.M. crops.
5. A food guarantee act should be formulated and enacted farmer who are also the largest consumer group will produce more if there is greater consumption.
6. Well defined, pro farmer and pro resource poor consumer food security policy is an urgent necessity.
7. The time has come for the government to setup a multitask holder National Food Security and Sovereign Board chaired by P.M. to develop and implement a socially and economically sustainable food security programme catering to the interest of all regions of the country and all sections of overpopulation.
8. Farm families should be protected from HIV/ADS/TB.
9. There is need for Rural Non Farm Livelihood Initiative for rural areas.
10. An Indian Trade Organization (ITO) should be established to help the govt. to operate a livelihood security box and link global policies.
11. Since agriculture is a state subject every state govt. should setup a state farmer's commission with an eminent farmer as chairperson.
12. Agriculture progress should be measured by the growth in the net income of farm families. Along with production growth rates.
13. Panchayats have to be empower with the needed information, training and tools for discharging their responsibility.
14. Elected members of panchayats can play a pivotal role in improving sanitation, quality of life, drinking water, child care early childhood education and nutrition security.
15. Mass media constitutes an important pillar of our democratic system of governance they can work as warning system to check serious famines. To establish Regional Media Resource Centre.

16. Agriculture should be included under the concurrent list in schedule vii, article 246 of constitution.

Farmer of the future

1. Cooperative farming and service cooperative
2. Group farming by self helps groups
3. Small holder's estate
4. Contract farming
5. Farmers companies
6. State farms

Provisions for Assured Remunerative Marketing Opportunities

Important steps needed are :

1. MSP : Mechanism has to be developed protected and implemented effectively across the country MSP of crops needs to keep pace with the rising input costs.
2. The market Intervention scheme MIS should respond speedily to exigencies especially in the case of sensitive crops in the rain fed areas.
3. The establishment of community food grain banks would help in the marketing of under utilized crops.
4. Indian farmers can produce a wide range of health foods and herbal medicines and market them strict quality control and certification procedures.
5. P.D.S. should be universal and should undertake the task of enlarging the food security basket by storing and selling nutritious millets and other under utilized crops.
6. Indian Trade Organization setup.

Livestock and Livelihoods

1. Setup a livestock Food Corporation of India.
2. Promote Community Fodder Banks in scarcity area.
3. In case of poultry production.
   a. Poultry estates may be organized with help from national egg coordination committee and the organized poultry

industry to help SC/ST and other resource poor committee take to economically viable poultry farming.

b. Quarantine arrangements may be strengthened to avoid the outbreak of epidemics eased by invasive alien species.

c. Poultry farming should be regarded as an agriculture enterprise and should be given special attention from the point of view of bioseurity.

Some of the important new initiative to mitigate the agrarian crisis

1. Bharat Nirman era now deal for rural finding
2. National Rural Employment Guarantee Programme
3. National Horticulture Mission
4. Expansion of Agril credit
5. Lowering of interest rate
6. NRAA National Rain fed Area Authority
7. National Fisheries' Development Board
8. Changes in the Agril produce marketing committee act, to make it farmer friendly
9. Integrated food law
10. Warehouse Receipt Act, making warehouse receipts a negotiable instrument, thereby helping to prevent distress rates
11. Knowledge connectivity through 1,00,000 rural common service centers (CSC)

National Agricultural Innovation Project (NAIP)

Launched July 2006

Specific objectives

1. To build critical capacity of the ICAR as a catalyzing agent for management of change of the NARS
2. To promote research in the production to consumption system mode in the priority areas/themes to enhance agricultural production
3. To improve livelihood security of the rural people living in selected disadvantaged regions through technology led innovation systems

4. To build capacity to undertake basic and strategic research in strategic areas of agricultural sciences to meet the technology development challenges in the immediate and predictable future.

Thrust areas for research

1. Production to consumption system development.
2. Research on sustainable rural livelihood improvement.
3. Integrated agricultural diversification.
4. Livestock and fisheries production.
5. Genetic resources management and exploitation.
6. Bioprospecting, natural resource management, integrated pest management.
7. Post harvest processing and value addition.

Components of the NAIP

The NAIP comprises four components.

1. The ICAR as the catalyzing agent for the management of change in the Indian NARS.
2. Research on production to consumption systems on end to end basis.
3. Research on sustainable rural livelihood security.
4. Basic and strategic research in the frontier areas of agricultural sciences.

"SHGs" Self Help Groups

Concepts: - The self help groups (SHGs) are informal voluntary associations of people formed to attain a collective goal. People who are homogenous with respect to social background heritage caste or traditional occupations come together for a common cause to raise and manage resources for the benefits of group members. Usually, the focus is on poor and that too on women.

The process through which selected people, usually poor, representing a specific economic state of society, with some specific identity are encouraged to form small homogenous groups for the purpose of participating in development activities i.e. savings, credit, income generation etc is called "Group Formation" or group promotion or group evaluation.

The SHGs can be formed for any common cause or development activity.

Types of possible groups

1. Savings and credit groups
2. Joint farming groups
3. Social Forestry Groups
4. Water Users Groups
5. Watershed Development Groups
6. Farmers Interest Groups, etc.

Concept of SHG is much more beyond a credit and thrift group, advocates a strategy for empowerment through organization to act an agent for structural changes at the policy level and also raise survival need as the poverty development issues. SHG are helpful in creating confidence awareness and sharing information resource or knowledge.

Objectives of SHG

1. Saving and credit facilities.
2. Production and action of social issues.
3. To impart economic social status of its members in terms of its needs and priorities.
4. Activities multipurpose.
5. Capacity building, networking and convergence of services and activities.

Principles of SHGs

1. Homogeneity in nature.
2. Mutual help.
3. Common purpose/interest.
4. Regular meeting and savings.
5. Rotation of saving among members as per needs.

6. Regular repayments and more rotation.
7. Transparency in the dealings everybody knows.
8. Sharing of leadership.
9. Decision making democratic and need.
10. Self discipline

Formation of group

Three stages

1. Motivation
2. Group formation
3. Initiation of activities for consolidation

Stages according to Fisher

1. Orientation
2. Conflict
3. Emergence
4. Reinforcement

Organization of Self Help Groups

In order to facilitate the initial task of organizing the poor into groups, attention was first focused by the NGO on some specific issues of immediate concern to the poor, like exploitation of vested interest, seasoned migration for employment, small credit needs to meet emerging situations, raw supply etc. Initial efforts were directed towards bringing about now awakening among the target groups i.e. the small and marginal farmers, landless agricultural labors, backward classes, women and SC/ST.

The group strategy i.e. formation of groups or identification of existing natural/social groups met with several obstacles. The groups which developed into a savings and credit management institutions meeting the credit needs of their members created resentment among the money lenders. In pockets money lending is the sale occupation, the money lenders organized opposition to group formation. The group strategy did not harmonize with anti poverty programmes of the govt.

which were individual oriented. The banks operatives where all the documents were in the name of individuals.Resistence from men wherever women were involved.

Non availability of common place for holding the meeting. With the entry of other institutions, like NABARD banks, govt. agencies etc. groups meant different people. The differences lay in (i) Understanding the structure of the group i.e. the difference between cooperative heterogeneous large group and homogeneous socially viable groups and (ii) The manner in which the people are approached and groups formed. Certain common features when other agencies have been involved one. The pressure to achieve annual targets. The performance of officials was assessed on the basis of targets achieved. The dominant criteria to select "eligible" beneficiaries such as non-defaulters or even to fix the number of group members were extraneous to and often destructive to the process of group identification, formation and functioning.

As result of these dominant pressures, the staff first identified the eligible families and brought that family together as a 'group'. In most cases these groupings were artificial though the members had a degree of economic and gender homogeneity. A large no of groups were not socially homogeneous, the members of the group linked only by the desire to become "beneficiaries" of the programme. The group formed in this fashion will not scrutinize the options of each member nor will apply sanctions throughout the process of management of the asset and recovery.

Mobilization

The mobilization essentially emirates on the skill development by involving them at every stage of economic and social development to solve their actual major problems like unemployment and very low standard of living.

Mobilization is the process of bringing together all fusible and practical inter sectorial social allies to raise people's awareness and demand for a particular development programme to assist in the delivery of resources services and to strengthen community, participation for sustainability and self reliance. The goal of social mobilization is to raise the post status in society through capacity built down for self - awareness, self - development and self realization.

The philosophy lies in the assumption that poor are creative and have the capacity to take action for self improvement. There is a need to mobilize them to form group and take initiative for their development. Such groups are a moving unit of interacting personalities and are constituted by the fact that there is some interest which holds its members together.

There is also a need to understand internal nature of the group as to how they are formed, what their in strucutre and resources are, how they function and effect individual members, other groups and organization. This understanding will help in mobilizing SHG in the society for their sustained development.

SHG mobilization process

There are basically three phases to mobilize and sustain SHGs

I. Group formation phase (0-4 months).

II. Building up or stabilization phase (4-15 months).

III. Self helping phase (15-36) formation

I. Group formation phase (0-4 months).

1. Rapport building - visit to panchayat.
2. Interaction with farmers' together information.
3. Awareness creation about the programme activities.
4. Identification of farmer's interest in the programme during initial phase.
5. Conduct of general meeting of farmers for locating specific problems, their solutions and cushion plans.
6. Fixing the date and purpose of next meeting to one or 2 or 3 selected farmers who showed genuine interest and actively participated in general meeting.
7. Assigning the responsibility or organizing the next meeting to one or 2 or 3 selected farmers.
8. Recording and documentation of deliberators in Ist meeting by selected farmers and extension staff.
9. Assigning the selected farmers to mobilize genuine farmers for second meeting.

10. Conducting second meeting by the extension staff on analysis of problems, solutions and action plans.
11. Recording and documentation of deliberations in IInd Meeting by selected farmers.
12. Preparation of action plan supplies programme and instructions etc.
13. Execution of various activities as per action plan.
14. Conduct of fortnightly/monthly meetings of farmers to review progress of action plan and selection of members for SHG.
15. Selection of leaders in respect of production credit and marketing aspects. SHG should nominate these leaders based on involvement, interaction, experience, and acceptance.
16. These are to act as master farmers to guide and help others. Completion of collection of basic details about group members in data sheet by extension staff.
17. Imitation of some group activities by extension staff such as campaign GD.

II Building up/stabilization phase (4-15 month)

1. Regular fortnightly group meeting on need-based manner for carrying group activities with participation of extension staff. Attendance is to be maintained.
2. Training of master farmer a production, credit and marketing aspects.
3. Master farmers must project their leadership qualities and have confidence to take group members along the programme. If they are not they can be changed.
4. Implementation of programme as per action plan as per schedule.
5. Identification and execution of group action programmes Soil testing, procurement of seed, method demonstrations, marketing IPM etc.
6. Development of procurement plant for extension by SHGs as per specific intends of procurement.

7. Market synchronized production planning at later half of building phase. By this time, SHGs should be.
8. A market synchronized production planning at later half of building phase. By this time, SHG should be sed upto field centers or consumer markets.
9. Master farmers should be trained on production, credit and marketing aspects. They should be given training kits so that they could guide farmers.
10. After a period of one year, change the 3 master farmers.

III Self-helping phase (15-35)

1. Development of fortnightly action programme by master farmers in consultation with the other members
2. Collection of fortnightly programmes of other SHGs by extension staff to develop site calendar.
3. Organization of group meeting for sharing experiences of group members for farmer technology transfer.
4. A love laisioning by extension staff with other collaborating institutions and master farmers
5. Felt information support and organizational support.
6. Development of new set of master farmers after tenure of 1 year term.
7. Organization of inter-SHG contests at site for competition of achievement of SHGs
8. Setting norms for exercising group pressure and group control by master farmers in case of defaults.
9. Ensuring active participation of members in each activity of SHG.

Microfinance and functioning of Self Help Group for empowerment and sustainability.

Microfinance is the provision of thrift, credit and other financial services and products of very small amounts to the poor in rural, semi urban, or urban areas for enabling them to raise their income levels and improve their living standards.

Microfinance through SHGs has reached a respectable position and it is being acknowledged as the biggest microfinance intervention in the world.Grammen Bank of Bangladesh model is being replicated in various parts of India.SIDBI is also promoting microfinance through NGOs RMK Rastriya Mahila Kashtri is also in the business of promoting microfinance in India through NGOs.

Scope

Availability of credit is the basic requirement to any citizen, more so, to a poor person, credit should be timely as well as easily available. It is often seen that formal institution including nationalized banks are unable to meet the micro credit requirements of the poor people. Therefore, microfinance programmmes have emerged as effective instruments of poverty alleviation in India.

Sustainability and SHG

1. Agro biodiversity conservation and effective management of the natural resources largely depends upon community organizations like SHGs where emphasis could be laid on participatory conservation systems and sustainable method. The members of the group could be made in charge of the efforts and drive to protect the diversity of natural resources around their habitations.

2. A provision of economic rewarding to the successful practitioner may be made so as to motivate the grass root level community organizations in generating on economic stack in the conservation of traditional genetic strains of crops. Through the establishment of joint forest management committees and van samrakshana samities, affectation and plantation activities could be carried out intensively to increase the forest cover in the country.

Gender issues

Women in agriculture and their empowerment

Women in Agriculture and their Main streaming

1. Gender concerns need to be main streamed in the agricultural extension process. Public extension systems, which must disseminate new technology and information, are still largely male dominated. Hence the necessity to target women in to ensure that they receive information relevant to their work, particularly with reference to crops and livestock.

2. Improving access to extension and training women farmers usually have been neglected in external efforts. Under these

innovations efforts will need to be made both by the central and state innovations efforts will need to be made both by the central and state govts to improve extension services through.

3. Extension policy reorientation
4. Training of men and women
5. Training of women in decision making
6. Training of women farmers on agril marketing particularly with respect to post-harvest processing, on farm value addition.
7. Redesign of extension services to reach women farmers.

Extension services are being redesigned to focus on women through

- Appropriate training
- Increasing the proportion of trained female extension workers to gradually ensure that at least on third of extension
- Sensitizing male extension worker to the needs, approaches and perspectives of women through appropriate training and orientation programmer.
- Thereby dispelling the notion that only women can address extension needs of farm women.
- Improving communication between women, researchers, marketing agencies.
- Developing appropriate extension methodologies.
- Establishing head of the farm families.
- Expanding the sphere of women extension workers.
- Reexamination of all service cadre rules for hidden gender biases.
- Improvement of female attendance at agril institutes.
- Building incentives such as scholarships and stipends for more women to like up undergraduate and post graduate courses in the agriculture and allied sciences.
- Redesigning of agricultural training curricula to include women's concerns.

- Ensuring that women are adequately represented in all training programms.
- Redesigning of training facilities to make them suitable for large numbers of female students and trainees.
- Inclusion in the teaching curricula for extension workers greater analysis and extension methods take into account women's time, mobility and cultural situation and exploring the specific role of farm women in the marketing of agril produce.

Empowerment of Women

UNDP report indicates that while 67% of the worlds work is done by women only 10% of global income is earning by women and mere 1% of global property is owned by women.Empoerment refers to increasing political, social and economic strength of individuals and communities. it enables person manage and came out of any difficult situation and perceive obstacles on a part of life and challenges. Empowerment of women involves many things, economic opportunity, property rights, political representation, social availability, personal rights.

The spectrum of some such roles

1. Family food security
2. Farm laborers
3. Post harvest processing of farm trade
4. Livestock
5. Fisheries
6. Group organizers

Different development programme for women empowerment

1. Empowering strategies

1. Indira Mahila Yojana (IMY)
2. Mahila Samridhi Yogana (MSY)

3. The rural women's development and empowerment project (RWDED)
4. Development of women and child in rural area (DWCRA)

2. Employment and income generation

1. TRYSEM
2. Swayamsidha
3. Swalamban
4. Support for training and employment programme (STEP)
5. Norwegian agency for international development (MORAD)
6. Socio Economic Programme (SEP)
7. Condensed courses of education and vocational training (CCEVT)
8. SGSY
9. Women's vocational training programme.
10. Development of Women and Children in Urban Area (DWCUA)

3. Welfare and support services

1. Hostels for Working Women (HWW)
2. Short Story Homes (SSH)
3. Swadhar

4. Other enabling measures

1. Rashtriya Mahila Kastra (RMK)
2. National Credit Fund for Women (NCFW)

5. NGOs

1. Bihar Rural Livelihood Promotion Project (BRLP)
2. National Alliance of Young Entire Premiers (NAYE)
3. Self Employment Women and Association (SEWA)
4. Association of Women Entrepreneurs of Karnataka (AWAKE)

6. Financial Institution

1. Development Bank for Women Entrepreneurs

Approaches for women empowerment

1. Educational Empowerment
2. Social Empowerment
   I. Status
   II. Gender bias
   III. Heath nutrition
3. Economic empowerment
   I. Access to resources
   II. Micro credit performance
   III. Access to cooperatives and local women's organization
4. Technological empowerment
   I. Capacity building
   II. Exposure to mass media
   III. Appropriate training programme
   IV. Tech for woman

Measures to strengthen women's empowerment and rural development

- Identification and organization innovative and high income generating activities for woman.
- Development of entrepreneurial ability of women by organization special types of training.
- Empowering women to take up part time jobs while being on the house itself to earn additional income to support the family.
- Future planning at micro level.
- Adequate representation of women experts in the case of women related planning.
- Central and State govt. to place more emphasis on untouched area in the service sector to in value women.
- Govt can establish a separate mechanism.

- At the C, state district level to look after women's employment and their problem.
- Of infrastructure facilities and supportive services like land, building transport and also to look after the children.
- Encouraging women through formal and non formal education to involve themselves in the service sector.
- Organizing women in the service through women's association, cooperation societies or Mahila Mandals SHG etc.
- Establishment of supervisory bodies to monitor the implmentator at institutional provisions selected to women.
- Step to make women aware to technical and financial assistance available to women entrepreneurs and to encourage them through concession and centers to enter into service sector related to women.
- Govt should encourage research and development in the service sector to find out high income generating activities suitable for women.
- Wide publicity to training programme in the service sector.
- Creating awareness among educated and uneducated women about availability of facilities and concessions for entrepreneurs.

Women empowerment

1. Power
2. Assets
3. Knowledge
4. Attitude

Strategies for empowerment of rural women

1. Short-term
2. Mid-term
3. Long-term

Concepts of Yield Gaps

Yield gaps : Difference between the actual Yield and potential yield calculated

Strategies for closing the yield gaps

1. Policy support
2. Identifying and classifying yield gaps
   i Exploitable
   ii Less Exploitable
   iii Unexploitable
3. Promotion of integrated crop management
4. Deployment of new technologies
5. Adequate input and farm credit supply
6. Post production
7. Linkage of research and extension

Factors causing yield gaps/constraints

1. Physical factors problem soils, poor water management drought, flash floods and temperature stresses.
2. Biophysical factors: - Varieties, seeds, weeds, insects, diseases, and other pests.
3. Socio economic factors: - Labor shortage cost benefit, farmers knowledge, skills and welfare conditions.
4. Institutional factors: - Govts policies, rice price, agriculture credit and input supply, land tenure agriculture research and extension.

Benefits of Yield Gap Bridging

1. Solution to global food problem
2. Increased production
3. Cost reduction
4. Poverty alleviation
5. Social justice and equity

Overcoming constraints

1. Physical
2. Socio-economic Risk, cost and return, credit, traditions, attitude, input availability, institutions

3. Biological constraints Stable performing varieties, weeds, biotic and abiotic stress, soil problems, soil fertility and fertilizers, water and irrigation integrated crop management, Post Harvest.

Classification of Yield Gaps / Types

A. According to constraints

- Agronomic gaps - due to biological = physical constraints
- Socio-economic gaps
- Institutional gaps
- Mixed crops

B. Classification of yield gaps according practical purpose

- Unexploitable
- Less exploitable
- Exploitable

C. Swaminathan classification

- Yield gap research GAP between the yield possible on theoretical consideration
- Gap II/Research cum management gap is the gap between the best yields obtained in the research formed by a good farmer.
- Yield gap III/extension gap difference between the best average yields realised by a farmer in a state and the state average yield.

Components of yield gaps

There are mainly two components of yield gaps

- *1st Component*: The first component is mainly due to factors which are generally not transferable such as environmental conditions and some built in component technologies available at research station this component therefore cannot be narrowed or is not exploitable.
- *IInd Component* : Due to differences in management practices this gap exist as farmers use suboptimal doses of inputs and cultural practices Gap II is manageable and can be narrowed by deploying more efforts in research and extension services as well as govts appropriate intervention particularly on the institutional issues.

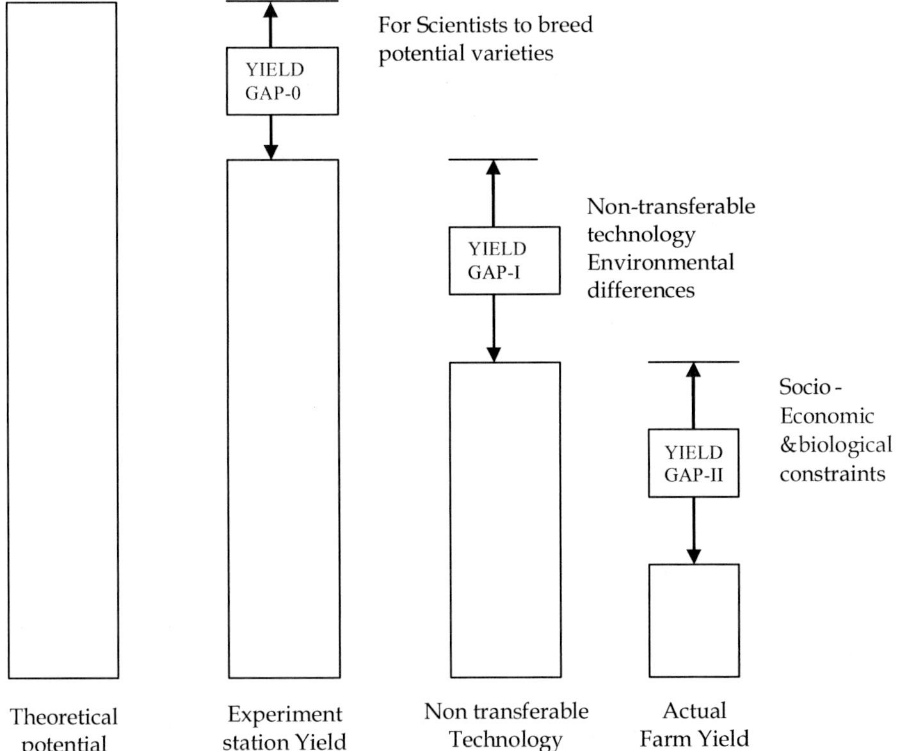

Information Revolution

The term information revolution (sometimes called also the "informational revolution") describes current economic, social and technological trends beyond the Industrial Revolution.

Many competing terms have been proposed that focus on different aspects of this societal development. The British polymath crystallographer J. D. Bernal (1939) introduced the term "*scientific and technical revolution*" in his book *The Social Function of Science* in order to describe the new role that science and technology are coming to play within society. He asserted that science is becoming a "productive force", using the Marxist Theory of Productive Forces. After some controversy, the term was taken up by authors and institutions of the then-Soviet Block. Their aim was to show that socialism was a safe home for the scientific and technical ("technological" for some authors) revolution, referred to by the acronym STR. The book *Civilization at the Crossroads*, edited by the Czech philosopher Radovan Richter (1969), became a standard reference for this topic.

Daniel Bell (1980) challenged this theory and advocated *Post Industrial Society*, which would lead to a service economy rather than socialism. Many other authors presented their views, including Zbigniew Brzezinski (1976) with his "Technetronic Society".

[edit] Information in social and economic activities

The main feature of the information revolution is the growing economic, social and technological role of information. Information-related activities did not come up with the Information Revolution. They existed, in one form or the other, in all human societies, and eventually developed into institutions, such as the Platonic Academy, Aristotle's Peripatetic school in the Lyceum, the Musaeum and the Library of Alexandria, or the schools of Babylonian astronomy. The Agricultural Revolution and the Industrial Revolution came up when new informational inputs were produced by individual innovators, or by scientific and technical institutions. During the Information Revolution all these activities are experiencing continuous growth, while other information-oriented activities are emerging.

Information is the central theme of several new sciences, which emerged in the 1940s, including Shannon's (1949) *Information Theory* and Wiener's (1948) *Cybernetics*. Wiener (1948, p. 155) stated also: "information is information not matter or energy". This aphorism suggests that information should be considered along with matter and energy as the third constituent part of the Universe; information is carried by matter or by energy.

We can distinguish between information, data, knowledge, and wisdom. Data comes through research and collection. Information is organized data. Knowledge is built upon information. Wisdom knows what to do with the knowledge—knowing how to synthesize the knowledge and make it useful to humankind.

Data and information are easily transferable; knowledge built by a person is not certain that it can be transferred to another. Following this, the notion of a "knowledge society" cannot be defined cogently. A *wisdom society* would have to be a futuristic society in which all human beings were educated to the highest level.

Information is then further considered as an economic activity, since firms and institutions are involved in its production, collection, exchange, distribution, circulation, processing, transmission, and control. Labor is also divided into physical labor (use of muscle power) and informational labor (use of intellectual power). A new economic sector is thereby identified, the Information Sector, which amalgamates information-related labor activities.

The Theory of Information Revolution

The term *information revolution* may relate to, or contrast with, such widely used terms as Industrial Revolution and Agricultural Revolution. Note, however, that you may prefer mentalist to materialist paradigm. The following fundamental aspects of the theory of information revolution can be given (Veneris 1984, 1990):

1. The object of economic activities can be conceptualized according to the fundamental distinction between matter, energy, and information. These apply both to the object of each economic activity, as well as within each economic activity or enterprise. For instance, an industry may process matter (e.g. iron) using energy and information (production and process technologies, management, etc.).

2. Information is a factor of production (along with capital, labor, land (economics)), as well as a product sold in the market, that is, a commodity. As such, it acquires use value and exchange, and therefore a price.

3. All products have use value, exchange value, and informational value. The latter can be measured by the information content of the product, in terms of innovation, design, etc.

4. Industries develop information-generating activities, the so-called Research and Development (R&D) functions.

5. Enterprises, and society at large, develop the information control and processing functions, in the form of management structures; these are also called "white-collar workers", "bureaucracy", "managerial functions", etc.

6. Labor can be classified according to the object of labor, into information labor and non-information labor.

7. Information activities constitute a large, new economic sector, the information sector along with the traditional primary sector, secondary sector, and tertiary sector, according to the three-sector hypothesis. These should be restated because they are based on the ambiguous definitions made by Colin

Clark (1940), who included in the tertiary sector all activities that have not been included in the primary (agriculture, forestry, etc.) and secondary (manufacturing) sectors. The quaternary sector and the quandary attempt to classify these new activities, but their definitions are not based on a clear conceptual scheme, although the latter is considered by some as equivalent with the information sector.

8. From a strategic point of view, sectors can be defined as information sector, means of production, means of consumption, thus extending the classical Ricardo-Marx model of the Capitalist (see Influences on Karl Marx). Marx stressed in many occasions the role of the "intellectual element" in production, but failed to find a place for it into his model.
9. Innovations are the result of the production of new information, as new products, new methods of production, patents, etc. Diffusion of innovations manifests saturation effects (related term: market saturation), following certain cyclical patterns and creating "economic waves", also referred to as "business cycles". There are various types of waves, such as Kondratiev wave (54 years), Kuznets swing (18 years), Juglar cycle (9 years) and Kitchin (about 4 years, see also Joseph Schumpeter) distinguished by their nature, duration, and, thus, economic impact.
10. Diffusion of innovations causes structural-sect oral shifts in the economy, which can be smooth or can create crisis and renewal, a process which Joseph Schumpeter called vividly "creative destruction".
11. From a different perspective, Irving E. Fang (1997) identified six 'Information Revolutions': writing, printing, mass media, entertainment, the 'tool shed' (which we call 'home' now), and the Information Highway. In this work the term 'information revolution' is used in a narrow sense, to describe trends in communication media.
12. Measuring and modeling the Information Revolution
13. Portal (1976) measured the Information Sector in the US using the input-output analysis; OECD has included statistics on the Information Sector in the economic reports of its member countries.

Modeling the Informational Revolution

Veneris (1984, 1990) explored the theoretical, economic and regional aspects of the Informational Revolution and developed a systems dynamics simulation computer model.

These works can be seen as following the path originated with the work of Fritz Machlup who in his (1962) book "The Production and Distribution of Knowledge in the United States", claimed that the "knowledge industry represented 29% of the US gross national product", which he saw as evidence that the Information Age has begun. He defines knowledge as a commodity and attempts to measure the magnitude of the production and distribution of this commodity within a modern economy. Machlup divided information use into three classes: instrumental, intellectual, and pastime knowledge. He identified also five types of knowledge: practical knowledge; intellectual knowledge, that is, general culture and the satisfying of intellectual curiosity; pastime knowledge, that is, knowledge satisfying non-intellectual curiosity or the desire for light entertainment and emotional stimulation; spiritual or religious knowledge; unwanted knowledge, accidentally acquired and aimlessly retained.

More recent estimates have reached the following results

- The world's technological capacity to receive information through one-way broadcast networks grew at a sustained compound annual growth rate of 7 % between 1986 and 2007;
- The world's technological capacity to store information grew at a sustained compound annual growth rate of 25 % between 1986 and 2007;
- The world's effective capacity to exchange information through two-way telecommunication networks grew at a sustained compound annual growth rate of 30 % during the same two decades;
- The world's technological capacity to compute information with the help of humanly guided general-purpose computers grew at a sustained compound annual growth rate of 61 % during the same period.

❒❒❒

Chapter 8

Recent Developments in Agriculture and Extension

The Organic Movement

The organic movement began in the early 1900s in response to the shift towards synthetic nitrogen fertilizers and pesticides in the early days of industrial agriculture. It lay dormant for many years, kept alive by a relatively small group of ecologically minded farmers. These farmers came together in various associations : Demeter International of Germany, which encouraged biodynamic farming and began the first certification program, the Australian Organic Farming and Gardening Society, the Soil Association of the United Kingdom, and Rodale Press in the United States, along with others. In 1972 these organizations joined to form the International Federation of Organic Agriculture Movements (IFOAM). In recent years, environmental awareness has driven demand and conversion to organic farming. Some governments, including the European Union, have begun to support organic farming through agricultural subsidy reform. Organic production and marketing have grown at a fast pace.

The term "organic" can be broadly described as food grown without the assistance of man-made chemicals (businesssweek.com). The beginnings of the organic movement can be traced back to the beginning of the 1800s. In 1840 Justus Von Liebig developed a theory of mineral plant nutrition. Liebig believed that manure could be directly substituted for mineral salts. Many years later in 1910, preceding the First World War, chemists Fritz Haber and Carl Bosh developed an ammonia synthesis process, making use of nitrogen from the atmosphere. This form of ammonia had already been used to manufacture explosives, so after the war, it was implemented into the fertilization of agriculture.

Organic food was initially seen as a fad observed by the eccentric few, however today it has become more widespread. "Organics have come to represent a safe house in a disturbing world where food quality and safety are constantly under siege" (Blythman). Today, whole foods stores have captured a significant share of the grocery shopping market, specifically, [Whole Foods], [Wild Oats], and others.

Structure

Specifications for what may be classified as "organic" food may vary by location. For all intents and purposes, the following quote from USDA. Govt. presents a fair structure for the movement.

"What is organic food? Organic food is produced by farmers who emphasize the use of renewable resources and the conservation of soil and water to enhance environmental quality for future generations. Organic meat, poultry, eggs, and dairy products come from animals that are given no antibiotics or growth hormones. Organic food is produced without using most conventional pesticides; fertilizers made with synthetic ingredients or sewage sludge; bioengineering; or ionizing radiation. Before a product can be labeled 'organic,' a Government-approved certifier inspects the farm where the food is grown to make sure the farmer is following all the rules necessary to meet USDA organic standards. Companies that handle or process organic food before it gets to your local supermarket or restaurant must be certified, too."

People

According to ResearchWikis.com, the purchasing of organic food stuffs has risen every year since 1990 to latest record breaking year.

Purchasing continues to gain among all income groups and races, and is gaining the most ground with women ages 25-35, white, & married demographic.

One of the pioneers was John Battendieri.

Battendieri pressed his first batch of "Mr. Natural" apple juice in 1972, making it one of the first packaged organic products.

Organic companies

The recent interest in the organic industry has sparked the interest of many business. Among these businesses are large companies that seek to distribute many products in large amounts to the American public. The organic market is now a 14 billion dollar a year industry, that continues to grow especially from large corporations such as Wal-Mart that are now offering organic choices to their customers.[3] Other companies leading the organic revolution include food companies such as; General Mills and Kraft that are making the switch to offer more organic option for customers. Along with major companies now offering more organic choices many large companies are buy up smaller already established organic companies such as Earth's Best, Rice Dream soy milk, Garden of Eaten', Celestial Seasonings and Health Valley. By having large companies buy up the already established smaller organic companies it exposes these companies to the popular culture, therefore making them popular because they are in large quantities.

Evergreen revolution

Dr. M.S. Swaminathan along with Nobel Laureate Normal Borlaug, is an honorary advisor to the US-India joint knowledge initiative in agriculture started last year when Prime Minister Manmohan Singh met US President George Bush in America. American Foreign Agricultural Service Administrator Ellen Torstar co-chairs the board, and the Indian counterpart was Mangala Rai, then Chief of the Indian Council for Agricultural Research, New Delhi.

Complete coverage

- Agriculture is becoming more knowledge intensive, not chemical intensive, and that is what we want for our small farmers too. In a sample survey, more than 40 per cent of our farmers said they wanted to quit farming. If this is true with older farmers, the younger farmers won't stay in the farms at all. That is what is happening already. Even graduates in agriculture do not go back to farming; they like to take up jobs in banks or elsewhere. So, how do we attract and retain the youth in farming? Only if it can become intellectually satisfying and economically rewarding. This means you have to bring down the cost of cultivation and improve productivity.

- If you look at the economic survey of 2006, you find that we have 100 million farms in India but over 60 per cent of them are less than one hectare. 40 per cent are below one acre — what we call sub-marginal. So, where is the source of income unless you have more productivity? If I produce just one tonne of rice, I may not be able to sell much because I need it for myself. If you produce four tonnes from the same area, I can sell three tonnes. So, the knowledge initiative is about improving productivity, but on an environmentally sustainable basis, without ecological harm. That is what I call an 'evergreen revolution'.

In what way is the US going to help Indian farmers improve productivity?

- They have a large number of universities, a large number of departments, and certain expertise that we do not have. For example, in areas like bio-fuels, bio-diesels, etc, they have much more expertise than we do. Therefore, it is up to us to choose what we want.
- There are four broad areas identified. One is biotechnology as applied to crops, as applied to animals and as applied to fish. The second area is post-harvest technology, in which agro-processing and small farmers' agro business will be given emphasis. We keep saying that only 7 per cent of our fruit and vegetables are processed while the rest is either consumed or lost in rotting. In the US, it is the reverse — just 10 per cent is consumed and 90 per cent is processed. They have a lot of knowledge on the area of food processing, quality control and packaging, selling, etc. If we want to export, we have to follow international standards.

When the real problem lies in India, in what way will the expertise and knowledge from the US benefit our farmers?

1. Sometimes, foreigners tell you point-blank what the problem is! They can tell us how they succeeded. You should never stop an exchange of knowledge and information. Let us follow Mahatma Gandhi's principle : 'Keep your doors and windows open; let fresh air and ideas come from all sides. Keep, however, your feet firmly on the ground and do not get blown off by untested ideas.'

2. Let knowledge come from everywhere. They have all the Nobel Prize [Images] winners there. They have invested a lot of money in developing ideas, and they have human resources. At the Rashtrapati Bhavan function, George Bush said there are two million Indians bringing about great changes to the US in the fields of science and technology. He acknowledged it.

Apart from the media, even President Bush in his address described the initiative bringing about a second green revolution.

I have never supported that term. The term 'Green Revolution' was coined by William Gaud in the US, in the '60s. It meant production to productivity improvement. For example, 20th century science has helped us produce more from 50 per cent of the land used, because of the vertical growth in productivity. I coined the term 'Evergreen revolution', which means improvement of productivity without associated ecological or environmental harm. Use of pesticides, etc will lead to a Greed Revolution. There is a distinction between Green Revolution and Greed Revolution. What we need is an Evergreen revolution!

In an earlier interview, you said it was not productivity, but accessibility to food that was India's problem. When we have situations where grain rots in godowns, how will this knowledge initiative help?

Yes, our poor do not have purchasing power. That is why I say we need more income and more employment orientation. This is where the knowledge initiative is going to help us. The situation now is different. We are already lagging behind in food production. It is less than the population growth. We are importing wheat now. We should not go back to the situation we faced in the '60s. What has gone wrong largely is the public policy. Rather, you don't have policies.

Are you confident about achieving good results from this partnership?

Yes. In the '60s the partnership was mutually beneficial, and we took the best advantage. All our agricultural universities were developed then with their help.

America accounts for 30 per cent of the world's food production. Is it knowledge that helped them achieve this?

They are supported by three things – technology which is knowledge, very high subsidy and capital driven large farms. One farm in Iowa may have a couple of million-worth equipment. Subsidy to one farmer alone in the US is one crore. But, in India, agriculture is a way of life and a means to livelihood. Our production is by masses; their production is mass production. 50 million people in our country produce 90 million tonnes of milk. 100,000 people in the US produce 70 million tonnes of milk. Where is 100,000 and where is 50 million?

Organic Farming

Organic farming methods have proven to be more effective than many conventional farmers realized, but they are still placed at a relative disadvantage by their lack of synthetic nitrogen. Thus they rely on price premiums, but these price premiums often allow them to reap higher profits than conventional farms.

Criticisms of the Organic Movement

1. The organic movement has transformed the mentality that the general public held about food. However, not everyone is a fan of organic products, higher prices for instances are one of the first things critics attack. Originally, organic farming was characterized by family farming, which, contributed to its wholesome image. Initially, organic products were for a small portion of the population, the portion that is left over after what society determined was popular culture. Over the years the idea of organic food and products caught on, becoming more popular and widespread. The event of organics becoming more normative marked the transition from high culture to popular culture. Today, organic farming is not representative of quality family farming, it is dominated by big corporations and has become big business. "A single company controls 70 percent of the market in organic milk, California grows about 400 million dollars per

year in organic produce and the Organic Trade Association includes the biggest names in agribusiness, such as Archer Daniels Midland, Gerber, and Heinz " Despite the modest beginnings of the organic movement, organic farming is now a lucrative source of revenue, and its humble beginnings are no longer representative of the organic products on the market today.

2. Some organic food advocates claim that organic food tastes better, and often it does. The organic movement emphasizes eating local produce that is in season, which in turn, produces fresher food. Essentially, the food being tastier is not attributable to the fact that it's organic. "Organic food may be fresh, but it's fresh from the manure fields"

3. Critics further protest against the idea that organic food is healthier. Miller, the author of the Organic Myth asserts, "organic food is no more nutritious than food fueled by industrial fertilizers that are sprayed with synthetic pesticides and genetically altered in science labs." He goes on to say that bacteria are a source of disease in organic food and organic farming is incredibly inefficient

Supporters of the Organic Movement

- Proponents of organic agriculture point to the fact that organically-grown foods contain higher nutrient levels than those grown with synthetic pesticides and herbicides, and the environmental benefits of not using these often persistent and toxic chemicals. Organic farming also promotes greater biodiversity on farms, as the lack of poison allows species other than those being cultivated to inhabit the farms.

- Groups such as IFOAM are active supporters of the organic movement throughout the world. They readily state their goals and are taking their idea to the world.

- "Agriculture is one of humankind's most basic activities because all people need to nourish themselves daily. History, culture and community values are embedded in agriculture. The Principles apply to agriculture in the broadest sense, including the way people tend soils, water, plants and animals in order to produce, prepare and distribute food and other goods. They concern the way people interact with living landscapes, relate to one another and shape the legacy of future generations." IFOAM.com

- This organization and others like it come under attack from political right in the US and other countries, but it has been accredited by many United Nations organizations.
- "Several of the management practices evolved by the organic farming movement have a direct relevance to the battle to achieve food security. Examples include diversification and crop rotation, as well as the use of natural means to combat pests."
- Quote from Jacque Diouf, Director General of the FAO in IFOAM's 25th anniversary issue of its magazine Ecology & Farming.
- IFOAM has been an FAO-accredited international organization since 1997.
- In March 2002, IFOAM obtained observer status with UNCTAD, and in addition received classification in a special category to participate in the meetings of the Commission on Trade on Goods and Services and Commodities, the Commission on Enterprise, Business Facilitation and Development and the meetings of the UNCTAD Board.
- "Organic agriculture can play an important role to achieve the goals of the Convention on Biodiversity, in the global context of the Millennium Development Goals, by significantly reducing the current rate of biodiversity loss. In addition, organic farming offers direct economic benefits to resource-poor farmers and thereby makes an important contribution to both poverty alleviation and sustainable development."
- Statement made by UNEP Executive Director Klaus Toepfer at the IFOAM Conference on Biodiversity and Organic Agriculture that UNEP co-hosted September 24–26, 2004 in Nairobi, Kenya.
- IFOAM is an accredited international organization with UNEP, but that only one aspect of a more complex relationship between the two organizations.
- In September 2004, UNEP hosted and jointly organized the Third International IFOAM Conference on Biodiversity and Organic Agriculture at the headquarters of UNEP in Nairobi, Kenya.
- IFOAM and the UNEP are in the process of developing a three-year joint work program for biodiversity and organic agriculture

in order to implement the recommendations resulting from the conference.

- With organizations such as these gaining in their presence worldwide organic farming practices will likely gain ground in the foreseeable future.

Timeline

- The first use of the term "organic farming" is by **Lord Northbound** (aka Walter James, 4th Baron Northbourne). The term derives from his concept of "the farm as organism", which he expounded in his book, Look to the Land (1940), and in which he described a holistic, ecologically balanced approach to farming. Northbound wrote of "chemical farming versus organic farming".
- Sir Albert Howard's 1940 book, *An Agricultural Testament*, was influential in promoting organic techniques, and his 1947 book "The Soil and Health, A Study of Organic Agriculture" adopted Northbourne's terminology and was the first book to include "organic" agriculture or farming in its title.
- In 1939, strongly influenced by Sir Howard's work, Lady Eve Balfour launched the Haughley Experiment on farmland in England. It was the first, side-by-side comparison of organic and conventional farming. Four years later, she published *The Living Soil*, based on the initial findings of the Haughley Experiment. It was widely read, and lead to the formation of a key international organic advocacy group, the Soil Association.
- During the 1950s, sustainable agriculture was a research topic of interest. The science tended to concentrate on the new chemical approaches. In the U.S., J.I. Rodale began to popularize the term and methods of organic growing. In addition to agricultural research, Rodale's publications through the Rodale Press helped to promote organic gardening to the general public.
- In 1962, Rachel Carson, a prominent scientist and naturalist, published *Silent Spring*, chronicling the effects of DDT and other pesticides on the environment. A bestseller in many countries, including the US, and widely read around the world, *Silent Spring* was instrumental in the US government's 1972 banning of DDT. The book and its author are often credited with launching the environmental movement.

- In the 1970s, worldwide movements concerned with environmental pollution caused by persistent agrichemical increased attention on organic farming. One goal of the *organic movement* was to promote consumption of locally grown food, which was promoted through slogans such as *"Know Your Farmer, Know Your Food"*.
- In 1972, the International Federation of Organic Agriculture Movements (IFOAM), was founded in Versailles, France. IFOAM was dedicated to the diffusion of information on the principles and practices of organic agriculture across national and linguistic boundaries.
- In the 1980s, around the world, various farming and consumer groups began seriously pressuring for government regulation of organic production to ensure standards of production. This led to various legislation and certification standards being enacted through the 1990s and to date. Currently, most aspects of organic food production are government-regulated in the US and the European Union.
- In the 2000s, the worldwide market for organic products (including food, beauty, health, bodycare, and household products, and fabrics) has grown rapidly. More countries are establishing formal, government-regulated certification of organic food : in 2002 in the US, in 2005 in China and projected for 2006 in Canada, among others. Monitoring and challenging certification rules and decisions have become a regular, high profile aspect of activists in the organic movement.

Permaculture

- Permaculture is an approach to designing human settlements and agricultural systems that is modeled on the relationships found in nature. It is based on the ecology of how things interrelate rather than on the strictly biological concerns that form the foundation of modern agriculture. Permaculture aims to create stable, productive systems that provide for human needs; it's a system of design where each element supports and feeds other elements, ultimately aiming at systems that are virtually self-sustaining and into which humans fit as an integral part.
- Permaculture as a systematic method was developed by Australians Bill Mollison and David Holmgren during the 1970s.

The word "permaculture" originally referred to "permanent agriculture", but was expanded also to stand for "permanent culture" as it was seen that social aspects were integral to a truly sustainable system. Mollison has described permaculture as "a philosophy of working with, rather than against nature; of protracted and thoughtful observation rather than protracted and thoughtless labor; and of looking at plants and animals in all their functions, rather than treating any area as a single project system."

- Permaculture draws from several other disciplines including organic farming, agroforestry, sustainable development, and applied ecology. "The primary agenda of the movement has been to assist people to become more self reliant through the design and development of productive and sustainable gardens and farms. The design principles which are the conceptual foundation of permaculture were derived from the science of systems ecology and study of pre-industrial examples of *sustainable land use.*"

Biodynamic Agriculture

- Biodynamic agriculture is a method of organic farming that treats farms as unified and individual organisms, emphasizing balancing the holistic development and interrelationship of the soil, plants and animals as a self-nourishing system without external inputs. Insofar as this is possible given the loss of nutrients due to the export of food. As in other forms of organic agriculture, artificial fertilizers and toxic pesticides and herbicides are strictly avoided. There are independent certification agencies for biodynamic products, most of which are members of the international biodynamic standards group Demeter International.
- Regarded by some as the first modern ecological farming system and one of the most sustainable, biodynamic farming has much in common with other organic approaches, such as emphasizing the use of manures and composts and excluding of the use of artificial chemicals on soil and plants. Methods unique to the biodynamic approach include the use of fermented herbal and mineral preparations as compost additives and field sprays and the use of an astronomical sowing and planting calendar. Biodynamic originated out of the work of Rudolf Steiner, the founder of anthroposophy.

- The development of biodynamic agriculture began in 1924 with a series of eight lectures on agriculture given by philosopher Rudolf Steiner at Schloss Koberwitz in Silesia, Germany, (now Kobierzyce in Poland east of Wroc³aw). The course was held in response to a request by farmers who noticed degraded soil conditions and deterioration in the health and quality of crops and livestock resulting from the use of chemical fertilizers. At the course of eight lectures there were 111 attendees coming from six countries. An agricultural research group was subsequently formed to test the effects of biodynamic methods on the life and health of soil, plants and animals.
- Today biodynamic is practiced in more than 50 countries worldwide. Demeter International is the primary certification agency for farms and gardens using the methods.

Organic Farming

- Organic farming is the form of agriculture that relies on techniques such as crop rotation, green manure, compost and biological pest control to maintain soil productivity and control pests on a farm. Organic farming uses fertilizers and pesticides but excludes or strictly limits the use of manufactured (synthetic) fertilizers, pesticides (which include herbicides, insecticides and fungicides), plant growth regulators such as hormones, livestock antibiotics, food additives, and genetically modified organisms.
- Organic agricultural methods are internationally regulated and legally enforced by many nations, based in large part on the standards set by the International Federation of Organic Agriculture Movements (IFOAM), an international umbrella organization for organic farming organizations established in 1972. IFOAM defines the overarching goal of organic farming as:
- "Organic agriculture is a production system that sustains the health of soils, ecosystems and people. It relies on ecological processes, biodiversity and cycles adapted to local conditions, rather than the use of inputs with adverse effects. Organic agriculture combines tradition, innovation and science to benefit the shared environment and promote fair relationships and a good quality of life for all involved."
- International Federation of Organic Agriculture Movements.

- Since 1990, the market for organic products has grown from nothing, reaching $55 billion in 2009 according to Organic Monitor. This demand has driven a similar increase in organically managed farmland. Approximately 37,000,000 hectares (91,000,000 acres) worldwide are now farmed organically, representing approximately 0.9 percent of total world farmland (2009).

Sustainable Agriculture

Sustainable agriculture is a way of raising food that is healthy for consumers and animals, does not harm the environment, is humane for workers, respects animals, provides a fair wage to the farmer, and supports and enhances rural communities.

Characteristics of this type of agriculture include :

- *Conservation and preservation* : What is taken out of the environment is put back in, so land and resources such as water, soil and air can be replenished and are available to future generations. The waste from sustainable farming stays within the farm's ecosystem and cannot cause build-up or pollution. In addition, sustainable agriculture seeks to minimize transportation costs and fossil fuel use, and is as locally-based as possible.
- *Biodiversity* : Farms raise different types of plants and animals, which are rotated around the fields to enrich the soil and help prevent disease and pest outbreaks. Chemical pesticides are used minimally and only when necessary; many sustainable farms do not use any form of chemicals.
- *Animal welfare* : Animals are treated humanely and with respect, and are well cared for. They are permitted to carry out their natural behaviours, such as grazing, rooting or pecking, and are fed a natural diet appropriate for their species.
- *Economically viable* : Farmers are paid a fair wage and are not dependent on subsidies from

Climate Change

What is climate change?

Climate change can manifest itself in gradual changes in temperature, precipitation and a rise in sea level, resulting in changes in frequency, intensity and duration of extreme events. Global warming means earth getting warmer and resulting in an ecological imbalance.

Impact of Climate Change

The major challenge faced by the entire world today is Climate Change or Global Warming. Every country is taking various measures to reduce the global warming. The climate change causes a greater impact on the agriculture. Particularly, Indian agriculture is hardly affected due to the climate change or global warming. The success of Indian agriculture much depends on the normal monsoon as also on favorable weather conditions. The fact is that all over the world, wherever, whenever the farmers meet, the crops and the prevailing prices. Aberrant weather conditions worry them. During recent years, burning of fossil fuels by the vehicles, coal burning by the power plants, emission from the industrial factories as also as a result of the large scale deforestation, the earth's surface temperature is increasing, rainfall pattern is shifting causing loss of moisture, occurrence of cyclones, thunderstorms, floods as also the rising of the sea-level, which may ultimately see many coastal cities and towns inundated. As far as food grains production is concerned, even a slight rise in the earth's surface temperature could cause drop in the country's wheat production substantially as also it could much impair the quality of rice (specially of the basmati rice), fruits, vegetables and medicinal plants, products now being much valued for their export. Despite these limitations, much useful work has been done on estimating the impacts of different climate change scenarios. In this paper climate change impacts in agriculture are examined with particular to Indian scenario.

Agriculture is one of the most weather-dependent of all human activities. In India, agriculture and allied activities constitute the single largest component of Gross Domestic Product (GDP) contributing nearly 25% of the total. The tremendous importance of this sector to the Indian economy can be ganged by the fact that it provides employment to two-thirds of the total workforce. The share of agricultural products in exports is also substantial, with agriculture accounting for 15% of export earnings. Agricultural growth also has a direct impact on poverty eradication, and is an important factor in employment generation. Further, Indian agriculture is fundamentally dependent on weather for higher productivity.

How Climate Change Affects Agriculture ?

- Climate change can affect agriculture in a variety of ways. Beyond a certain range of temperatures, warming tends to reduce yields because crops speed through their development, producing less grain in the process. Evaporation from the soil accelerates when temperatures rise and plants increase transpiration—that is, lose more moisture from their leaves. The combined effect is called "evapotranspiration." Because global warming is likely to increase rainfall, the net impact of higher temperatures on water availability is a race between higher evapotranspiration and higher precipitation. Typically, that race is won by higher evapotranspiration. But a key culprit in climate change—carbon emissions— can also help agriculture by enhancing photosynthesis in many important, so-called C3, crops (such as wheat, rice, and soybeans). The science, however, is far from certain on the benefits of carbon fertilization. But we do know that this phenomenon does not much help C4 crops (such as sugarcane and maize), which account for about one-fourth of all crops by value.

- Potential direct effects on agricultural systems : Seasonal changes in rainfall and temperature could impact agro-climatic conditions, altering growing seasons, planting and harvesting calendars, water availability, pest, weed and disease populations, etc. Evapotranspiration, photosynthesis and biomass production is altered. Land suitability is altered.

- India, as a developing country has reasons to be concerned about the adverse impact of climate change on its economy. A large part of its population depends on climate sensitive sectors for livelihoods which makes it highly vulnerable to climate change. Climate change can have serious impact on its crops, forests, coastal regions, etc. which can in turn affect the achievement of its important national development goals. The issue of climate change cannot however be taken up without linking it to developmental needs such as poverty, health, energy access and education. Efforts to address climate change adaptation and mitigation needs should not take resources away from the core development needs and growth objectives of the developing countries. Climate Change mitigation and poverty reduction should be addressed simultaneously.

Effect of Global Warming

- Global warming is increasing the earth's average temperature. The green house gases are the main culprits of the global warming. The green house gases like carbon dioxide, methane and nitrous oxide are playing hazards in the present time. These green house gases trap heat in the earth's atmosphere and thus result in increasing the temperature of the earth. The excessive emission of these gases is the major cause of global warming.
- The major source of carbon dioxide is the power plants. These power plants emit large amounts of carbon dioxide produced from burning of fossil fuels for the purpose of electricity generations. Coal is the major fuel that is burnt in these power plants. Coal produces around 1.7 times as much carbon dioxide per unit of energy when flamed as does natural gas and 1.25 times as much as oil. Coal gives out 80% more carbon per unit of energy it produces as compared to natural gas. Another major source of carbon dioxide in the atmosphere is the emission from the cars and other vehicles. About 20% of carbon dioxide emitted in the atmosphere comes from burning of gasoline in the engines of the vehicles.
- Buildings, both commercial and residential represent a large source of global warming pollution than cars and trucks. Building of these structures requires a lot of fuel to be burnt which emits a large amount of carbon dioxide in the atmosphere.
- The second major greenhouse gas after CO_2 which causes global warming is Methane. Methane is obtained from resources such as rice paddies, Bovine Flatulence, bacteria in bogs and fossil fuel manufacture. Almost in all parts of the world, rice is grown on flooded fields. When fields are flooded, anaerobic situation built up and the organic matter in the soil decays, releasing methane to the atmosphere. Nitrous oxide which is a colorless gas with sweet odour is another green house gas. The main sources of nitrous oxide include nylon and nitric acid production, cars with catalytic converters, the use of fertilizers in agriculture and burning of organic matter. A greater emission of nitrous oxides in the recent decades is leading global warming.
- Another major cause of global warming is deforestation. Deforestation is to be blamed for 25% of all carbon dioxide release entering the atmosphere, by the cutting and burning of about 34

million acres of trees each year. Trees collect CO_2 that we breathe out and they give back oxygen that we breathe in. Thus, cutting of trees is leading to greater concentration of CO_2 in the atmosphere. Greater urbanization, requirement of land for factories and buildings, requirement of timber are all reasons that are leading to deforestation which in turn is leading to global warming.

- The global warming has led to increase in mean earth surface temperature and thus melting of the polar ice. There are frequent meltdown of glaciers that result in floods and other natural calamities. The melting of ice at the poles has led the increase in sea level. And further increase in temperature may melt the ice and lead to increase in mean sea level which will engulf low lying countries.
- The effect of global warming is very evident on the animal kingdom. Some animals have become extinct due to loss of their natural habitat or their inability to evolve the rapid change in their life style because of the changes in the seasons. The migrating birds have changed their time to travel and also their place of migration.
- The effects of global warming can be felt on the seasons too. There is shift in season cycle, as the summers are getting longer than the winters. This has affected the animals and made them to change their life style accordingly and those who failed to do so have perished or on the verge of extinction.
- The global warming is also responsible for the introduction of some new diseases. The bacteria are more effective and multiply much faster in warmer temperatures compared to cold temperatures. The increase in temperature has led to increase in the microbes that cause diseases.
- Climate change is also effecting the crop production, as the crops are getting destroyed due to the sudden change in temperatures or sudden onset of rains. Also, the flash floods and other natural calamities affect the crop.
- As the matter of fact, because of global warming, the earth's atmosphere is getting more unpredictable with heavy rains in the areas which have scantly rainfall or drought in the areas which received good annual rainfall. The months of rainfall are also getting affected.

Impact of Climate Change in Agriculture

- Agriculture is the mainstay of Indian economy and provides food and livelihood security to a substantial section of the Indian population. The impact of climate change as witnessed in recent times has immense potential to adversely affect agriculture in this country in a variety of ways. As a large part of the arable land in India is rain-fed, the productivity of agriculture depends on the rainfall and its pattern. Agriculture will be adversely affected not only by an increase or decrease in the overall amounts of rainfall but also by shifts in the timing of the rainfall. Any change in rainfall patterns poses a serious threat to agriculture, and therefore to the economy and food security. Summer rainfall accounts for almost 70% of the total annual rainfall over India and is crucial to Indian agriculture.

- Climate Change is projected to have significant impacts on conditions affecting agriculture, including temperature, precipitation and glacial run off. It affects agriculture in more ways than one. It can affect crop yield as well as the types of crops that can be grown in certain areas, by impacting agricultural inputs such as water for irrigation, amounts of solar radiation that affect plant growth, as well as the prevalence of pests.

- Rise in temperatures caused by increasing green house gases is likely to affect crops differently from region to region. As a result of climate change the amount of arable land in high-latitude region is likely to increase by reduction of the amount of frozen lands. At the same time arable land along the coast lines are bound to be reduced as a result of rising sea level. Erosion, submergence of shorelines, salinity of the water table due to the increased sea levels, could mainly affect agriculture through inundation of low lying lands.

- The impact of climate change on agriculture could result in water shortages and drought, new diseases, heat stress for animals and risks liked to extreme weather events. Scientists are studying very closely the effect of climate change in different fields including agriculture. Various studies have indicated that rise in temperature is going to affect the agricultural production though its contribution to green house gas is only 28%. Climate change is already upon us. Indian agriculture, despite being a relatively minor contributor to greenhouse gas (GHG) emissions, will be a big loser if global temperatures rise.

- Agriculture will be adversely affected not only by an increase or decrease in the overall amounts of rainfall, but also by shifts in the timing of the rainfall. Higher temperatures reduce the total duration of a crop cycle, leading to a lower yield per unit area, especially for India's wheat and paddy crops. Soil erosion, increased numbers of pests and weeds brought by climate change will also affect agriculture in India. For instance, the amount of moisture in the soil will be affected by changes in factors such as rainfall, runoff and evaporation. Global warming will seriously affect agriculture. Some species of agricultural crops will be adversely affected by higher temperatures, increased weeds and harmful insects. It is also possible that global warming will lead to global food shortages.

Climate Change in India

- The vulnerability of Indian agriculture to climate change is well acknowledged. But what is not fully appreciated is the impact this will have on rain-fed (non-irrigated) agriculture, practiced mostly by small and marginal farmers who will suffer the most.
- The crops that may be hit include pulses and oilseeds, among others. These are already in short supply and are consequently high-priced. Nearly 80 million hectares, out of the country's net sown area of around 143 million hectares, lack irrigation facilities and hence, rely wholly on rain water for crop growth. Over 85% of the pulses and cereals, more than 75% of the oilseeds and nearly 65% of cotton are produced from such lands. The crop yields are quiet low.
- According to A K Singh, Deputy Director-General (Natural Resource Management) of the Indian Council of Agricultural Research (ICAR), medium-term climate change predictions have projected the likely reduction in crop yields due to climate change at between 4.5% and 9% by 2039.
- The long run predictions paint a scarier picture with the crop yields anticipated to fall by 25 per or more by 2099. This will have a detrimental effect on farmers' income and purchasing power, with obvious down-the-line repercussions. Though the rainfall records available with the Indian Meteorological

Department (IMD) do not indicate any perceptible trend of change in overall annual monsoon rainfall in the country, noticeable changes have been observed within certain distinct regions.

- At least three meteorological sub-divisions - Jharkhand, Chhattisgarh and Kerala have shown significant decrease in seasonal rainfall though some others have recorded an uptrend in precipitation as well. Since, rain-fed crops like coarse grains, pulses and oilseeds are grown mostly during the kharif season, these are impacted by both low as well as excess rainfall.
- India is among countries most threatened by climate change with experts warning that rising temperatures will lead to more floods, heat waves, storms, rising sea levels and unpredictable farm yields. Here are the main potential effects of climate change on a country which is the world's seventh largest in area and is home to 1.1 billion people, a sixth of humanity.

Impact of Climate Change in Indian Agriculture

The Food and Agricultural Organization (FAO) has warned that India could lose upto 125 million tones of cereals. The Inter-governmental Panel on Climate Change (IPCC) which was awarded Nobel Prize in the year 2008 has warned that due to global warming, the sea levels in Asia will rise by atleast 40cm by 2100 flooding vast areas on the coastline, including some of the most densely populated cities, whose population will be forced to migrate inland (The Hindu Business Line dated 03-11-2007). The first to be affected by the climate change will be those living in the fringe zones, working outdoors or whose livelihoods depend directly on the physical environment. Viewed from these perspectives, the main sectors facing risks from direct impact could be agriculture, water resources and health.

The IPCC report indicates high probability of crop losses with increase in temperature in the tropical regions. Studies do confirm this trend. Among cereal crops important for food security, wheat is most sensitive to even small increase in temperature. Relatively, rice has greater tolerance to temperature increases. Increasing climate variability could result in considerable season/ annual fluctuation in food production. All agricultural commodities are subject to such variability.

Solutions for Global Warming

The most important contributor to global warming is the increase in atmospheric CO_2 levels due to human activity. The following are the some of the suggestions to prevent global warming.

- CO_2 emission can be cut by reducing the use of fossil fuels by cutting back on car use, investing in energy efficiency, implementing energy conservation measures and utilizing renewable resources such as wind, solar and hydropower. Reducing fossil fuel user will, coincidentally, also reduce emissions of ethane, nitrous oxide and ozone.
- Since CO_2 is consumed by plants and trees (known as carbon-sinks), reversing deforestation and implementing reforestation programmes may reduce levels in the atmosphere. The need to take such measures was first recognized at the RIO climate change convention in 1992, after which over 160 countries pledged to limit emissions of CO_2 and to protect and enhance natural 'sinks' of CO_2.
- Trees play a unique role in the global carbon cycle. They are the largest land-based mechanism for removing CO_2 from the air. Trees are able to store a large amount of CO_2 in their structure. An acre of forest will absorb about ten times the CO_2 amount absorbed by an acre of crop land. Planting more trees and reducing more trees and reducing timber cut worldwide will help to restore the balance.
- Try to cut down demand for electricity when we reduce electricity power use, we save money, breathe cleaner air and help to reduce the global warming problem.
- Government should also encourage the use of mass transit; provide tax rebates for people who buy low and no-pollution vehicles and subsidies to fossil fuel and the nuclear industries.
- Green house gas emissions be managed using an incentive based policy, such as market based approach to capping and reducing such emissions. This type of strategy provides clear incentives for changes in business practices and the development of new technologies. It assures that economic forces are directed to keeping the cost of reducing emissions as low as they can be.
- Many industrial nations have now adopted policies intended to limit green house gases. Try to follow the environment-friendly

policies like 'Reduce', 'Reuse' and 'Recycle'. Save containers, bags, everything that can be used in the future. Also we use cloth towels and napkins instead of paper ones and use rechargeable batteries instead of disposable ones.

- Try to minimize the usage of air conditioners, refrigerators and other electronic goods which cause hazardous effects to the mother earth and can maximize the usage of the solar energy and wind energy for the power generation.
- Scientists are convinced that human actions are causing global warming. If this is so, it stands to reason that our own actions can also help to reduce this threat. Today global warming poses an extraordinary challenge. The world's leading atmospheric scientists tell us that a gradual warming of our climate is under way and will continue. This long-term warning trend poses even greater risks to many other nations, particularly poorer countries that will be far less able to cope with a changing climate and low-lying countries where sea-level rise will cause significant damage.
- Meeting the challenge of global warming will require sustained effort over decades on the part of Governments, who must establish the rules and modify them. Although much is being done to reduce the emission of green house gases, the efforts are still not enough. The Indian policy makers must remember that climate change has a creeping effect on the economy and can further hurt the already fragile agricultural system.

Farm Mechanization

The process of developing agricultural machines and substituting this machine power for human and animal power in agricultural production practices is farm mechanization.

Mechanized agriculture is the process of using agricultural machinery to mechanize the work of agriculture, massively increasing farm output and farm worker productivity. In modern times, powered machinery has replaced many jobs formerly carried out by men or animals such as oxen, horses and mules.

- The history of agriculture contains many examples of tool use, but only in recent time has the high rate of machine use been at such a level.

- The first pervasive mechanization of agriculture came with the introduction of the plough, usually powered by animals. It was invented in ancient Mesopotamia.
- Current mechanized agriculture includes the use of Tractors, trucks, combine harvesters, airplanes (crop dusters), helicopters, and other vehicles. Modern farms even sometimes use computers in conjunction with satellite imagery and GPS guidance to increase yields. See : List of agricultural machinery.
- Mechanization was one of the factors responsible for urbanization and industrial economies. Besides improving production efficiency, mechanization encourages large scale production and improves the quality of farm produce. On the other hand, it displaces unskilled farm labor, causes environmental pollution, deforestation and erosion.

Jethro Tull's seed drill (ca. 1701) was a mechanical seed spacing and depth placing device that increased crop yields and saved seed. It was an important factor in the British Agricultural Revolution.

- Since the beginning of agriculture threshing was done by hand with a flail, requiring a great deal of labor. The threshing machine, which was invented 1794 but not widely used for several more decades, simplified the operation and allowed the use of animal power.
- Before the invention of the grain cradle an able bodied laborer could reap about one quarter acre of wheat in a day using a sickle. It was estimated that for each of **Cyrus McCormick's** horse pulled reapers freed up five men for military service in the U.S. Civil War. Later innovations included raking and binding machines. By 1890 two men and two horses could cut, rake and bind 20 acres of wheat per day.
- In the 1880s the reaper and threshing machine were combined into the combine harvester. These machines required large teams of horses or mules to pull.
- Steam power was applied to threshing machines in the late 19th century. There were steam engines that moved around on wheels under their own power for supplying temporary power to stationary threshing machines. These were called *road engines,* and Henry Ford seeing one as a boy was inspired to build an automobile.

- With internal combustion came the first modern tractors in the early 1900s, becoming more popular after the Fordson tractor **(ca. 1917).** At first reapers and combine harvesters were pulled by tractors, but in the 1930s self powered combines were developed.[5] (*Link to a chapter on agricultural mechanization in the 20th Century at reference*)
- The horse population in the U.S. began to decline in the 1920s after the conversion of agriculture and transportation to internal combustion. Peak tractor sales in the U.S. were around 1950. In addition to saving labor, this freed up much land previously used for supporting draft animals.[7]
- The greatest period of growth in agricultural productivity in the U.S. was from the 1940s to the 1970s, during which time agriculture was benefiting from internal combustion powered tractors and harvesting machines, chemical fertilizers and the green revolution.
- The human society around the world, over a period of time, has established greater contact, but the pace has increased rapidly since the mid 1980's.The term globalization means international integration. It includes an array of social, political and economic changes. Unimaginable progress in modes of communications, transportation and computer technology have given the process a new lease of life.
- The world is more interdependent now than ever before Multinational companies manufacture products across many countries and sell to consumers across the globe. Money, technology and raw materials have broken the International barriers. Not only products and finances, but also ideas and cultures have breached the national boundaries.
- Laws, economies and social movements have become international in nature and not only the Globalization of the Economy but also the Globalization of Politics, Culture and Law is the order of the day. The formation of General Agreement on Tariffs and Trade (GATT), International Monetary Fund and the concept of free trade has boosted globalization.

Globalization in India

In early 1990s the Indian economy had witnessed dramatic policy changes. The idea behind the new economic model known as

Liberalization, Privatization and Globalization in India (LPG), was to make the Indian economy one of the fastest growing economies in the world. An array of reforms was initiated with regard to industrial, trade and social sector to make the economy more competitive. The economic changes initiated have had a dramatic effect on the overall growth of the economy. It also heralded the integration of the Indian economy into the global economy. The Indian economy was in major crisis in 1991 when foreign currency reserves went down to $1 billion and inflation was as high as 17%. Fiscal deficit was also high and NRI's were not interested in investing in India. Then the following measures were taken to liberalize and globalize the economy.

Steps Taken to Globalize Indian Economy

Some of the steps taken to liberalize and globalize our economy were :

1. *Devaluation* : To solve the balance of payment problem Indian currency were devaluated by 18 to 19%.
2. *Disinvestment* : To make the LPG model smooth many of the public sectors were sold to the private sector
3. *Allowing Foreign Direct Investment (FDI)* : FDI was allowed in a wide range of sectors such as Insurance (26%), defense industries (26%) etc.
4. *NRI Scheme* : The facilities which were available to foreign investors were also given to NRI's.

The Merits of Globalization are as follows :

- There is an International market for companies and for consumers there is a wider range of products to choose from.
- Increase in flow of investments from developed countries to developing countries, which can be used for economic reconstruction.
- Greater and faster flow of information between countries and greater cultural interaction has helped to overcome cultural barriers.
- Technological development has resulted in reverse brain drain in developing countries.

The Demerits of Globalization are as follows :

- The outsourcing of jobs to developing countries has resulted in loss of jobs in developed countries.
- There is a greater threat of spread of communicable diseases.
- There is an underlying threat of multinational corporations with immense power ruling the globe.
- For smaller developing nations at the receiving end, it could indirectly lead to a subtle form of colonization.

India gained highly from the LPG model as its GDP increased to 9.7% in 2007-2008. In respect of market capitalization, India ranks fourth in the world. But even after globalization, condition of agriculture has not improved. The share of agriculture in the GDP is only 17%. The number of landless families has increased and farmers are still committing suicide. But seeing the positive effects of globalization, it can be said that very soon India will overcome these hurdles too and march strongly on its path of development.

Impact of Globalization on Indian Agriculture

The liberalization of India's economy was adopted by India in 1991. Facing a severe economic crisis, India approached the IMF for a loan, and the IMF granted what is called a 'structural adjustment' loan, which is a loan with certain conditions attached which relate to a structural change in the economy. The government ushered in a new era of economic reforms based on these conditions. These reforms (broadly called Liberalization by the Indian media) can be broadly classified into three areas : Liberalization, privatization and globalization. Essentially, the reforms sought to gradually phase out government control of the market (liberalization), privatize public sector organizations (privatization), and reduce export subsidies and import barriers to enable free trade (globalization). There was a considerable amount of debate in India at the time of the introduction of the reforms, it being a dramatic departure from the protectionist, socialist nature of the Indian economy up until then. However, reforms in the agricultural sector in particular came under severe criticism in the late 1990s, when 221 farmers in the south Indian state of Andhra Pradesh committed suicide. (The damage done, 2005) The trend was noticed in several other states, and the figure today, according to a leading journalist and activist, P. Sainath1, stands at 100,000 across the country. (Senath, 2006) Coupled with this was a sharp drop in agricultural growth from 4.69% in 1991 to 2.06% in 1997 (Agriculture Statistics at a Glance, 2006). This paper seeks to look into these and other similar negative trends in Indian agriculture today, and in analyzing the causes, will look at the extent to which liberalisation reforms have contributed to its current condition. It will look at supporting data from three Indian states which have been badly affected by the crisis : Andhra Pradesh, Maharashtra and Kerala. Andhra Pradesh's (AP's) experience is particularly critical in this debate because it was headed by Chief Minister Chandrababu Naidu, who pursued liberalization with enthusiasm. Hence liberalization in AP has been faster than other states, and the extent of its impact has been wider and deeper. (Sainath, 2005).

Indian Agriculture Today : A Snapshot

Agriculture employs 60% of the Indian population today, yet it contributes only 20.6% to the GDP. (Isaac, 2005) Agricultural production fell by 12.6% in 2003, one of the sharpest drops in independent India's history. Agricultural growth slowed from 4.69% in 1991 to 2.6% in 1997-1998 and to 1.1% in 2002-2003. (Agricultural Statistics at a Glance, 2006) This slowdown in agriculture is in contrast to the 6% growth rate of the Indian economy for almost the whole of the past decade. Farmer suicides were 12% of the total suicides in the country in 2000, the highest ever in independent India's history. (Unofficial estimates put them as high as 100,000 across the country, while government estimates are much lower at 25,000. This is largely because only those who hold the title of land in their names are considered farmers, and this ignores women farmers who rarely hold land titles, and other family members who run the farms.) (Sainath, P) Agricultural wages even today are $1.5 - $2.0 a day, some of the lowest in the world. (Isaac, 2005) Institutional credit (or regulated credit) accounts for only 20% of credit taken among small and marginal farmers in rural areas, with the remaining being provided by private moneylenders who charge interest rates as high as 24% a month. (Sainath, 2005) An NSSO2 survey in 2005 found that 66% of all farm households own less than one hectare of land. It also found that 48.6% of all farmer households are in debt. The same year, a report by the Commission of Farmer's welfare in Andhra Pradesh concluded that agriculture in the state was in 'an advanced stage of crisis', the most extreme manifestation of which was the rise in suicides among farmers. Given the performance of agriculture and figures of farmer suicides across the country, this can be said to apply to Indian agriculture as a whole.

The Crisis Facing Indian Agriculture

The biggest problem Indian agriculture faces today and the number one cause of farmer suicides is debt. Forcing farmers into a debt trap are soaring input costs, the plummeting price of produce and a lack of proper credit facilities, which makes farmers turn to private moneylenders who charge exorbitant rates of interest. In order to repay these debts, farmers borrow again and get caught in a debt trap. I will examine each one these causes which led to the current crisis in Andhra Pradesh, Kerala and Maharashtra, and analyses the role that liberalization policies have played.

As was mentioned earlier, AP's experience is particularly relevant in this analysis because of its leadership. Chandrababu Naidu, Chief Minister of Andhra Pradesh from 1995-2004, was an IT savvy neo-liberal, and believed that the way to lead Andhra Pradesh into the future was through technology and an IT revolution. His zeal led to the first ever state level (as opposed to national level) agreement with the World Bank, which entailed a loan of USD 830 million (AUD 1 billion) in exchange to a series of reforms in AP's industry and government. Naidu envisaged corporate style agriculture in AP, and implemented World Bank liberalization policies with great enthusiasm and gusto. He drew severe criticism from opponents, saying he was using AP as a laboratory for extreme neo-liberal experiments. Hence, AP's experience with liberalization is critical.

The Debt Trap and the Role of Liberalization

The biggest input for farmers is seeds. Before liberalisation, farmers across the country had access to seeds from state government institutions. For example, AP's APSSDC3 produced its own seeds, was responsible for their quality and price, and had a statutory duty to ensure seeds were supplied to all regions in the state, no matter how remote. The seed market was well regulated, and this ensured quality in privately sold seeds too. (The damage done, 2005) With liberalization, India's seed market was opened up to global agribusinesses like Monsanto, Cargill and Syn Genta. Also, following the deregulation guidelines of the IMF, 14 of the 24 units of the APSSDC's seed processing units were closed down in 2003, with similar closures in other states. This hit farmers doubly hard : in an unregulated market, seed prices shot up, and fake seeds made an appearance in a big way. Seed cost per acre in 1991 was Rs. 70 (AUD 2) but in 2005, after the dismantling of APSSDC and other similar organizations, the price jumped to Rs. 1000 (AUD 28), a hike of 1428%, with the cost of genetically modified pest resistant seeds like Monsanto's BT Cotton costing Rs. 3200 or more per acre, (AUD 91) a hike of 3555%. (Sainath, 2005) BT Cotton is cotton seed that is genetically modified to resist pests, the success of which is disputed : farmers in Andhra Pradesh and Maharashtra now claim that yields are far lower than promised by Monsanto, and there are fears that pests are developing resistance to the seeds. (Menon & Jayaraman, 2002) Expecting high yields, farmers invest heavily in such seeds. Also BT Cotton and other new seeds guarantee a much lower germination rate of 65% as opposed to a 90% rate of state certified seeds. Hence 35% of the farmer's investment in seeds is a waste. (Sainath, 2004) Output is not commensurate with the heavy investment in the seeds, and farmers are pushed into debt. The abundant availability of spurious seeds is another problem which leads to crop failures. Either tempted by their lower price, or unable to discern the difference, farmers invest heavily in these seeds, and again, low output pushes them into debt. Earlier, farmers could save a part of the harvest and use the seeds for the next cultivation, but some genetically modified seeds, known as Terminator, prevent harvested seeds from germinating, hence forcing the farmers to invest in them every season.

Fertilizer and Pesticide

One measure of the liberalisation policy which had an immediate adverse effect on farmers was the devaluation of the Indian Rupee in 1991 by 25% (an explicit condition of the IMF loan). Indian crops became very cheap and attractive in the global market, and led to an export drive. Farmers were encouraged to shift from growing a mixture of traditional crops to export oriented 'cash crops' like chilli, cotton and tobacco. (The damage done, 2005) These need far more inputs of pesticide, fertilizer and water than traditional crops. Liberalisation policies reduced pesticide subsidy (another explicit condition of the IMF agreement) by two thirds by 2000. Farmers in Maharashtra who spent Rs. 90 an acre (AUD 2.5) now spend between Rs. 1000 and 3000 (AUD 28.5 - 85) representing a hike of 1000% to 3333%. Fertilizer prices have increased 300% (Sainath, 2005) Electricity tariffs have also been increased : in Andhra Pradesh tariff was increased 5 times between 1998 and 2003. (Seeds of ruin, 2005) Pre-liberalisation, subsidised electricity was a success, allowing farmers to keep costs of production low. These costs increased dramatically when farmers turned to cultivation of cash crops, needing more water, hence more water pumps and higher consumption of electricity. Andhra Pradesh being traditionally drought prone worsened the situation. This caused huge, unsustainable losses for the Andhra Pradesh State Electricity Board, which increased tariffs. (This was initiated by Chandrababu Naidu in partnership with Britain's DFID4 and the World Bank.) Also, the fact that only 39% of India's cultivable land is irrigated makes cultivation of cash crops largely unviable, but export oriented liberalisation policies and seed companies looking for profits continue to push farmers in that direction. (Isaac, 2005).

The Debt Trap : Low Price of Output

With a view to open India's markets, the liberalization reforms also withdrew tariffs and duties on imports, which protect and encourage domestic industry. By 2001, India completely removed restrictions on imports of almost 1,500 items including food (The damage done, 2005). As a result, cheap imports flooded the market, pushing prices of crops like cotton and pepper down. Import tariffs on cotton now stand between 0 - 10%, encouraging imports into the country. This excess supply of cotton in the market led cotton prices to crash more than 60% since 1995. As a result, most of the farmer suicides in Maharashtra were concentrated in the cotton belt till 2003 (after which paddy farmers followed the suicide trend). (Hardikar, 2006). Similarly, Kerala, which is world renowned for pepper, has suffered as a result of 0% duty on imports of pepper from SAARC5 countries. Pepper, which sold at Rs. 27,000 a quintal (AUD 771) in 1998, crashed to Rs. 5000 (AUD 142) in 2004, a decline of 81%. As a result, Indian exports of pepper fell 31% in 2003 from the previous year. (Sainath, 2005) Combined with this, drought and crop failure has hit the pepper farmers of Kerala hard, and have forced them into a debt trap. Close to 50% of suicides among Kerala's farmers have been in pepper producing districts. (Mohankumar & Sharma, 2006).

The Debt Trap : Lack of credit facilities and dependence on private money lenders

In 1969, major Indian banks were nationalized, and priority was given to agrarian credit which was hitherto severely neglected.

However, with liberalisation, efficiency being of utmost importance, such lending was deemed as being low-profit and inefficient, and credit extended to farmers was reduced dramatically, falling to 10.3% in 2001 against a recommended target of 18%. (Seeds of ruin, 2005) A lack of rural infrastructure deters private banks from setting up rural branches, with the responsibility falling on the government, which has reduced rural spending as a result of its liberalization policies. Rural development expenditure, which averaged 14.5% of GDP during 1985 - 1990 was reduced to 8% by 1998, and further to 6% since then. This at a time when agriculture was going through a crisis proved disastrous for farmers, who turned to private money lenders who charge exorbitant rates of interest, sometimes up to 24% a month. (Seeds of Suicide, 2005) With input costs and output prices being what they are, coupled with crop failures and drought, they are pushed into debt which is impossible to repay. 12 out of India's 28 states have 50% and higher indebtedness among farm households. Andhra Pradesh has the highest percentage of indebted farm households - 82%. 64.4% of Kerala's farm households and 54.8% of Maharashtra's farm households are indebted (NSSO, 2003) Indebtedness has been identified as the single major cause of suicides in both Andhra Pradesh, Kerala and Maharashtra. (Analysis of Farmer Suicides in Kerala, 2006, Who's suicide is it anyway, 2005, Saving small farmers, 2005).

Liberalization and how it Failed

- **Branco Milanovic**, a World **Bank economist** describes how he believes liberalisation helps developing countries achieve growth, when a country lowers trade barriers, reduces government intervention in the market in order to allow market forces to operate freely, increases competition and attracts foreign investment, it will increase productivity and reduce inefficiency, which will lead to economic growth, and in a few generations, if not less, the poor will become rich, illiteracy will disappear, and poor countries will catch up with the rich.' (The damage done, 2005) This argument is an economic rationalist one, which views government intervention with profound suspicion, and has equally profound faith in unfettered market forces. (Whitwell quoting Robert Manne, 1998) What Mr. Milanovic neglects to mention, though, is that rich countries which now preach liberalization protected their 'infant industries' at the time they began to industrialize, till they were strong enough to compete globally. The US government, for example, had a protectionist trade policy in the late nineteenth century to help US companies become competitive in the world. Besides, apart from wool, the US, Germany, Britain and France were all almost self-sufficient in the raw materials that they needed for industrialization, and took off from that platform, a luxury that India and other developing countries do not have. (Issac, 2005) As German economist Friedrich List says, the adoption of these values (of liberalization) assumes that all countries are at the same starting place, which (as we have seen above) is clearly not the case. (Issac, 2005) In fact, it is this very reason that has brought about the crisis that Indian agriculture is facing today. Most farmers in India were already in a position of minimum security, with no education system, credit facilities, access to alternative employment, or efficient technology. Their only support was government subsidy and regulation. Liberalization policies came in and dismantled their only support structure. It halted the sharp reduction in rural poverty from 55% in the 1970s to 34% in the 1980s. Not only has the incidence of poverty in rural areas not gone lower than 34% in the 1990s, it has gone to higher levels of 42% in individual years (Ghosh, 2000).

- The second most popular argument of the economic rationalists in favour of liberalisation is that competition will weed out the inefficient, and in the growth that ensues, employment will be provided in other areas of the economy, thus lifting the poor out of poverty. This argument however assumes that the poor will be able to take advantage of the opportunities presented to them. As Robert Isaac says in 'The Globalization Gap', "Globalization encourages the well positioned to use tools of economics and politics to exploit market opportunities, boost technical productivity, and maximize short-term material interests." This is compounded in India, where the gap between one who is 'well positioned' and one who is not can be extreme. With a lack of investment, chances of generation of rural employment are slim. Unemployment and underemployment are chronic problems in India, with the rate of unemployment being close to 10% in 2004. (Sainath, 2005) Primary education in rural areas is mismanaged and bad quality, and there is no system which helps agricultural workers find alternate employment, or develop alternate skills. (Chossudovskly, 1997) In the face of such obstacles, it is nearly impossible to expect agricultural workers to shift to alternate fields. Coming back to AP, the IT Revolution spearheaded by Chandrababu Naidu attracted companies like Google, Amazon, Microsoft and Dell and created thousands of jobs. However, given the skills and education of most farmers, it is obvious that none of this translated into job opportunities for them.

- The final argument that supporters of globalization have is the much touted 10% reduction in poverty (60 million declines in poor) in India in the year 2000. However, this figure was challenged by experts. Poverty is defined according to how many people consume less than the nutritional minimum prescribed. (2400 calories for rural areas, and 2100 for urban areas) Major changes in survey design in 1999-2000 not only made the resultant estimates incomparable to previous years' estimates, but an over-estimation of consumption (meaning people were getting enough food, hence were not considered poor) meant a sharp reduction in poverty figures. After experts challenged it, the Planning Commission of India accepted that the figure was inaccurate, and could not be compared to previous years' estimates, hence the 10% drop in poverty is incorrect. With adjusted figures, experts have determined that the decrease in poverty was a mere 2.3%, and that the number of poor increased by nine million in 2002 as compared to 1999.

Liberalization and 'Growth'

Many economists now concede that the relationship between liberalisation and growth are 'uncertain at best'. According to the Center for Economic and Policy research, which studied impact of liberalisation reforms on the developing world, key economic and social indicators such as increases in life expectancy, infant and child mortality, education and literacy levels slowed down in the 20 years between 1980 and 2000 when liberalisation policies were implemented, compared to the 20 years leading to 1980. (The damage done, 2005) This defeats the economic rationalist argument of free trade eliminating poverty, since the 20 years leading up to 1980 witnessed high protectionist policies and trade barriers. Following the suicides in 2000, the World Bank and Britain's DFID abandoned power reforms in Andhra Pradesh four years before schedule. It admitted that it had 'substantially underestimated' the 'complexity of the process' and that there must be 'increased consultation with the farmers to get their acceptance' of any further reform. (The damage done, 2005).

The Andhra Pradesh government sponsored report by the Commission of Farmer's Welfare squarely laid the blame for its agrarian crisis on the state and central government's policies : "While the causes of this crisis are complex and manifold, they are they are dominantly related to public policy. The economic strategy of the past decade at both central government and state government levels has systematically reduced the protection afforded to farmers and exposed them to market volatility and private profiteering without adequate regulation; has reduced critical forms of public expenditure; has destroyed important public institutions, and has not adequately generated other non-agricultural economic activities." A report on suicides in Kerala similarly held the liberalization policies of the government responsible. (Mohankumar & Sharma, 2006).

It is clear that the liberalisation policies adopted by the government of India played a dominant role in the agrarian crisis that is now being played out. However, this is not to say that privatisation, liberalisation and globalization are per say bad, or inherently inimical to an economy. It is the 'one size fits all' brand of liberalisation adopted by the IMF and the World Bank which forces countries to privatize, liberalise and globalize without exception which has failed. Without taking into account the state of an economy, and in this case, the state and nature of the agricultural sector in India, the IMF and the World Bank, with the cooperation of the Indian government, embarked on mismatched reforms, which have caused misery and despair among millions of Indian farmers, driving large numbers of them to suicide. It is also essential to break the link between aid and liberalization, which caused India in the first place to accept the conditions of the IMF. Remember that India was on the brink of a financial crisis in 1991 when it applied for the IMF loan and accepted its conditions-perhaps the course of economic reform in India would have taken a very different course if there was no urgent need to borrow from the IMF. The start to this process may have already occurred : recognizing the failure of its liberalization policies, (and perhaps also the failure of DFID with AP's power reforms) the Blair government of Britain announced in 2004 that it will no longer make liberalization and privatization conditions of aid. In another blow to the neo-liberal lobby, Chandrababu Naidu suffered the worst ever defeat in the 2004 state elections in his party's history, with rural AP clearly rejecting his brand of World Bank sponsored liberalization. The battle, however, has not yet been won. It is essential for the rest of the G8 to follow Britain's example in order to influence World Bank and IMF policy towards India to ensure blind liberalization is not pursued, and so that countries like India can adopt tailor-made reforms to suit their economy.

The boxes

In WTO terminology, subsidies in general are identified by "boxes" which are given the colours of traffic lights : green (permitted), amber (slow down – i.e. be reduced), red (forbidden). In agriculture, things are, as usual, more complicated. The Agriculture Agreement has no red box, although domestic support exceeding the reduction commitment levels in the amber box is prohibited; and there is a blue box for subsidies that are tied to programmers that limit production. There are also exemptions for developing countries (sometimes called an "S&D box", including.

Amber box

- All domestic support measures considered to distort production and trade (with some exceptions) fall into the amber box, which is defined in Article 6 of the Agriculture Agreement as all domestic supports except those in the blue and green boxes. These include measures to support prices, or subsidies directly related to production quantities.
- These supports are subject to limits : "de minimize" minimal supports are allowed (5% of agricultural production for developed countries, 10% for developing countries); the 30 WTO members that had larger subsidies than the de minimize levels at the beginning of the post-Uruguay Round reform period are committed to reduce these subsidies.
- The reduction commitments are expressed in terms of a "Total Aggregate Measurement of Support" (Total AMS) which includes all supports for specified products together with supports that are not for specific products, in one single figure. In the current negotiations, various proposals deal with how much further these subsidies should be reduced, and whether limits should be set for specific products rather than continuing with the single overall "aggregate" limits.

Blue box

- This is the "amber box with conditions" – conditions designed to reduce distortion. Any support that would normally be in the amber box is placed in the blue box if the support also requires farmers to limit production (details set out in Paragraph 5 of Article 6 of the Agriculture Agreement).

- At present there are no limits on spending on blue box subsidies. In the current negotiations, some countries want to keep the blue box as it is because they see it as a crucial means of moving away from distorting amber box subsidies without causing too much hardship. Others wanted to set limits or reduction commitments, some advocating moving these supports into the amber box.

Green box

- The green box is defined in Annex 2 of the Agriculture Agreement.
- In order to qualify, green box subsidies must not distort trade, or at most cause minimal distortion They have to be government-funded (not by charging consumers higher prices) and must not involve price support.
- They tend to be programmer that are not targeted at particular products, and include direct income supports for farmers that are not related to (are "decoupled" from) current production levels or prices. They also include environmental protection and regional development programmers. "Green box" subsidies are therefore allowed without limits,
- In the current negotiations, some countries argue that some of the subsidies listed in Annex 2 might not meet the criteria of the annex's first paragraph – because of the large amounts paid, or because of the nature of these subsidies, the trade distortion they cause might be more than minimal. Among the subsidies under discussion here are : direct payments to producers, including decoupled income support and government financial support for income insurance and income safety-net programmers and other paragraphs. Some other countries take the opposite view – that the current criteria are adequate, and might even need to be made more flexible to take better account of non-trade concerns such as environmental protection and animal welfare.

Reforming the Agricultural Extension System in India

In order to realize agricultural potential and to increase agricultural yields, India's extension system has experienced major conceptual, structural, and institutional changes since the late 1990s.

This paper reviews existing reform programs and strategies currently existing in agricultural extension in India. It distinguishes strategies that have been employed to strengthen both the supply and demand sides of service provision in the area of agricultural extension, and it reviews the effects of the demand- and supply-side strategies on the access to and the quality of agricultural extension services. The ultimate objectives are (1) to gain a view on what works where and why in improving the effectiveness of agricultural extension in a decentralized environment; (2) to identify measures that strengthen and improve agricultural extension service provision; and (3) to reveal existing knowledge gaps.

Although the range of extension reform approaches is wide, this paper shows that an answer to the question of what works where and why is complicated by the absence of sound and comprehensive qualitative and quantitative impact and evaluation assessment studies. Even evidence from the National Agricultural Technology Project and the Diversified Agricultural Support Project of the World Bank, the women empowerment programs of the Danish International Development Agency, the Andhra Pradesh Tribal Development Project, and the e-Choupal program of the Indian Tobacco Company is subject to methodological and identification problems. Conclusions regarding the importance

- of implementing both decentralized, participatory, adaptive, and pluralistic demand- and supply-side extension approaches;
- of involving the public, private, and third (civil society) sectors in extension service provision and funding; and
- of strengthening the capacity of and the collaboration between farmers, researchers, and extension workers are necessarily tentative and require further quantification. The paper seeks to inform policymakers and providers of extension services from all sectors about the need to make performance assessments and impact evaluations inherent components of any extension program so as to increase the effectiveness of extension service reforms.

Policy Framework for Agricultural Extension

Introduction

Rapid agricultural growth continues to be the key to poverty alleviation and overall economic development. Agriculture accounts for about one-fourth of the Gross Domestic product and is the source of livelihood or nearly two-thirds of the population. The agriculture sector in India has been successful in keeping pace with the rising food demand of a growing population, expected to cross one billion. Food grains production more than quadrupled since the early 1950s from 51 million tons to over 209 million tons in 2000, while population nearly tripled from 350 million to one billion during the same period.

The Green Revolution has been the cornerstone of India's agricultural achievement, transforming the country from one of food deficiency to self-sufficiency. While recognizing the impact of the Green Revolution in imparting dynamism to the agriculture sector, it must be recognized that the Green Revolution remained restricted to the well-endowed irrigated areas of the country. Of late, deceleration in production and of factor productivity growth in some of the major irrigated production systems, especially in the North and North-West regions, has been recorded. Potentially high production areas (Eastern and Central States) are still lagging behind in productivity increases. Moreover, in the area of agricultural research, the success has been restricted to selected crops. Even in this arena, a growing disparity between the actual and the potential yields points to a crucial gap between research and extension.

Public research and extension played a major role in bringing about the Green Revolution. In the post-Green Revolution era, however, extension faces important challenges in the areas of relevance, accountability and sustainability. The changing economic scenario in India and the need for appropriate agricultural technologies and agro-management practices to respond to food and nutritional security, poverty alleviation, diversifying market demands, export opportunities, and environmental concerns is posing new challenges to the technology dissemination systems. It is expected that future agricultural growth would largely accrue from improvements in productivity of diversified farming systems with regional specialisation and sustainable management of natural resources, especially land and water. Effective linkages of production systems with agro-processing and other value added activities including marketing would play an increasingly important role in the diversification of agriculture.

It is becoming increasingly evident that public extension by itself can no longer respond to the multifarious demands of farming systems. There is need for reappraisal of the capacity of agricultural extension to address, effectively, contemporary and future needs of the farming community. Public funding for sustaining the vast extension infrastructure is also under considerable strain. Meanwhile, in response to market demand the existing public extension network is inexorably being complemented, supplemented and being replaced by private extension. As the nature and scope of agricultural extension undergoes fundamental changes, the outlook is for a whole new policy mix nurturing a plurality of institutions.

Changing Role of Agricultural Extension

Community Development Approach to Extension. Public sector extension has undergone several changes since the early 1950s. Beginning with the Community Development Programme in 1952 through the National Extension Service in 1953, the focus of extension was on human and community development, but there has been a steady progression towards technology transfer, with the Intensive Agriculture District Programme started in 1961-62, followed by the Intensive Agriculture Area Programme in 1964-65, the High Yielding Varieties Programme 1966-67, the Farmers Training and Education Programme 1966-67 and the Small and Marginal Farmers Development Programme in 1969-70.

Transfer of Technology Approach through T & V. The most significant recent development was the introduction of the Training and Visit (T&V) extension management system, starting in the mid-1970s. T&V extension was well suited to the rapid dissemination of broad-based crop management practices for the high yielding wheat and rice varieties that were released since the mid-1960s. The T&V system profoundly influenced extension practices and registered impressive gains in irrigated areas, because of the similarity between the agro-ecological conditions where technologies were generated and where they were ultimately used, and the favourable socio-economic situations and developmental infrastructure for their wider uptake. Indeed, the T&V system played an important role in ushering in the Green Revolution.

Limitations of the T & V System. The focus of the T & V system being on disseminating Green Revolution technology for major cereal crops, extension activities have been largely carried out by State Departments of Agriculture (DOA). The other line departments, such as the Departments of Animal Husbandry (DAH), Horticulture (DOH), and Fisheries (DOF), have not been able to focus on extension due to lack of infrastructure, trained personnel and resources. The T&V system operated largely in the inter-personnel mode without planned and optimum utilisation of information support and with low level of involvement of farmers. The "top-down" approach generated uniformity rather than specificity and lacked focus on location specific needs of regions, disadvantaged areas, target groups enterprises etc. Individual and institutional issues under Human Resource Development, the training of researchers, subject matter specialists and extension functionaries and farmers had not been addressed adequately. The linkages between research-extension and farmer remained weak or non-existent. Media and information management largely remained in the public sector and characterized by centralized operations. Farmer driven and farmer accountable feedback systems were not adequately developed.

Post-Green Revolution Period. The transfer of T&V extension approach to rainfed farming areas where fundamentally different production systems predominate and more importantly, local conditions vary widely, resulted in serious limitations and failures. The system well suited to the rapid dissemination of pre-set agronomic practices for the high yielding wheat and rice varieties, failed to respond to the more location-specific, risk-prone agriculture of the unirrigated tracts. Similarly, extending the system to programmer for natural resource management, sustainable agricultural practices such as integrated pest management, integrated nutrient management and to diversified agriculture such as high value horticulture, livestock activities and fisheries did not meet with success. Nor could the T&V system adapt to the more holistic Farming Systems Approach towards which the new thrust of both research & extension had begun to focus.

Towards a Farming Systems Approach. The extension approaches of the 1950s and 1960s centered around 'farmers' ignorance' as an explanation of non-adoption of agricultural technology and therefore the extension policies remained confined to "extension education" with the key activities related to "teaching". In the decade of the 1970s and 1980s, farm level constraints were considered to be the explanation of non-adoption. The key activities of extension system were confined to 'input supply' for removal of such farm level constraints. The basic philosophy of these extension approaches centered on "technology transfer". By the early 1990's and the completion of the third National Agricultural Extension Project (NAEP), there was growing recognition that the T&V extension approach needed to be overhauled in meeting the technology needs of farmers during the 21st century. First, it was recognised that extension should begin to broad base its programmes, by utilising a Farming Systems approach. For example, attention should be given to the needs of farmers in rainfed areas, and to diversifying extension programmes into livestock, horticulture, and other high value commodities that would increase farm incomes. Secondly, to support and strengthen the Farming Systems approach, issues of financial sustainability, farmer participation in programme planning, and research-extension linkages, marketing and value addition would have to be concurrently addressed. Present day agriculture is defined by key concepts of stability, sustainability, diversification and commercialization. There is need for reorientation of the philosophy of extension from technology transfer mode to technology application.

Proposed Extension Policy Framework

In the context of meeting the holistic needs of increasing agricultural production, yet do so in a sustainable manner, agricultural extension has a crucial role to play. Reforms in the system envisage an extension service more broad-based and holistic in content and scope, thus beyond agricultural technology transfer. Its normal task of transferring and disseminating appropriate technologies and agronomic practices would not be sufficient. Extension agencies, services and workers will need to exercise a more proactive and participatory role, serve as knowledge/ information agents, initiating and facilitating mutually meaningful and equitable knowledge based transactions among agricultural researchers, trainers and primary producers. All this needs to be done in an effective and cost efficient manner.

Technology generation and its application will have to focus more strongly than before on the themes of optimization by producers of their available resources, sustainability, coping with diversity by adapting technology more specifically to agro-ecological or social circumstances aimed at creation of a policy environment that promotes profitable, productive and sustainable farming.

Reforms in agricultural extension already initiated and proposed to be undertaken on wider scale are discussed under the following sub-heads :

Policy Reforms

- Institutional Restructuring
- Management Reforms
- Strengthening Research -Extension Linkages
- Capacity building & Skill Up-gradation
- Empowerment of Farmers
- Mainstreaming of Women in Agriculture
- Use of Media & Information Technology
- Financial Sustainability
- Changing Role of Government

Policy Reforms

Farming Systems Approach. Policy reforms in Agricultural Extension envisage the replacement of the old single-discipline based, commodity-oriented approach of the T & V system by the Farming Systems (FS) approach. The FS approach considers the farm, the farm household and off-farm activities in a holistic way to take care not only of farming but also aspects of nutrition, food security, sustainability, risk minimisation, income and employment generation which make up the multiple objectives of farm households. FS considers interdependencies of the components under the control of members of the household as well as how these components interact with the physical, biological and socio-economic factors not under the household's control. The FS approach emphasises that research and extension agendas should be determined by explicitly defined farmers' needs through an understanding of the existing farming systems rather than perceptions by research scientists or extension functionaries.

Multi-Agency Extension Service. For many years agricultural extension was considered the monopoly of the public sector. However, with the wide range of demands for agricultural technology in the changing scenario there is growing recognition that public extension by itself cannot meet the specific needs of various regions and different classes of farmers. The new extension regime recognizes the role of a multi-agency dispensation comprising different strengths. Policy environment will promote private extension to operate in roles that complement, supplement, work in partnership and even substitute for public extension. The three arms of the agricultural extension network are :

Public extension services

- State government line departments operated extension (Departments of Agriculture, Horticulture & Livestock development).
- State agriculture universities based extension (Directorates of Extension, Krishi Vigyan Kendras (KVKs) and Krishi Gyan Kendras (KGKs)
- ICAR extension (Zonal Research Stations/ Krishi Vigyan Kendras, Agriculture Technology Information Centres (ATICs), Institute Village Linkage Programme (IVLP) etc.)

Private extension services

- Community Based Organization (Farmers' Organizations, Farmers' Cooperatives, Self Help Groups etc.)
- Para Extension Workers (contact farmers, link farmers, *gopals, mitra kisans, mahila mitra kisans, etc.*)
- Agri-clinics & Agribusinesses
- Input Suppliers/Dealers (Pesticides, Seeds, Nutrients, Farm Implements, etc.)
- Corporate Sector (Commercial Crops - tobacco, tea, coffee, oilseeds (sunflower), vegetables, Seeds, Farm Implements - tractors, threshers, sprinklers, drip irrigation, etc.)

Mass Media & Information Technology

- Print Media - Vernacular Press
- Radio, Television, Private Cable Channels, etc.
- Electronic Connectivity through Computers, NICNET, Internet, V-SAT, etc.
- Farm Information and Advisory Centres (FIACs)
- Private Portals
- Public & Private Information Shops

Public extension services. Despite the rise of the private sector in the provision of agro-services, such extension will gravitate towards selected regions, crops and sectors where gains are to be appropriated.

Pure public goods, economically backward regions, small, marginal farmers and landless labourers will not attract the for-profit private sector. Public Extension will therefore continue to play a central role in technology dissemination. For example, public extension should focus its efforts on those *knowledge-based* technologies that are central to farmers' concerns and that will maintain the natural resource base. These are subject matter areas that are not likely to be taken by the private sector. Examples include dissemination of production management technologies that are specific to different crops and livestock systems; natural resource management technologies, such as soil and water management, integrated pest management, agro-forestry and other technologies associated with sustainable development; and farming systems technologies, including farm management skills that will enable farmers to improve their efficiency, increase cropping intensities and to diversify into more high value commodities in conformity with marketing trends.

Promotion of farmer participatory approach. There is, therefore, need for a more farmer participatory approach in working out the system description, problem diagnosis, search for appropriate technology, designing the process of implementation, monitoring and evaluation, and feedback. The extension agent is no longer seen as the expert who has all the useful information and technical solutions; the indigenous technical knowledge of farmers and their ingenuity, individually and collectively, are recognized as a major source; and solution to local problems are to be developed in partnership between the extension agent and farmers. Extension workers therefore need new skills of negotiations, conflict resolution and mobilizing and nurturing community organizations.

Promotion of demand-driven and farmer-accountable extension. Under the T&V system the technology dissemination regime was more supply-driven. Research and extension agendas were pre-set based on technologies for high-yielding varieties of wheat and rice. An important reason why research and extension organisations have not focused on farmer problems is due to the lack of an effective feedback system. The vast majority of small and marginal farmers in India, especially women, lack an effective voice in influencing, research and extension priorities. Under the new policy agenda a demand-driven extension system will be created, by providing farmers with access to linkage mechanisms through which they would be provided all relevant information/data to help them articulate their problems and needs with reference to their production & marketing plans. A key factor in improving these feedback systems is to organise farmers into functional groups, such as Self-Help Groups (SHGs), Farmer Interest Groups (FIGs), Commodity Associations (CAs), and/or other types of farmer organisations (FOs). These FOs can provide, an effective channel for both the dissemination of technology to large number of small and marginal farmers and feedback to research and extension. Linkage mechanisms would also ensure meaningful farmer representation in the governing bodies of public and private extension services, farmer influence on decisions on the planning, implementation and monitoring of public extension at local, regional and national levels and farmer influence on the incentives of extension staff, including supervisors and subject matter specialists.

Thrust on Marketing Extension. Farmers have increasingly begun to perceive marketing rather than production as the major constraint to enhancing farm incomes. With major thrust of extension agencies on production techniques, marketing extension so far has not received the attention it deserves. This assumes greater significance in the light of the new international trading regime under the WTO and the export opportunities being opened up. Public extension functionaries are presently ill equipped to deal with marketing extension. The multi-agency extension service will need to address these issues through strengthening capacity of the public agency, supporting private sector in marketing extension and making extensive use of media and IT in information and technology dissemination. Marketing extension so far a peripheral issue in the extension scenario will need to be brought centre-stage. Indeed, production will now need to be significantly dictated by market requirements.

Enabling Farmers for Problem Solving Skills. Under the new dispensation there will be a paradigm shift from top-down blanket dissemination of technological packages, towards providing producers with the knowledge and understanding with which to solve their own location specific problems. This means that the existing public organizations should improve their efficiency and effectiveness in research and technology application. This will call for interdisciplinary approach aiming at location-specificity of technical solution.

Encouraging Private Sector Involvement in Technology Transfer. A significant deterrent to expansion of private sector involvement in technology transfer is the provision of subsidized agro goods and services by public agencies. This often leads to the creation of an uneven playing field and discourages market entry by private providers. Wherever possible such subsidies will be phased out, in order to stimulate emergence of a private input supply network to provide hybrid seeds, artificial insemination services, fertilisers, agro-chemicals, animal feed, machinery and equipment, and other agricultural supplies and services to farmers on a full cost recovery basis. Generally, the costs associated with the research, development and transfer of these material technologies are embodied in the prices of these products, therefore, these costs are passed along to farmers, making this component of the Agricultural Technology System (ATS) financially sustainable. Targeted subsidies may be retained to protect the interest of the poor and vulnerable sections. In the field of *material technology* dissemination, which includes distribution of inputs such as fertilizer, seed, planting material, chemicals for plant protection, agricultural implements etc. a competitive private sector has developed in almost all states except for the north-eastern states. The new policy agenda envisages withdrawal of the public sector from areas where agro-services can be effectively and competitively provided by the private sector. In such cases the role of the public sector becomes one of facilitator and enabler. This implies moving towards a realistic cost recovery of agro-services by the State. For, if the public sector continues to subsidize the services, this will prevent a "level playing field" to the private sector, which will ultimately get crowded out. There will need to be a re-examination of existing Rules, Regulations & Acts to abolish provisions, which constrain private investment in delivery of agro-services.

Public funds for private extension services. Promotion of private extension needs to be matched by corresponding shifts in the allocation of public resources. Public funds would be made available to NGOs, Farmer Associations, Para-professionals or private foundations for extension work. An environment in which private investment in technology generation and transfer is more attractive will, therefore, have to be created.

Charging for extension services. The emergence of a market for private extension advice or consultancy services will be encouraged. Processors with contracted producers, also commercial suppliers of seed, agro-chemicals, machinery, vaccines, artificial insemination and the like should recover the costs of providing advice to their clients out of profit margins. However, the vulnerable group will need to be protected through targeted subsidies and safety nets.

Institutional Restructuring

It is clear that no one uniform extension system will serve as a panacea to all States. Even within States there will be a combination of various agencies and different institutional arrangements to address needs of differing agro-climatic zones as well as different sections of farmers. A menu *of various models* will be available to the States to select and adapt to their own requirements.

Restructuring Public Extension Public Extension will continue to remain central to Technology Dissemination, small & marginal farmers & economically backward regions will need to be serviced by it. This implies that public extension functionaries (including VEWs & SMSs) will have to be placed in new decentralized institutional arrangements which are demand-driven, farmer-accountable, bottom-up and have a Farming Systems Approach (broad-based). States have before them several models namely,

- the ATMA model (7 States)
- Single Window - Broad Based extension model (Maharashtra),
- Panchayati Raj Institutions (Kerala, West Bengal, Madhya Pradesh) and
- the SAU - Farmer Direct Contact (Punjab). With supplementation from the private sector, media and Information Technology the public extension service would be made leaner and professional. It is envisaged that the approximately 100, 000 public extension functionaries will be gradually reduced to be supported by the other two arms of services providers.

District Level Agriculture Technology Management Agency (ATMA) model. A key concept is to *decentralize decision-making* to the district level through the creation of the ATMA as a registered society. A second goal is to increase farmer input into program planning and resource allocation, especially at the block level, and to increase accountability to stakeholders. A third major goal is to increase program coordination and integration between departments so that the following program thrusts can be more effectively and efficiently implemented.

- *Farming System Innovations*—especially the intensification and/or diversification into high value commodities and/or value-added marketing and processing activities,
- *Farmer Organizations*—especially for high value commodities and resource poor farmers,
- *Technology Gaps* in both crops and livestock production systems.
- *Natural Resource Management*—especially soils and waters management and to reduce pesticide use through integrated pest management (IPM) programs.
- *Marketing and Agro-processing Linkages* between farmers' groups, markets and private processors.

Strategic Research and Extension Plans (SREPs) through Participatory Rural Appraisal (PRA). In the process of creating a more bottom-up extension system, PRA procedures would be introduced across all system levels (district, block, mandal, and village), and across each participating line department (DOA, DOH, DAH and Department of Marketing) and research institutions (ZRS and KVKs) within the district. On the basis of PRA, Strategic Research Extension Plans would be prepared for the districts. The district SREP must be *grounded* at the block or mandal level, where extension programs can be fine-tuned to the needs of farmers and more effectively implemented. The SREP would take account of the research, training and extension requirements for production as well as marketing activities. The rural periodic markets and wholesale assembling markets where farmers visit regularly would be used as important locations for dissemination of market and production technologies.

Block/ Mandal Level Technology Centre for Single Window Extension System. Concept of a Block (or Mandal) Technology Centre (BTC) has emerged wherein a multidisciplinary technology team (comprising block level agriculture, horticulture, soil & water conservation, agricultural marketing and livestock extension officers) would be assigned to organise and implement extension programs within the block. Other line department units and personnel would continue to provide essential service in developmental activities. In effect, the BTC would result in the functional integration of extension activities within the block or mandal and, in effect, become the operational arm of the ATMA. This Centre would become the common meeting point for extension personnel from the line departments to prepare integrated work plans (WPs) and to co-ordinate their implementation. It would also be the level where farmer input could be more effectively mobilized through a single Farmer Advisory Committee (FAC). The FAC would include 10-12 members representing all major stakeholders within the block. The FAC would help set block extension priorities and recommend resource allocation across program areas. The block technology team would be responsible for operationalizing the SREP in each block and moving toward a *single window extension system*.

Upgrading and Restructuring the Extension Staff – Field Extension Functionaries as Farm Advisors. The DOA's extension field staff would be restructured and upgraded to create a professional cadre of Farm Advisors. In the process, the village extension worker (VEW) cadre would be incrementally phased out through reassignment and normal attrition. Eventually, these farm advisors would be in charge of all extension activities within the block and they would all be required to meet a minimum educational requirement for service entry i.e. B.Sc. (Agri.). In addition, the project would provide in-service training in new planning, diagnostic and technical skills. By the end of this project, this new cadre of extension professionals should be able to identify and provide useful advice for most farmer problems (i.e. become more demand-driven). First, they should be able to carry out a systematic need assessment programme to prioritize farmer problems. Then, by utilizing the strengthened cadre of research and extension specialists (SMSs) within the district, they would be expected to organize and deliver a broader range of extension and farmer training programmes. In addition, these upgraded farm advisors would be expected to formulate and target location specific recommendations, including systems-based technologies, to reflect the needs of different socio-economic groups of farmers within their block or service area. Besides, the SMS cadre would be expanded and strengthened in the district to support the primary production and farming systems by supplying market related information to the producers. To facilitate collaboration between the line departments, district-level SMS positions would remain within each development department, but their extension activities would be coordinated under the ATMA framework.

Group Approach to Extension : The *contact farmer* approach to extension popularised by the T & V is to be replaced by the group approach. Formation and mobilisation of Farmer Interest Groups

(FIGs), Farmers Co-operatives and Self-Help Groups will be encouraged with the support of NGOs. Group extension will help to replace the top down approach with bottom up approach in technology transfer, as FIGs will first generate a demand for information, technology and management techniques. The extension workers would then respond to the group demand. This would lead to a farmer-extension worker participatory process with emphasis on problem solving rather than disseminating routine messages. The group approach in extension would also be in line with the Self-Help Groups of rural credit delivery, water user associations, and co-operatives.

Strengthening research-extension-farmer linkages. There is a need for close interaction between farmers, extensionists and production systems researchers in diagnosing problems together and working out location specific recommendations emphasizing participation education rather than prescription and joint actions in the field. Accepted to be more knowledge intensive, these new recommendations will require greater skills both to develop and to apply. There will be strengthening of research-extension-farmer linkages not only at the state and SAU levels but also at the district level; not only between the DOA and the SAUs, but between DOH, DOS and DAH for horticulture, fodder, agro-forestry and silvi-pasture as well as on-farm land and water management in a farming systems approach with due coverage of agricultural marketing concerns. The research-extension interface at all levels from district to national level will be supported

Promotion of Multi-Agency Extension Services. Widening the range of extension delivery agencies for the resource poor farmers and those residing in the hilly, tribal and remote areas, the public system will have to remain as the chief extension mechanism with NGOs possibly being able to play a significant role.

ICAR Role in Extension – Transfer of KVKs to DAC. The primary mandate of ICAR is research. Its extension programmes should be limited to reinforcing the research activities to make them more demand-driven and farmer centric. At present the extension activities through 261 KVKs, are drawing away about 12% of the ICAR Plan funds. This is likely to increase further as the ICAR strives to establish one KVK in every district by the end of the 9th Plan. Efforts were made towards the end of the 8th Plan to transfer the financial and administrative responsibilities of KVKs to their respective state governments, which could not materialize because of the severe resource constraints faced by the states. Hence, Krishi Vigyan Kendras would continue to operate in pro-active mode, retaining their allegiance with ICAR, for project implementation activities. Apart from focusing on production related issues, ICAR research would adequately address different components of marketing and make available need based packages in consonance with the changed/changing agricultural marketing scenario. Links with KVKs will be strengthened at the district level through institutions such as ATMAs.

Management Reforms in Agricultural Extension

Central Government Support to State Governments for Extension Services on their Undertaking Policy and Institutional Reforms. After the close of the World Bank supported NAEP, central support to the state extension services dried up, leaving them with the operation and maintenance of personnel and infrastructure created under T&V. States have barely been able to pay the salaries of extension personnel. Less than 10% of the budget is available for operational expenses, which has practically immobilized the service with scarcely any technology dissemination in the field. It is proposed to support the state extension services provided policy reforms and institutional restructuring is undertaken with demonstrated ability to be demand-driven, farmer-accountable, sustainable & farming systems based with broad-based integrated delivery. While funding for salaries of public functionaries will continue to be the responsibility of the State Governments, funds for technology dissemination and application (operation & management) would be shared between States & Central Government.

Central Government Funds to be Pooled at ATMA or ATMA like Registered Agencies at District Level. Funds from the central government together with state share for all technology transfer and extension activities would be pooled at these district level agencies and released for various activities according to the Strategic Research and Extension Plan prepared for the district. At present, annually about Rs. 200 crores worth of funds are released to the states under 100 centrally sponsored schemes (crops, horticulture, inputs, soil & water management) for the purpose of *transfer of technology.* Wherever ATMAs have been established they should be the conduits of these funds.

Central Government Assistance to State Agriculture Universities for Expanded Role in Field Extension. On the pattern of the successful scientist-farmer-extensionist model developed by the Punjab Agriculture University, the Directorates of Extension of SAUs would be supported to play a larger role in providing extension services in their service-areas.

Central Government Assistance to KVKs. Under the present arrangement the ownership and mainstreaming of KVKs with the state extension mechanisms has been weak. KVKs, set up as Centres for location specific, adaptive research, if effectively organized to achieve their primary objective of refinement and validation of local technologies could play a strategic role in linking the research and extension systems particularly in the area of farming systems based technologies. It is likely that State Governments will be more willing to own and mainstream KVKs once their relevance as district level technology refinement institutions integrated with the extension machinery is demonstrated rather than as just another vocational training organization, which they are largely perceived as at present and of which there are several others at the district level.

Promotion of Community -Based Private Extension Services. Group approach is the cornerstone of the restructured extension mechanism. A major component of extension services will be the mobilization of the community into farmers groups — FIGs, FOs, and SHGs. Farmers' Organizations will be linked with Panchayats through existing statutory institutional arrangements such as the Land Management Committees, Development Committees etc. FOs will be supported directly through public funds and will be involved in the planning, implementation, monitoring and feedback of programmes. FOs at the village level would be federated at higher levels. Representatives of FOs would be members of decision making bodies such as ATMAs, Block level Farmer Advisory Committees, Watershed Associations. Ultimate aim is for FOs to internalize extension services for its members and provide backward (inputs, credit, technology) and forward linkages (post-harvest facilities, markets, value addition) in a vertically integrated arrangement.

Promotion of NGOs based private extension services. Strength of NGOs is in their ability to mobilize communities into Farmers Organizations/Farmer Interest Groups/ Watershed Associations/ Market Associations. As such NGOs complement the public extension effort in several centrally sponsored programmes. Also extension services are contracted out and out-sourced to NGOs at the Block level in some states. In such cases the NGOs substitute for public extension. Public funds are used to support NGOs and are usually met from the provision of administrative expenses built into the Project Costs. NGOs are also supported directly by the central government in undertaking extension work. Of the 261 KVKs in the country, 86 are operated by NGOs. A systematic training, capacity building and technical backstopping mechanism, supported through public funds is to be developed for NGOs involved in providing extension services.

Promotion of para-professional based private extension. Para-extension workers normally supplement public extension in a relatively cost-effective manner and overcome constraints of absentee public extension functionaries (Gopals for AI services, Mitra Kisan for agri-services such as soil testing etc.). Under the new policy agenda para-extension workers at grassroot level will be supported through publicly funded training and capacity building and payment of honorarium in the early years. The honorarium will be routed through the Farmer Organizations/ Farmer Groups serviced by the para-extension workers to ensure accountability to the client group. Once the para-worker is able to demonstrate his/ her usefulness to the client group the honorarium provided through public funds will be phased out and the client group would take on the onus of paying for the services of the para extension worker. The public extension machinery will also assist para-workers in procuring loans from credit institutions for equipment, mobility and linkages with SMSs in Line Departments and SAUs. There will be an element of partial/ full cost recovery of services provided by para-workers who must ultimately become economically viable units except in the case of vulnerable clients where the State may continue the targeted subsidy.

Panchayati Raj Institutions and extension services. After the 73rd Amendment most states are conducting regular elections to the Panchayats. Some states have also delegated suitable administrative and financial powers to the three tier Panchayati Raj institutions. In these states the extension personnel are placed under the administrative control of the panchayats, whereas for technical guidance they remain under the control of their respective technical line departments. Since the panchayat systems are evolving in different states and are currently in a state of flux, the ATMA model at the district, the BTCs and FACs at the Block and the FOs at the village level may be organized as conceived, and suitable linkages be forged with the Panchayati Raj Institutions, e.g. the CEO of the Zila Panchayat is the Vice-Chairman of ATMA, the Chairman of the FAC at the Block is the elected chairman of the Development Committee of the Block Panchayat and suitable linkages will be established between various FOs at the village level and the Village Panchayats through the Land Management Committees, Development Committees etc.

Competitive Agriculture Extension Grant Fund. Similar to the Competitive Agriculture Research Grant Fund set up in ICAR and several state governments, wherein both public & private sector research institutions compete for funds to address specific research problems, it is proposed to set up a Competitive Agriculture Extension Grant Fund. Resources under this fund could be accessed through a competitive bidding process. Contracting out extension services to private sector, community-based organizations or NGOs in selected geographical areas (e.g.. a village, cluster of villages, Block) would be done through a transparent, laid out procedure under this Fund. This would also imply a strict monitoring and evaluation process.

Linkage of performance with funding for public sector. In a manner similar to the private extension agencies who must compete with one another to access funds and whose subsequent eligibility to compete for funds will depend upon their performance as indicated by an independent impact evaluation, it is proposed that on a pilot basis Public extension agencies also be made to compete with private extension agencies for operational funds under Competitive Agriculture Extension Grant Fund (CAEGF).

Contracting out extension support services. Wherever possible extension services in whole or in part could be contracted out for greater cost effectiveness. This applies, in addition, to administrative services such as security, mobility, computer and secretarial services, participatory planning to NGOs (being done in watershed management), staff training to a University/Institute, monitoring to a Farmer Organizations/IIM/Other Institutions.

Improving Research-Extension Linkages

Promotion of direct interface between farmers and scientists. The direct interface between scientists and farmers is the most ideal and should be undertaken wherever possible. It is an often repeated refrain that farmers learn best from scientists or other successful farmers. Moreover, transmission losses are minimized in the direct interface. However, there are relatively high costs attached to this direct mode of technology transfer and the outreach of scientists is limited. Punjab Agriculture University has achieved significant success through this system. It must be noted however, that Punjab is a small state geographically and what is applicable to Punjab may not be possible in large states such as Uttar Pradesh, Madhya Pradesh and Maharashtra.

Activating Existing Interface Mechanisms. Regional Committees of the ICAR, Zonal interfaces initiated by DAC, national level pre-kharif and pre-rabi DAC-ICAR interface, state level bi-annual meetings between line departments and SAUs are all formally instituted mechanisms for improving research-extension linkages. Several of these mechanisms have fallen into disuse or are conducted in a perfunctory manner. As a result the desired results are not being achieved. These will be activated.

Research priority setting based on SREP. Micro-level extension strategies reflected in the Strategic Research and Extension Plans (SREPs) based on PRA and developed jointly by the district technology teams including the marketing department officials and scientists of the KVKs/ ZRS or SAUs should be formally feedback into the research systems through a research priority setting mechanism in the ICAR.

Capacity Building of Extension Functionaries

Formulation of HRD Policy by States. Central Government support for HRD in Agricultural Extension would be available to the states only after the formulation and adoption of a HRD Policy and Action Plan through a systematic skill-gap analysis (Such a policy would incorporate compulsory training and skill upgradation of all extension functionaries). It would also build in an effective system of rewards and incentives for public extension functionaries.

Formulation of Training Plan for Extension functionaries. A long-term training plan should be developed by each state based on a thorough skill gap analysis. A massive campaign will need to be launched for skill upgradation and capacity building of extension functionaries using resources of all training institutes. The training be divided into *Foundation Courses* comprising skill up gradation in (i) need assessment techniques including the role of participatory rural appraisal (ii) group formation (iii) development of entrepreneurial skills for agri-business (iv) agri-business management (v) WTO and its implications (vi) marketing of agricultural produce (vii) post harvest

Mainstreaming Women in Agriculture

Mainstreaming women in agriculture. Gender concerns need to be mainstreamed in the agricultural extension process. Public extension systems, which must disseminate new technology and information, are still largely male dominated. Hence the necessity to target women is to ensure that they receive information relevant to their work, particularly, with reference to crops and livestock.

Improving access to extension and training : Women farmers usually have been neglected in extension efforts. Gender inequality had so far not been challenged by the agricultural extension system in the past. However, with the changing scenario, the need for innovating changes in extension approaches has assumed centre-stage. Under these innovations efforts will need to be made both by the central and state governments to improve extension services to reach farm women through (i) extension policy reorientation that explicitly recognizes farm women as agricultural extension clientele; (ii) training of men and women extension staff on women's role in agriculture and rural development and how agricultural extension work could be organized and conducted to meet women's needs in agriculture and rural development activities; (iii) training of women on decision-making in the context of farm and home management (iv) training of women farmers on agricultural marketing, particularly with respect to post-harvest processing, on farm value addition and market requirements/ demand.

Redesign of extension services to reach women farmers. Extension services are being redesigned to focus on women through (i) appropriate training/sensitization of extension personnel towards the role and contribution of women in the total agriculture system; (ii) increasing the proportion of trained female extension workers to gradually ensure that at least one-third of all extension workers are women (iii) sensitising male extension workers to the needs, approaches and perspectives of women through appropriate training and orientation programmes thereby dispelling the notion that only women can address extension needs of farm women (iv) improving communication between women, researchers, marketing agencies and extension workers required for the development of technology suitable for women; (v) developing appropriate extension methodologies that recognize the multi-dimensional role of women and the socio-cultural barriers, in which women farmers operate in a rural society; (vi) establishing Head of the Farming family as the target group, for extension services and assuming that the information will automatically trickle down to women farmers.

Expanding the sphere of women extension workers. The number of female agricultural extension workers would be increased through (i) re-examination of all service cadre rules for hidden gender biases (ii) improvement of female attendance at agricultural institutes and school (iii) building incentives such as scholarships and stipends for more women to take up under-graduate and post graduate courses in the agricultural and allied sciences (iv) redesigning of agricultural training curricula to include women's concerns: (v) ensuring that women are adequately represented in all training programmes whether domestic or overseas (vi) redesigning of training facilities to make them suitable for large numbers of female students and trainees; (vii) inclusion in the teaching curricula for extension workers, greater analysis and extension methods that take into account women's time, mobility and cultural situation; and (viii) exploring the specific role of farm women in the marketing of agricultural produce.

Use of Information Technology

Information Technology revolution is unfolding, and has very high visibility. However, its benefits have remained confined primarily to the urban areas. Rural communities have not been able to gain to the same extent from IT. As a means of agricultural technology transfer to farmers, information technology, had a limited impact. Even the vast potential of the broadcasting network has been tapped only minimally for extension.

Increasing Use of Information Technologies. Harnessing Information Technology for agricultural extension will receive high priority in the new policy agenda. Extensive use of modern information technology will be promoted for communication between researchers, extension workers and their farmer clients to transfer technologies and information more cost effectively. Information technology should be made available, particularly, to those with specific inquiries to guide them in adopting the more knowledge intensive forms of agriculture, which will expand in future.

IT Application in Agriculture Marketing. Agriculture produce marketing requires connectivity between the market and exporters/ growers/ traders, industry consumers, through wide area network of national and international linkages in order to provide day-to-day information with regard to commodity arrivals and prevailing rates etc. to provide links for on-line International market information; to provide export related documentation, to inform about the latest research in agricultural marketing, packaging, storage related information and to provide connectivity with lead international and national market organizations.

Wider Use of Electronic Mass Media for Agricultural Extension. Radio and TV have vastly increased their reach, as also reception facilities. "Local" radio and new FM transmitters open up possibilities of area-specific broadcasts. In communicating with an audience with low literacy skills, an audio-visual medium like TV has advantages. Today Doordarshan covers the entire population. Much wider and creative use of the mass media - All India Radio, private FM, Doordarshan, private cable network will be promoted for more rapid and effective dissemination of general information and advice to farming communities. This will include market information; market led production planning, on farm and post harvest management/value addition, e-contracting, market networks, market intelligence and wider application of World Wide Web. Face to face contacts should follow and back up these methods of information dissemination, not precede or substitute for them. Central government will support states in their effort to make fuller use of electronic media. Central government would also consider supporting an exclusive Agriculture Channel on Television.

Farmer Participation in IT Programmes. In developing any system of IT for agriculture technology transfer; the farmer is to be kept centre-stage. She/he is not to be treated as a passive recipient but rather as a player, a generator and user of knowledge. The upgradation of his/her skills and knowledge is therefore a crucial part of the process. The farmer will be an effective participant in the process.

Support to States for Information Technology Increased use of information technology at State/district and block levels would be promoted. This would include electronic access, through NICNET, to technical and administrative information; [e.g. electronic mail (EM) access to ICAR and SAU researchers, plus state and national administrative offices, and electronic access to ARIS system databases and, eventually, to the World Wide Web]. Under the NATP all the Block level Farm Advisory Information Centres in 28 project districts would be electronically linked to district, state and national institutions. Central Government will support States in the use of electronic linkages and computerization so that marketing, research, extension and farming communities are linked to each other, and into local, national and global networks. The objective is to link all 5200 Blocks to the state and national network and the Internet in a phased manner.

Private Information Shops/ Kiosks. The ultimate aim is to promote private Information shops/kiosks franchised out to private sector especially unemployed rural educated youth, in the manner of PCOs/ STD shops. Private sector will be encouraged to establish information shops at Block/Mandal/Village level. A major programme for development of software will need to be mounted so that Information Shops could have access to suitable material. Electronic connectivity and access to e-mail would put the franchisees in contact with district KVKs, Line Departments, markets and other sources of information. Such information could be dispensed to farmers, farmers groups on payment. Credit facilities for purchase of equipment for setting up such Information Shops would be permissible under the micro-credit programme for agriculture and allied activities.

Capacity Building for use of IT. Application of IT is constrained by lack of or inadequacy of complementary inputs (equipment, power, etc), appropriate organisational and institutional structures, information management and skills development. A major training programme for developing capacity for IT usage will be promoted. Training Institutes will run suitable courses for the purpose.

Financial Sustainability & Resource Mobilization

Publicly funded extension will continue to play a predominant role in technology dissemination firstly because the large numbers of small disadvantaged farmers may not have access to or be able to afford any other kind, and secondly, because much of the new technology will not be commercially marketable for instance watershed management, land capability assessment and land use planning, breaking of yield ceilings sustainable management of natural resources and socio- economic research. But pressures on government expenditure mean that public funds will have to be more carefully targeted and more efficiently used.

Cost-cutting mechanisms for extension services. Cost effectiveness may be improved by relying on fewer but better qualified (graduate or post-graduate) field advisers who interact directly with researchers for subject matter advice and then multiply their impact in the field by working with farmer groups rather than individual contact farmers. Cost cutting mechanisms, including the exploitation of mass media, encouragement of NGO and private sector involvement in extension, or needs-based coverage

Efficient use of available resources. Optimum fund utilization will be achieved with better matching the farmers needs with extension delivery, a stronger focus on the economics of farming, and the use of participatory methods to assess needs, create commitment to action, and monitor impact.

Privatization of agro-services. An environment in which private investment in technology generation and transfer is more attractive will be created. Product diversification both horizontal and vertical shall be promoted to not only improve profitability sustainability and more efficient use of production resources but also to encourage greater involvement of the private sector. Where opportunities exist to contract out publicly-funded services, or to transfer costs to the corporate sector or to users themselves, these opportunities should be exploited-for instance for diversification into higher-value or export crops, or to develop new commercial inputs or machines. Privatization of selected "private goods" and agro services wherever a competitive market exists, such as AI services, soil testing, fertilizer advice, farm improvement plans or breeding plans would be undertaken. Wherever feasible contract farming through the involvement of private sector would be taken up, particularly, in the area of high value/export oriented agriculture.

Towards a realistic cost recovery of agro-services. Wherever farmers have the capacity to pay for public services, which are in the nature of private goods, realistic cost of such services should be recovered. However, provision is made for targeted subsidies to protect the vulnerable class of users.

Co-financing of public extension. Co-financing of public extension services by farmers and farmers' associations to reduce pressure on public finances and to improve the accountability and responsiveness of extension to farmers.

Initiating new financial systems. Modification in rules and regulations and innovations in financial institutions will also be required to allow for arrangements such as "revolving funds" for government farms, nurseries, etc. While budgetary support to these units may continue to meet pay and allowances of government staff, the funds for recurring expenditure and operational costs could be generated by these units from the commercial activities undertaken by them. All efforts would be made to develop credit linkages of farmers and farmers groups with credit institutions.

Changing Role of Government

Role of State in Effective Regulation & Enforcement. As a multi-agency extension regime proliferates, the responsibility of the State for effective enforcement of legislation, which ensures quality control of inputs such as seed, pesticides, fertilizers etc. will increase. State's role as arbitrator of conflicts between various private sector extension agents will also increase and systems to address grievances will need to be developed. This role will increase as the number of private extension agencies grows. Guidelines for private agencies would be required. However, in the emerging pluralistic scenario the role of public extension would need to be redefined from one of solely a provider of services to become increasingly an appropriate mix of provider, coordination, facilitator and regulator. The large section of small and marginal farmers and landless labourers as well as remote and backward regions would continue to need the services of the public extension functionaries, as they are not likely to be serviced by a competitive private sector in the near future. Public Extension's role would increase in arbitration of conflicts, assuring accountability of all service providers to the farmers and ensure transparency through provision of information. The overall environment of private provision of extension services deserves to be encouraged through policy reforms and institutional changes so that rural people's needs serviced more efficiently.

Creating an Enabling Environment. Generally, this implies appropriate legislation, rules and regulations, and application of the rule of law. In particular, it implies that private contracts and property are protected and a judiciary exists to enforce contracts without partiality and undue delay. Where many individual smallholders are concerned a legal course of action may, however, not be practical or politically expedient for handling conflicts and disputes. Government can instead support institutions like an independent arbiter, an ombudsman, or a small farmer reference service that would certify bona fide borrowers or agricultural producers. Governments can also set minimum standards and norms for commodities such as food, pesticides, and packaging materials when it is in the interest of public health. To protect the weaker of the contracting parties, governments can propose minimum standard contract clauses and guidelines for small farmer/agribusiness transactions. It is essential that such proposals be seen as recommendations, not prescriptions. An important condition for lasting farmer-agribusiness linkages is security of tenure. Activities supported by international donors in many countries include land titling, formal transfer of public land to *de facto* users, and outright land reform.

Enhancing Competition. The enhancement of competition is another government contribution too improving the institutional environment. It involves all measures to ensure open, fair and transparent competition and to facilitate entry of newcomers. It may include breaking up of monopolies and cartels, ensuring minimum professional standards of business conduct, and resisting demands for non-technical obstacles to official licensing by rent-seeking lobbies. A lack of financial means is frequently the reason that prevents newcomers from establishing new businesses. Governments can assist young entrepreneurs to access credit and venture capital by providing technical assistance to prepare business plans, conduct market surveys and hire help to resolve special engineering or legal problems and through this to improve confidence of funding sources in new ventures. Part of creating an enabling environment would also be to address the downside of privatization and liberalization. This might include the provision of safety nets and skills for alternative livelihood to those who are unable to survive the competition in deregulated markets.

Strengthening Farmers' Associations. Government services can help identify existing associations or cooperatives of farmers and support them to develop their organization. The aim must be to assist the groups to define their objectives, such as savings mobilization or specific post-harvest operations, to ensure group coherence and continuity, and to assist them with setting up group structures and organization. Over time such groups can establish a track record of organizational maturity that will allow, possibly after joining with other groups for economies of scale, to engage in their own business activities and to gain access to formal credit. Government extension services and NGO staff need to receive suitable training to act as group facilitators. Training would be centred on group dynamics, record keeping and accountancy, financial management of savings and credit, identification of group business opportunities and producing business plans, and on imparting technical skills to undertake new ventures. Support to farmers' organizations is perhaps the main single input that governments can provide for the promotion of farmer integration with agribusiness.

Strengthening Physical Infrastructure. The government's role would increasingly be in the area of physical infrastructure provision, including communications and utilities, farm-to-market roads, and rural and urban markets. Promotion of private sector would be through making available sites with road and electricity connections to attract enterprises that may set up marketing or processing facilities as pert of an industrial estate. Rural or farm-to-market roads also facilitate linkages between farmers and private service providers. Similarly development of wholesale market yards is also supported by governments at given stages of marketing development. At more advanced stages such facilities tend to lose their functions, as alternative forms of marketing develop for a variety of agricultural products that rely on direct producer/agribusiness/consumer linkages and by-pass traditional markets.

Improving Information. Another way of leveling the playing field for private sector is the improvement of information. Information gathering and analysis is costly. Compared to commercial business, farmers are at a disadvantage on knowledge about prices, volumes, qualities, alternative marketing channels and other feature governing market transactions. Government can improve the communications flow and the quality of information to farmers through training workshops and publications and by his improve transparency and facilitate transactions. Government can also sponsor market-matching exercises, that is, sponsor meetings and workshops involving farmers and agribusiness enterprises to improve mutual understanding of constraints and requirements, and promote concrete business deals.

Through their sheer numbers and outreach the public extension system would continue to play a prominent role in technology dissemination. The large section of small and marginal farmers and landless labourers would need to be serviced by the public extension systems. The other actors involved in the extension/ transfer of technologies such as NGOs, Farmers Organizations, Private Sector (both corporate & informal), para-workers etc. would actively complement/ supplement the efforts of the public extension agency and wherever possible replace it. Extension mechanisms will have to be driven by farmers' needs, be location specific and address diversification demands. Technologies required to address total farming systems are knowledge intensive. Public extension system will need to be redefined with focus on knowledge-based technologies to upgrade and improve the skills of the farmers.

As agricultural extension transforms itself into a more diversified farming systems approach from its present simplistic accent on yield enhancement by increasing some limited inputs, farmers will be required to adopt a wider range of inputs and practices and develop skills in their more efficient use. The task of extension will become more challenging in the wake of post WTO era, which demands a system of market led extension with specific focus on diversification, post harvest management and export orientation. This will present a more complex role, but simultaneously requiring a flexible approach allowing specific information to be customized for different farmer-groups. A strategy of institutional innovations in extension will be evolved which optimizes the strengths of the public-private sectors to service the needs of the farming community.

New Agriculture Policy

- Agriculture is a way of life, a tradition, which, for centuries, has shaped the thought, the outlook, the culture and the economic life of the people of India. Agriculture, therefore, is and will continue to be central to all strategies for planned socio-economic development of the country. Rapid growth of agriculture is essential not only to achieve self-reliance at national level but also for household food security and to bring about equity in distribution of income and wealth resulting in rapid reduction in poverty levels.

- Indian agriculture has, since Independence, made rapid strides. In taking the annual food grains production from 51 million tonnes in early fifties to 206 million tonnes at the turn of the century, it has contributed significantly in achieving self-sufficiency in food and in avoiding food shortages.

- Over 200 million Indian farmers and farm workers have been the backbone of India's agriculture. Despite having achieved national food security the well being of the farming community continues to be a matter of grave concern for planners and policy makers. The establishment of an agrarian economy which ensures food and nutrition to India's billion people, raw materials for its expanding industrial base and surpluses for exports, and a fair and equitable reward system for the farming community for the services they provide to the society, will be the mainstay of reforms in the agriculture sector.

- The National Policy on Agriculture seeks to actualise the vast untapped growth potential of Indian agriculture, strengthen rural infrastructure to support faster agricultural development, promote value addition, accelerate the growth of agro business, create employment in rural areas, secure a fair standard of living for the farmers and agricultural workers and their families, discourage migration to urban areas and face the challenges arising out of economic liberalization and globalisation. Over the next two decades, it aims to attain :

The salient features of the new agricultural policy are :

1. Over 4 per cent annual growth rate aimed over next two decades..
2. Greater private sector participation through contract farming.
3. Price protection for farmers.
4. National agricultural insurance scheme to be launched.
5. Dismantling of restrictions on movement of agricultural commodities throughout the country.
6. Rational utilization of country's water resources for optimum use of irrigation potential.
7. High priority to development of animal husbandry, poultry, dairy and aquaculture.
8. Capital inflow and assured markets for crop production.
9. Exemption from payment of capital gains tax on compulsory acquisition of agricultural land.
10. Minimize fluctuations in commodity prices.
11. Continuous monitoring of international prices.
12. Plant varieties to be protected through a legislation.
13. Adequate and timely supply of quality inputs to farmers.

14. High priority to rural electrification.
15. Setting up of agro-processing units and creation of off-farm employment in rural areas.

Sustainable Agriculture

- The policy will seek to promote technically sound, economically viable, environmentally non-degrading, and socially acceptable use of country's natural resources - land, water and genetic endowment to promote sustainable development of agriculture. Measures will be taken to contain biotic pressures on land and to control indiscriminate diversion of agricultural lands for non-agricultural purposes. The unutilized wastelands will be put to use for agriculture and forestation. Particular attention will be given for increasing cropping intensity through multiple-cropping and inter-cropping.
- Rational utilization and conservation of the country's abundant water resources will be promoted. Conjunctive use of surface and ground water will receive highest priority. Special attention will be focused on water quality and the problem of receding ground-water levels in certain areas as a result of over-exploitation of underground aquifers. Proper on-farm management of water resources for the optimum use of irrigation potential will be promoted.
- Erosion and narrowing of the base of India's plant and animal genetic resources in the last few decades has been affecting the food security of the country. Survey and evaluation of genetic resources and safe conservation of both indigenous and exogenously introduced genetic variability in crop plants, animals and their wild relatives will receive particular attention. The use of bio-technologies will be promoted for evolving plants which consume less water, are drought resistant, pest resistant, contain more nutrition, give higher yields and are environmentally safe. Conservation of bio-resources through their *ex situ* preservation in Gene Banks, as also *in situ* conservation in their natural habitats through bio-diversity parks, etc., will receive a high priority to prevent their extinction. Specific measures will also be taken to conserve indigenous breeds facing extinction. There will be a time bound programme to list, catalogue and classify country's vast agro bio-diversity.

- Sensitization of the farming community with the environmental concerns will receive high priority. Balanced and conjunctive use of bio-mass, organic and inorganic fertilizers and controlled use of agro chemicals through integrated nutrients and pest management (INM & IPM) will be promoted to achieve the sustainable increases in agricultural production. A nation-wide programme for utilization of rural and urban garbage, farm residues and organic waste for organic matter repletion and pollution control will be worked out.
- Agro-forestry and social forestry are prime requisites for maintenance of ecological balance and augmentation of bio-mass production in agricultural systems. Agro-forestry will receive a major thrust for efficient nutrient cycling, nitrogen fixation, organic matter addition and for improving drainage. Farmers will be encouraged to take up farm/agro-forestry for higher income generation by evolving technology, extension and credit support packages and removing constraints to development of agro and farm forestry. Involvement of farmers and landless labourers will be sought in the development of pastures/forestry programmes on public wastelands by giving financial incentives and entitlements to the usufructs of trees and pastures.
- The history and traditional knowledge of agriculture, particularly of tribal communities, relating to organic farming and preservation and processing of food for nutritional and medicinal purposes is one of the oldest in the world. Concerted efforts will be made to pool, distil and evaluate traditional practices, knowledge and wisdom and to harness them for sustainable agricultural growth.

Food and Nutritional Security

- Special efforts will be made to raise the productivity and production of crops to meet the increasing demand for food generated by unabated demographic pressures and raw materials for expanding agro-based industries. A regionally differentiated strategy will be pursued, taking into account the agronomic, climatic and environmental conditions to realize the full growth potential of every region. Special attention will be given to development of new crop varieties, particularly of food crops, with higher nutritional value through adoption of bio-technology particularly genetic modification, while addressing bio-safety concerns.

- A major thrust will be given to development of rainfed and irrigated horticulture, floriculture, roots and tubers, plantation crops, aromatic and medicinal plants, bee-keeping and sericulture, for augmenting food supply, exports and generating employment in rural areas. Availability of hybrid seeds and disease-free planting materials of improved varieties, supported by a network of regional nurseries, tissue culture laboratories, seed farms will be promoted to support systematic development of horticulture having emphasis on increased production, post-harvest management, precision farming, bio-control of pests and quality regulation mechanism and exports.

- Animal husbandry and fisheries also generate wealth and employment in agriculture sector. Development of animal husbandry, poultry, dairying and aquaculture will receive a high priority in the efforts for diversifying agriculture, increasing animal protein availability in the food basket and for generating exportable surpluses. A national livestock breeding strategy will be evolved to meet the requirements of milk, meat, egg and livestock products and to enhance the role of draught animals as a source of energy for farming operations and transport. Major thrust will be on genetic upgradation of indigenous/native cattle and buffaloes using proven semen and high quality pedigreed bulls and by expanding artificial insemination network to provide services at the farmer's doorstep.

- Generation and dissemination of appropriate technologies in the field of animal production as also health care to enhance production and productivity levels will be given greater attention. Cultivation of fodder crops and fodder trees will be encouraged to meet the feed and fodder requirements and to improve animal nutrition and welfare. Priority will also be given to improve the processing, marketing and transport facilities, with emphasis on modernization of abattoirs, carcass utilization and value addition thereon. Since animal disease eradication and quarantine is critical to exports, animal health system will be strengthened and disease-free zones created. The involvement of cooperatives and private sector will be encouraged for development of animal husbandry, poultry and dairy. Incentives for livestock and fisheries production activities will be brought at par with incentives for crop production.

- An integrated approach to marine and inland fisheries, designed to promote sustainable aquaculture practices, will be adopted. Biotechnological application in the field of genetics and breeding, harmonal applications, immunology and disease control will receive particular attention for increased aquaculture production. Development of sustainable technologies for fin and shell fish culture as also pearl-culture, their yield optimization, harvest and post-harvest operations, mechanization of fishing boats, strengthening of infrastructure for production of fish seed, berthing and landing facilities for fishing vessels and development of marketing infrastructure will be accorded high priority. Deep sea fishing industry will be developed to take advantage of the vast potential of country's exclusive economic zone.

Generation and Transfer of Technology

- A very high priority will be accorded to evolving new location-specific and economically viable improved varieties of agricultural and horticultural crops, livestock species and aquaculture as also conservation and judicious use of germplasm and other bio-diversity resources. The regionalization of agricultural research, based on identified agro-climatic zones, will be accorded high priority. Application of frontier sciences like bio-technology, remote sensing technologies, pre and post-harvest technologies, energy saving technologies, technology for environmental protection through national research system as well as proprietary research will be encouraged. The Endeavour will be to build a well organized, efficient and result-oriented agriculture research and education system to introduce technological change in Indian agriculture. Up gradation of agricultural education and its orientation towards uniformity in education standards, women empowerment, user-orientation, vocationalization and promotion of excellence will be the hallmark of the new policy..

- The research and extension linkages will be strengthened to improve quality and effectiveness of research and extension system. The extension system will be broad-based and revitalized. Innovative and decentralized institutional changes will be introduced to make the extension system farmer-responsible and farmer-accountable. Role of Krishi Vigyan Kendras (KVKs), Non-

Governmental Organizations (NGOs), Farmers Organizations, Cooperatives, corporate sector and para-technicians in agricultural extension will be encouraged for organizing demand-driven production systems. Development of human resources through capacity building and skill upgradation of public extension functionaries and other extension functionaries will be accorded a high priority. The Government will endeavour to move towards a regime of financial sustainability of extension services through effecting in a phased manner, a more realistic cost recovery of extension services and inputs, while simultaneously safeguarding the interests of the poor and the vulnerable groups.

- Mainstreaming gender concerns in agriculture will receive particular attention. Appropriate structural, functional and institutional measures will be initiated to empower women and build their capabilities and improve their access to inputs, technology and other farming resources.

Inputs Management

- Adequate and timely supply of quality inputs such as seeds, fertilizers, plant protection chemicals, bio-pesticides, agricultural machinery and credit at reasonable rates to farmers will be the endeavour of the Government. Soil testing and quality testing of fertilisers and seeds will be ensured and supply of spurious inputs will be checked. Balanced and optimum use of fertilizers will be promoted together with use of organic manures and bio-fertilizers to optimize the efficiency of nutrient use.
- Development, production and distribution of improved varieties of seeds and planting materials and strengthening and expansion of seed and plant certification system with private sector participation will receive a high priority. A National Seed Grid will be established to ensure supply of seeds especially to areas affected by natural calamities. The National Seeds Corporation (NSC) and State Farms Corporation of India (SFCI) will be restructured for efficient utilization of investment and manpower.
- Protection to plant varieties through a *sui generis* legislation, will be granted to encourage research and breeding of new varieties particularly in the private sector in line with India's obligations under TRIPS Agreement. The farmers will, however, be allowed

their traditional rights to save, use, exchange, share and sell their farm saved seeds except as branded seeds of protected varieties for commercial purpose. The interests of the researchers will also be safeguarded in carrying out research on proprietary varieties to develop new varieties.

Integrated pest management and use of biotic agents in order to minimize the indiscriminate and injudicious use of chemical pesticides will be the cardinal principle covering plant protection. Selective and eco-friendly farm mechanization through appropriate technology will be promoted, with special reference to rainfed farming to reduce arduous work and to make agriculture efficient and competitive as also to increase crop productivity.

Incentives for Agriculture

- The Government will endeavour to create a favourable economic environment for increasing capital formation and farmer's own investments by removal of distortions in the incentive regime for agriculture, improving the terms of trade with manufacturing sectors and bringing about external and domestic market reforms backed by rationalization of domestic tax structure. It will seek to bestow on the agriculture sector in as many respects as possible benefits similar to those obtaining in the manufacturing sector, such as easy availability of credit and other inputs, and infrastructure facilities for development of agri-business industries and development of effective delivery systems and freed movement of agro produce.
- Consequent upon dismantling of Quantitative Restrictions on imports as per WTO Agreement on Agriculture, commodity-wise strategies and arrangements for protecting the grower from adverse impact of undue price fluctuations in world markets and for promoting exports will be formulated. Apart from price competition, other aspects of marketing such as quality, choice, health and bio-safety will be promoted. Exports of horticultural produce and marine products will receive particular emphasis. A two-fold long term strategy of diversification of agricultural produce and value addition enabling the production system to respond to external environment and creating export demand for the commodities produced in the country will be evolved with a view to providing the farmers incremental income from export earnings. A favourable economic environment and supportive

public management system will be created for promotion of agricultural exports. Quarantine, both of exports and imports, will be given particular attention so that Indian agriculture is protected from the ingress of exotic pests and diseases.

- In order to protect the interest of farmers in context of removal of Quantitative Restrictions, continuous monitoring of international prices will be undertaken and appropriate tariffs protection will be provided. Import duties on manufactured commodities used in agriculture will be rationalized. The domestic agricultural market will be liberalized and all controls and regulations hindering increase in farmers' income will be reviewed and abolished to ensure that agriculturists receive prices commensurate with their efforts, investment. Restrictions on the movement of agricultural commodities throughout the country will be progressively dismantled.
- The structure of taxes on foodgrains and other commercial crops will be reviewed and rationalized. Similarly, the excise duty on materials such as farm machinery and implements, fertilizers, etc., used as inputs in agricultural production, post harvest storage and processing will be reviewed. Appropriate measures will be adopted to ensure that agriculturists by and large remain outside the regulatory and tax collection systems. Farmers will be exempted from payment of capital gains tax on compulsory acquisition of agricultural land.

Investments in Agriculture

- The agriculture sector has been starved of capital. There has been a decline in the public sector investment in the agriculture sector. Public investment for narrowing regional imbalances, accelerating development of supportive infrastructure for agriculture and rural development particularly rural connectivity will be stepped up. A time-bound strategy for rationalisation and transparent pricing of inputs will be formulated to encourage judicious input use and to generate resources for agriculture. Input subsidy reforms will be pursued as a combination of price and institutional reforms to cut down costs of these inputs for agriculture. Resource allocation regime will be reviewed with a view to rechannelizing the available resources from support measures towards assets formation in rural sector.

- A conducive climate will be created through a favourable price and trade regime to promote farmers' own investments as also investments by industries producing inputs for agriculture and agro-based industries. Private sector investments in agriculture will also be encouraged more particularly in areas like agricultural research, human resource development, post-harvest management and marketing.
- Rural electrification will be given a high priority as the prime mover for agricultural development. The quality and availability of electricity supply will be improved and the demand of the agriculture sector will be met adequately in a reliable and cost effective manner. The use of new and renewable sources of energy for irrigation and other agricultural purposes will also be encouraged.
- Bridging the gap between irrigation potential created and utilized, completion of all on-going projects, restoration and modernization of irrigation infrastructure including drainage, evolving and implementing an integrated plan of augmentation and management of national water resources will receive special attention for augmenting the availability and use of irrigation water.
- Emphasis will be laid on development of marketing infrastructure and techniques of preservation, storage and transportation with a view to reducing post-harvest losses and ensuring a better return to the grower. The weekly periodic markets under the direct control of *Panchayat Raj* institutions will be upgraded and strengthened. Direct marketing and pledge financing will be promoted. Producers markets on the lines of Ryatu Bazars will be encouraged throughout the width and breadth of the country. Storage facilities for different kinds of agricultural products will be created in the production areas or nearby places particularly in the rural areas so that the farmers can transport their produce to these places immediately after harvest in shortest possible time. The establishment of cold chains, provision of pre-cooling facilities to farmers as a service and cold storage in the terminal markets and improving the retail marketing arrangements in urban areas, will be given priority. Upgradation and dissemination of market intelligence will receive particular attention.

- Setting up of agro-processing units in the producing areas to reduce wastage, especially of horticultural produce, increased value addition and creation of off-farm employment in rural areas will be encouraged. Collaboration between the producer cooperatives and the corporate sector will be encouraged to promote agro-processing industry. An interactive coupling between technology, economy, environment and society will be promoted for speedy development of food and agro-processing industries and building up a substantial base for production of value added agro-products for domestic and export markets with a strong emphasis on food safety and quality. The Small Farmers Agro Business Consortium (SFAC) will be energized to cater to the needs of farmer entrepreneurs and promote public and private investments in agri-business.

Institutional Structure

Indian agriculture is characterized by pre-dominance of small and marginal farmers. Institutional reforms will be so pursued as to channelize their energies for achieving greater productivity and production. The approach to rural development and land reforms will focus on the following areas :

- Consolidation of holdings all over the country on the pattern of north-western States;
- Redistribution of ceiling surplus lands and waste lands among the landless farmers, unemployed youth with initial start-up capital;
- Tenancy reforms to recognize the rights of the tenants and share croppers;
- Development of lease markets for increasing the size of holdings by making legal provisions for giving private lands on lease for cultivation and agri-business;
- Updating and improvement of land records, computerization and issue of land pass-books to the farmers, and
- Recognition of women's rights in land.
    - The rural poor will be increasingly involved in the implementation of land reforms with the help of Panchayati Raj Institutions, Voluntary Groups, Social Activists and Community Leaders.

- Private sector participation will be promoted through contract farming and land leasing arrangements to allow accelerated technology transfer, capital inflow and assured markets for crop production, especially of oilseeds, cotton and horticultural crops.
- Progressive institutionalization of rural and farm credit will be continued for providing timely and adequate credit to farmers. The rural credit institutions will be geared to promote savings, investments and risk management. Particular attention will be paid to removal of distortions in the priority sector lending by commercial banks for agriculture and rural sectors. Special measures will be taken for revamping of cooperatives to remove institutional and financial weaknesses and evolving simplified procedure for sanction and disbursement of agriculture credit. The endeavour will be to ensure distribution equity in the disbursement of credit. Micro-credit will be promoted as an effective tool for alleviating poverty. Self Help Group - Bank linkage system, suited to Indian rural sector, will be developed as a supplementary mechanism for bringing the rural poor into the formal banking system, thereby improving banks outreach and the credit flows to the poor in an effective and sustainable manner.
- The basic support to agriculture has been provided by cooperative sector assiduously built over the years. The Government will provide active support for promotion of cooperative-form of enterprise and ensure greater autonomy and operational freedom to them to improve their functioning. The thrust will be on :

- ❑ Structural reforms for promoting greater efficiency and viability by freeing them from excessive bureaucratic control and political interference;
- ❑ Creation of infrastructure and human resource development;
- ❑ Improvement in financial viability and organizational sustainability of cooperatives;
- ❑ Democratizations of management and increased professionalism in their operations, and
- ❑ Creating a viable inter-face with other grass-root Organizations.

The Legislative and regulatory framework will be appropriately amended and strengthened to achieve these objectives.

Risk Management

- Despite technological and economic advancements, the condition of farmers continues to be unstable due to natural calamities and price fluctuations. National Agriculture Insurance Scheme covering all farmers and all crops throughout the country with built-in provisions for insulating farmers from financial distress caused by natural disasters and making agriculture financially viable will be made more farmer-specific and effective. Endeavour will be made to provide a package insurance policy for farmers, right from sowing of crops to post-harvest operations, including market fluctuations in the prices of agricultural produce.

- In order to reduce risk in and impart greater resilience to Indian agriculture against droughts and floods, efforts will be made for achieving greater flood-proofing of flood prone agriculture and drought-proofing of rainfed agriculture for protecting farmers from vagaries of nature. For this purpose, contingency agriculture planning, development of drought and flood resistant crop varieties, watershed development programmer, drought prone areas and desert development programmes and rural infrastructure development programmes, will receive particular attention.

- The Central Government will continue to discharge its responsibility to ensure remunerative prices for agricultural produce through announcement of Minimum Support Prices policy for major agricultural commodities. The food, nutrition and other domestic and exports requirements of the country will be kept in view while determining the support prices of different commodities. The price structure and trade mechanism will be continuously reviewed to ensure a favourable economic environment for the agriculture sector and to bring about an equitable balance between rural and urban incomes. The methodology used by the Commission on Agricultural Costs and Prices (CACP) in arriving at estimates of cost of production will be periodically reviewed. The price structure of both inputs and outputs will be monitored to ensure higher returns to the farmers and bring about cost effectiveness throughout the economy. Domestic market prices will be closely monitored to prevent distress sales by farmers. Public and cooperative agencies undertaking marketing operations, will be strengthened.

- The Government will enlarge the coverage of futures markets to minimize the wide fluctuations in commodity prices as also for hedging their risks. The endeavour will be to cover all important agricultural products under futures trading in course of time.

Management Reforms

- Effective implementation of policy initiatives will call for comprehensive reforms in the management of agriculture by Central and State Governments. Central Government will supplement/complement the State Governments' through regionally differentiated Work Plans, comprising crop/area/target group efforts specific interventions, formulated in an inter-active mode and implemented in a spirit of partnership with States. Central Government will move away from schematic approach to Macro-Management mode and assume a role of advocacy, articulation and facilitation to help States in their efforts towards achieving accelerated agricultural development.
- The Government will focus on quality aspects at all stages of farm operations from sowing to primary processing. The quality of inputs and other support services to farmers will be improved. Quality consciousness amongst farmers and agro-processors will be created. Grading and standardization of agricultural products will be promoted for export enhancement. Application of science and technology in agriculture will be promoted through a regular system of interface between S&T institutions and users/potential users, to make the sector globally competitive.
- The database for agriculture sector will be strengthened to ensure greater reliability of estimates and forecasting which will help in the process of planning and policy making. Efforts will be made to significantly improve and harness latest remote sensing and information technology to capture data, collate it, add value and disseminate it to appropriate destinations for managing the risk and in accelerating the growth process. The objective will be to engage in a meaningful continuous dialogue with the external environment in the changing scenario and to have on-line and real time system of 'Agriculture on-line' capacity to analyze signals emanating from the farms and markets for the benefit of farmers.

Rural and Agricultural Finance

Who are the potential clients of rural and agricultural finance?

Rural and agricultural finance clients are a complex and overlapping blend of rural households, small farmers, agribusinesses, and off-farm enterprises. These can broadly be categorized as :

- Off farm microenterprises and rural households (non agricultural) - Households not directly related to agriculture, as well as non-agriculture related businesses
- Farm and agriculture-related enterprises – Input suppliers, farmers, producer groups, local traders and processors
- Agribusinesses (non-rural) – Agri-processors, distributors, and exporters located in urban and peri-urban areas

What are their financial service needs?

The financial service needs are broad and may include :

1. Short term credit/working capital - for inputs, cash flow management, income smoothing
2. Term credit - for fixed asset or land acquisition, leasing, inventory build-up
3. Deposit and transfer services including domestic payment services and remittances - for cash flow management, risk mitigation, investment and asset-building
4. Insurance products (health, life, weather)

   - Working capital is more readily available to rural enterprises than long-term credit or other financial services. It meets seasonal needs for inputs, labor, and production services. When timed correctly, working capital allows use of seed varieties, fertilizer, labor and other inputs that may lead to increased income. Working capital has limitations in that it is short-term financing for immediate needs of an enterprise. Also, very short-term loans (for instance, three months) for agricultural inputs are often too brief to allow repayment from the sale of seasonal crops. Without access to the flexibility of investment capital or other long- term sources of capital, enterprises cannot easily expand or upgrade their business or overcome unforeseen events.

- Term funds finance capital improvements such as a new barn, storage facility, equipment or livestock expansion. Investments in these assets help businesses to grow and jump to the next level. Long-term credit is not readily available in rural areas, at least to smaller farms and enterprises.
- Agriculture enterprises and rural households need other financial services. Deposit services enable rural households to manage crises (such as a sudden illness or a flood), to invest when opportunity strikes, or to pay for large expected expenses, such as school fees, a wedding, or a new roof. Cash flow management, money transfers and risk mitigation tools such as insurance are critical to weathering unforeseen costs like a family emergency, a natural disaster or crop failure.

What is the ability of rural and agricultural finance clients to repay loans?

Sources of repayment for a typical client may include :

- Crop harvest or sale of animals
- Sale of processed product to buyer/exporter
- Microenterprise income
- Household members' salaries and other earned income
- Remittances or domestic money transfers
- Sale of assets

Formal collateral (as opposed to collateral substitutes such as group guarantees) offered by RAF clients take varied forms, including :

- Land
- Crop harvest
- Animals
- Farming equipment
- House
- Household appliance or personal assets

There are many potential problems in using these types of collateral. In some countries, land may not constitute an effective guarantee, either due to lack of land titling or judicial or political reluctance to enforce legal contracts (e.g., claiming land in compensation of non-payment on a loan) that would drive poor farmers away from their means of livelihood. Banks cannot use a house as collateral if the house cannot be seized (primary residence laws). In some countries the lack of lien laws or movable property registries may prevent the use of equipment as collateral.

Who supplies rural and agricultural finance?

The main source of credit for many farmers and agribusinesses is other agribusinesses along the value chain including input suppliers, traders, and processors. Moreover, the clients' own savings and credit from financial institutions continues to play a role in agricultural production.

Suppliers of rural and agricultural finance can be broadly categorized as :

- Value Chain Actors : exporters/wholesalers; processors; local traders and processors; producer groups; farmers; and input suppliers.
- Financial Institutions :
- Banks (commercial, agricultural banks, state development banks)
- Non bank financial institutions (NBFIs, commercial MFIs, other non-bank lenders, and leasing companies
- Not-for-profit MFIs, which tend to work with poorer clients
- Credit unions and agricultural cooperatives

Why is it difficult for rural and agricultural businesses and households to get credit?

The development of microfinance and rural and agricultural financial markets often share similar environmental challenges such as an inhospitable policy, legal and regulatory framework, lack of adequate collateral, lack of registered credit history, lack of market information, and income variability among potential clients.

In addition, rural and agricultural markets have some unique characteristics that impede the supply of finance and the ability of rural households and firms to access the financial services they need.

- *Weather risk* : These are often correlated risks, affecting many people in an area by a single event, such as a drought, excessive rainfall, earthquake or other disaster. Rural households often rely on informal strategies to cope with risk, but these can break down when correlated catastrophic losses take place.

- *Commodity risk* : Agricultural enterprises are subject to uncertainties in the future market value of their produce and their future income, due to cost and price volatility in local, regional, and global markets.

- *Seasonality* : Rural farms and households are vulnerable to the cyclical or seasonal nature of the agricultural sector, even if their primary livelihood/economic base is not agriculture per se. Clients may face severe constraints in repaying a loan during certain times of the year (e.g., before the harvest comes in). Harvest season is a period of positive revenue while the planting season is characterized by a period of heavy spending and very little, if any, revenue generation. The toughest time is before the harvesting season, when liquidity is scarce and families in many regions need access to loans or other, non-agricultural sources of income.

- *Geographic dispersion or distance to urban center/financial institution*: Because clients live far from urban centers where financial institutions are often located, it increases the transaction cost for the borrower. These transaction costs include transportation costs and the opportunity cost of lost labor days.

- *Poor physical infrastructure* : Poorly developed roads or other public infrastructure can limit a borrower's access to financial institutions in urban centers or increase the time it takes to get there, thereby increasing transaction costs.

- *Social exclusion* : Ethnic, caste and gender divisions might be more pronounced in rural areas. Hill tribes in India, for example, are relegated to working on low productivity land thereby diminishing their attractiveness to lenders.

The impact of these constraints on lenders are increased operating costs, greater information costs (owing to the heterogeneity among communities and farms) and higher real and perceived risks.

The high transaction costs and risk associated with agricultural production prevent financial institutions from playing a more active a role in the rural context. In urban areas, financial institutions are the primary suppliers of agricultural finance. However, in rural areas, agribusiness enterprises are the primary suppliers of finance.

What macro-level constraints do financial institutions face in supplying rural and agricultural finance?

- *Lack of diversification of the rural economy and covariant risk* : Covariant risk arises when many farms or households in one area are adversely affected by a single phenomenon such as drought, flood, epidemic, unexpected changes in world prices, macroeconomic crisis or civil conflict. This is distinct from individual risks, which randomly affect individual households. Individual risks relate more to individual illness, predation on livestock, old age, job loss, crime, etc.
- *Policy and regulatory constraints* : An example of a policy and regulatory constraint that limits the recourse a lender has in the case of non-payment is a prohibition against repossessing real property from the poor. This is primarily a supply-side constraint.
- *Political meddling and crowding out* : Politicians often use rural development programs to garner political support. Often accomplished through subsidized credit programs, this political meddling results in market distortions and crowds out potential financial institutions in the area.
- The impacts of these constraints on lenders are higher operating costs, increased information costs (owing to the heterogeneity among communities and farms), and higher real and perceived risks.

How can technology be used to serve rural clients?

- Various technologies show promise for lowering the costs, managing the risks and increasing the efficiency of financial services in rural areas, including automated teller machines (ATMs), point-of-sale (POS) devices linked to smart cards, cellular phones, GIS mapping systems, and loan officers using personal digital assistants (PDAs).
- ATMs, smart cards, and debit cards can provide flexible payment options and more convenient access to client accounts. They can

also reduce branch infrastructure and employee costs, and facilitate financial services in areas with poor communications and electricity supplies.

- PDAs can streamline the work of loan officers and speed decision making—as long as the financial institution's loan analysis and client monitoring systems are sufficiently developed. The value of fast, field-level decisions can occasionally be enhanced by incorporating credit scoring into PDA systems.
- Cellular phones can be used to check loan balances and repayment schedules. They have the potential to be used for stored-value transactions and to facilitate remittance transfers and payments if linked to point-of-sale devices and other payment points. Cellular networks may also be used for low-cost deposits and withdrawals if they are linked to local merchants and agents.

How does rural and agricultural finance assist the rural poor?

Agriculture is still a leading economic sector, exporter, and source of livelihood in many developing countries, especially for the poor and for women. Rural livelihoods are often closely linked to agriculture. Financial services for the rural poor offer the same benefits of risk mitigation and income-smoothing as they do for urban and peri-urban microenterprises. Rural finance can also assist households in maintaining food security. Financial savings enable rural households to diversify and complement in-kind savings, while payment services enable them to cost-effectively send and receive money.

What roles do remittances and money transfers play in consumption smoothing and asset accumulation in rural areas?

- Income from national or international remittances is important for most developing economies, and it is disproportionately important for many rural areas where it may be the principal income source. Remittance monies can make significant contributions to consumption smoothing in poor households. Efficient mechanisms for money transfers are in large demand by the rural poor.
- MFIs engaged in remittances to and from rural areas include the National Microfinance Bank of Tanzania, Uganda Microfinance Union for domestic transfers, Equity Bank in Kenya and

Centenary Bank in Uganda for domestic transfers using Western Union as subagent, uBank for mineworkers in South Africa making transfers domestically and to Botswana, Lesotho, Mozambique, and Swaziland, Fonkoze in Haiti, Banco Solidario in Ecuador, PRODEM in Bolivia, and the ProCredit Bank in Kosovo

The Reserve Bank India had constituted an Advisory Committee on flow of credit to agriculture and related activities from the banking system under the chairmanship of noted economist, Prof. V.S. Vyas.

The Terms of Reference of the Committee were as under :

- To assess the progress made in implementation of the recommendations of the Expert Committee on Rural Credit (Vyas Committee) appointed by NABARD in August 2000.
- To suggest measures to reduce the rate of interest on agriculture credit given by Commercial, Cooperative and Regional Rural Banks;
- To examine the role of NABARD as the apex institution for providing and regulating credit for the promotion and development of agriculture and the role of Regional Rural Banks (RRBs) in purveying agricultural credit and suggest measures for improving the same without sacrificing overall viability considerations;
- To study the role and effectiveness of the RIDF mechanism and suggest ways to improve the same, or to suggest alternatives, with a view to increase direct agriculture lending;
- To identify the impediments in the flow of credit to the disadvantaged sections such as small and marginal farmers, tenant farmers, oral lessees and landless laborers and suggest measures to be taken by banks for providing financial assistance to them;
- To suggest short-term and medium-term measures to improve the flow of credit to agriculture, with particular emphasis on direct financing of farmers based on linkages for supply of inputs and sale of outputs and institutional and procedural arrangements required therefore;

- Scope for involving innovative location-specific catalytic agents to bridge the gap between the demand and supply of timely credit in rural areas;
- The problems faced by banks in extending their outreach; the need to modify the Service Area Approach; and feasibility for harnessing new technological developments in smoothening the process of credit delivery to the rural and agricultural sector.
- To study the role of micro finance in poverty alleviation and adoption of the Self Help Group (SHG) approach in extending banks' outreach to the disadvantaged sectors.
- To examine the need to regulate micro finance institutions and to suggest appropriate regulatory model.

To examine the norms relating to NPAs in cases of crop failure where seasonality and uncertainty are not captured.

Micro Finance : Issues & Recommendations

- Micro Credit/finance is defined as provision of thrift, credit and other financial services and products of very small amount to the poor in rural, semi-urban and urban areas for enabling them to raise their income levels and improve living standards. The loans and credit extended are typically small ("micro"). They are provided in varying contexts, either to individuals or groups, ranging from personal micro-credit, to small enterprise support and rural finance.

The institutions that provide micro-finance and credit services are diverse, including non-governmental organizations (NGOs), credit unions, non-bank financial intermediaries, and commercial banks.

Reserve Bank had in October 2002 constituted four informal groups to further stimulate the process of constructive dialogue amongst the major players in the area of micro finance. The four groups studied issues relating to (i) Structure and Sustainability; (ii) Funding; (iii) Regulations; and (iv) Capacity Building. The recommendations of these groups are now being circulated by RBI for greater public debate.

Recommendations

1. The group on structure and sustainability has recommended creation of an autonomous and professionally-managed National Micro Finance Equity Fund with an initial subscription of Rs 200

crore. Contributions made by banks to the fund are be treated as weaker section lending under the priority sector. The corpus of the fund is to be raised to Rs 500 crore over two to three years.

2. The group on funding issues has suggested for the creation of a separate category of non-banking finance companies (NBFCs) for attending to microfinance business with entry capital requirement of Rs 25 lakh. The amount of deposits mobilized by any such MFI should not exceed Rs 5,000 per depositor and all such deposits may be covered by the guarantee of Deposit Insurance Credit Guarantee Corporation.

3. The group on capacity building said the RBI should facilitate establishing micro-finance funds for capacity building. Besides the subsidy funds of Swarnajayanti Gram Swarozgar Yojana (SGSY), funds shall be mobilized from Rural Infrastructure Development Fund (RIDF), National Bank for Agriculture and Rural Development (NABARD) and also part of the profits of commercial banks.

   The group has recommended that the RBI should constitute a permanent working group on microfinance constituting members from formal financial institutions, government, apex development banks, non-government organisations (NGOs) and MFIs to monitor and review the progress on allocation of resources and undertaking of capacity building initiatives.

   The RBI shall facilitate setting up of business incubation fund through SGSY programme for providing venture capital support.

4. According to the group on regulatory issues, NGOs/ federation of self help groups (SHGs) which have already undertaken onward lending to transform themselves into mutually aided co-operative Societies /NBFCs within two years if they intend to continue mobilisation of savings and lending activity.

   If the NGOs intend to transform themselves into Section 25 companies, they should prepay the deposits collected and continue the lending activity only. RBI could permit 2-3 years as a transitional period to convert into either MACS or NBFCs.

   The size of the loan to individual members of SHGs should not exceed Rs 50,000. The interest charged by NGO to be made public and method of fixation be transparent, it could cover the capacity building and management costs as well. Managing agriculture knowledge : role of information and communication technology

The emergence of Information and Communication Technologies (ICT) in the last decade has opened new avenues in knowledge management that could play important roles in meeting the prevailing challenges related to sharing, exchanging and disseminating knowledge and technologies. ICT allows capitalizing to a greater extent on the wealth of information and knowledge available for Agriculture Knowledge, Science and Technology (AKST). The ultimate objectives of AKST activities are to come up with results that can advance research more in certain areas, and engender technologies that AKST stakeholders can use to increase production, conserve the environment, etc. The following paragraphs will discuss the knowledge management challenges, explain how ICT could play a role in addressing them, and highlight an example of an institution in the Egyptian Agriculture Research Center whose main function is to conduct applied research in ICT in Agriculture.

Challenges related to sharing, exchanging and disseminating knowledge and technologies

The first challenge is the poor mechanisms and infrastructure for *sharing and exchanging* agriculture knowledge generated from research at national and regional levels. Many research activities are repeated due to the lack of such mechanisms and infrastructure at the national level. Researchers can find research papers published in international journals and conferences more easily than finding research papers published nationally in local journals, conferences, theses and technical reports. The second challenge is the inefficient mechanisms and infrastructure for transferring technologies produced as the result of research to growers either directly or through intermediaries (extension subsystem). Knowledge and technologies fostering agricultural production and environment conservation are examples. Although many extension documents are produced by national agriculture research and extension systems to inform growers about the latest recommendations concerning different agricultural practices, these documents are not disseminated, updated or managed to respond to the needs of extension workers, advisers and farmers. This is also true for technical reports, books and research papers related to production. The third challenge is keeping the indigenous knowledge as a heritage for new generations. It is available through experienced growers and specialists in different commodities. These inherited agricultural practices are rarely documented, but they embody a wealth of knowledge that researchers need to examine thoroughly. The forth challenge is easily accessing and availing economic and social knowledge to different stakeholders at operational, management and decision-making levels, so that those responsible will be able to make appropriate decisions regarding the profit making of certain technologies and their effect on resource-poor farmers.

ICT Role in Agriculture Knowledge Management

Knowledge sharing, exchanging and dissemination are elements in a broader theme which is knowledge management. The central purpose of knowledge management is to transform information and intellectual assets into enduring value (Metcalfe, 2005). The basic idea is to strengthen, improve and propel the organization by using the wealth of information and knowledge that the organization and its members collectively possess (Milton, 2003). It has been pointed out that a large part of knowledge is not explicit but tacit (Schreiber et al., 1999). This is true for knowledge in agriculture where a lot of good practices are transferred without being well documented in books, papers or extension documents. To manage the knowledge properly, ICT is needed. In effect, there are many information technologies that can be used for knowledge management. The following paragraphs describe these technologies and emphasize their roles in agriculture knowledge management.

Content *management system in its wider sense including data bases and multimedia, is* the core technology of information and knowledge management. This technology can be used in different applications :

- Building a national agriculture research information system (NARIS) needs to include research outcomes, projects, institutions and researchers in every country, and a regional research information system that works as a portal for all the NARIS. An example NARIS has been developed at the agriculture research center in Egypt (ARC, 2007).
- Developing an information system of indigenous agricultural practices can enable researchers to examine this knowledge and decide on its usefulness for sustainable development. Such a system will also keep this knowledge for future generations before it disappears as a result of advanced technologies.

- Developing an information system recording matured technologies that on a trial basis have proven successful and success stories that have achieved economic growth will strengthen the interaction between inventors and innovators. This will lead to an innovation-driven economic growth paradigm.
- Storing and retrieving images, videotapes and audiotapes related to different agricultural activities.

Geographic information systems (GIS) are needed to store databases about natural resources with a graphical user interface that enables users to access these data easily using geographical maps.

Decision support system techniques are needed in many applications:

- Simulating and modeling methods can be used to build computer systems that can model and simulate the effect of different agricultural production policies on the economy and the environment to help top management make decisions.
- Using expert systems technology to improve crop management and track its effect on conserving natural resources is elaborated in Rafea (1999). This technology may also be appropriate for keeping indigenous knowledge (Rafea, 1995, 1998, 2000). Expediting the expert systems development by generating agriculture specific tools to overcome the well known problem of knowledge acquisition is addresses in Abdul-Hadi et. al. (2006). Enhancing the explanation capabilities of expert systems developed for agriculture was also investigated by Said et. al. (2009).

Modern ICT—Internet and Web technology—is needed to make these systems available regionally and globally. Accessing the Internet will bring a wealth of information to all agriculture stakeholders in rural and urban areas and will help in overcoming the digital divide. As most farmers in CWANA have no hands-on experience or access to digital networks, leaders of national agricultural research and extension systems should be encouraged to consider the ICT option. Training farmers and extension workers, including women, in ICT will help them access a lot of useful information if each country tries to develop contents in the language people are using.

Intellectual Property

Intellectual property (IP) is a term referring to a number of distinct types of creations of the mind for which a set of exclusive rights are recognized—and the corresponding fields of law.[1] Under intellectual property law, owners are granted certain exclusive rights to a variety of intangible assets, such as musical, literary, and artistic works; discoveries and inventions; and words, phrases, symbols, and designs. Common types of intellectual property rights include copyrights, trademarks, patents, industrial design rights and trade secrets in some jurisdictions.

Although many of the legal principles governing intellectual property have evolved over centuries, it was not until the 19th century that the term *intellectual property* began to be used, and not until the late 20th century that it became commonplace in the majority of the world. The British Statute of Anne 1710 and the Statute of Monopolies 1623 are now seen as the origins of copyright and patent law respectively.

Humanity has always relied on technological know-how derived from intellectual performance. The incentive for development may be viewed as an inherent characteristic of human nature. Where economic and financial interests interfere, this spirit for innovation is substantially conditioned. Where society is based on economic structures, financial success determines the main interest for the development of new technologies; the high costs of development, usually inherent to inventive success, make this inevitable. Where this is not valid for the inventor, it undoubtedly is for his promoters.

It is implied by this scenario that existing know-how is a valuable asset. Where there is no means of protection for innovating technology, it is exposed to exploitation by any third party. There is no incentive to invest funds and efforts for new technologies where there is the opportunity to exploit results elsewhere available. High costs of development are eventually deemed unnecessary where low-cost reproduction leads to the same results. This represents a substantial deterrent for research and development in a market economy. This is why Intellectual Property (IP) has been created as a means of protection and reward for inventiveness.

On the other hand, there are public use interests that should be considered in developing IP legislation. Innovation is essentially a public good. It is a key task for authorities establishing means of protection for IP, to find a sustainable balance between these aspects.

This module covers the following aspects :

- The definition of different types of intellectual property;
- Positive and negative aspects of IPR in relation to trade;
- The relationship between IPR and plant genetic resources; and
- Different perspectives of North and South in relation to protecting plant genetic resources

Means of Protection and Categories of IPR

- On the basis of the concepts set out above, IPR have been developed in several countries and in manifold forms. Rights are accredited to persons over the creations of their minds. IPR give the creator an exclusive right over the use of her/his creation for a certain period of time within a certain territory, normally that of the State recognizing the protection.
- Intellectual property legislation customarily recognizes two main categories of protection. One category of rights consists of ***copyright and rights related thereto***. These are mainly concerned with the rights over the creations of people's minds, over the expressions of their thoughts (i.e. the rights of authors over their literary and artistic works, the rights of programmers over their computer programmes, the rights of composers over their musical compositions, etc.).
- The development of new methods of propagation for intellectual output has led to the necessity to extend protection as to the subject-matter in question. Copyright and related rights are therefore likewise extended to producers of phonograms and broadcasting organizations and any similar products. In any case, the main social purpose of protection of copyright and related rights is to encourage and reward creative work and to assure its equitable and sustainable exploitation.

Industrial property including trademarks and geographical indications

The other category of IPR, namely ***industrial property,*** can usefully be divided into two main areas. The first area can be characterized as the protection of distinctive signs, in particular trademarks and geographical indications. The protection of such distinctive signs aims to stimulate and ensure fair competition and to protect consumers, by enabling them to make informed choices between various goods and services. Within a constantly expanding market where consumers are submitted to a broad quantity of different products, distinctive signs ensure distinctness of products.

A second area of industrial property is protected primarily to stimulate innovation, design and the creation of technology. Inventions (protected by patents), industrial designs and trade secrets fall into this category. The social purpose is to provide protection for the results of investment in the development of new technology, thus giving the incentive and means to finance research and development activities. Such intellectual property regimes also facilitate the transfer of technology in the form of foreign direct investment, joint ventures or licensing.

While the basic social objectives of intellectual property rights remain those outlined above, it should also be noted that the exclusive rights granted are generally subject to a number of limitations and exceptions, aimed at fine-tuning the balance that has to be found between the legitimate interests of right holders and users.

IPR and their Relation with Trade

International IPR policy coordination is likely to imply a change in global trade relations. The demand for general legal protection against unfair imitation of creations, inventions, marks and names has to be understood as a product of the commercial revolution. That demand has swollen immensely with the development of modern advertising and large-scale retailing.

Protection against counterfeit and pirated copies

- An adequate protection of IPR reduces trade distortions and barriers to international trade, in particular trade in counterfeit goods and pirated copyright goods. IPR give the framework for a free and fair competition. They also protect consumers since it is thanks to IPR that consumers are able to make informed choices between the various goods and services available.
- IPR support creativity by being the means through which imagination, ability and inventiveness are rewarded. IPR promote

the use of creations, inventions, expressions of ideas and distinctive signs to create wealth, through their application in trade, industry and culture development.

Encourages exports and technology transfer

- IPR favour exports and facilitate technology development and transfer since they encourage investment and at the international level they enforce the position of exporting industries.
- Measures and procedures to enforce IPR however should not themselves constitute barriers to legitimate trade through practices of unfair competition, restriction of trade and international transfer of technology, etc.

Positive implications

Provides incentive for private investment in innovation

- Knowledge that can be classified as innovative typically implies high research and development costs. The reason why industries undergo such investments is the expectation of a technological advantage in relation to other market competitors and, as a result thereof, higher profits.
- Given the ease of reproducing IP-protected goods, this objective might be severely endangered. If investors are not given a means to protect their results and recover their costs, the development of new technologies will be retarded.
- IPR are considered a reliable mechanism for recovering investment. The inventor/investor is thereby accredited a "monopoly-like" position as to the invention, permitting exclusive exploitation, i.e. precluding others from it. This results in the establishment of a safe market position, assuring the yield to its promoter and creating an incentive for private investment in innovative technology.
- IPR are a matter of national jurisdiction. A literary or artistic work, or an invention, is protected only in the country where the relevant IPR has been granted. If a company is doing business in many countries, it is compelled to apply for protection in all those countries. If this is not the case, each transaction, which crosses borders, bears the risk of counterfeiting or piracy in absence of remedies. This scenario represents a substantial

deterrent for any international trade involving innovative knowledge, eventually impeding a sustainable technology transfer between States.

Negative implications

Notwithstanding the positive connotations of IPR, some negative effects are inevitable. Particularly governments of developing countries express doubts about the introduction of a strong international IPR regime in the framework of negotiations on mutual IPR policy.

Raises costs of technology acquisition to developing countries

- Developing countries frequently base their economy and scientific research on the employment of foreign basic technology imported from industrialized countries. Both areas may be negatively affected by strong IP protection. The granting of the respective rights allows the inventor to receive high economic rents for the use of the invention. This results in higher end-product prices and leakages from the domestic economy in the form of royalty payments, license fees, etc.
- Likewise, subsequent innovation will be deterred by unaffordable foreign basic technology. Companies holding IPR will see their capital value rise and will be in a position to buy out competing firms to consolidate their market positions[1]. Industrial innovation eventually becomes a developed country monopoly countervailing the purpose of technology transfer, i.e. the widespread diffusion thereof.

Damages basic scientific research

Other problems may be encountered where scientific institutions (e.g. universities) are confronted with IPR regimes. In many countries, any advanced disclosure of an invention, e.g. by publication, precludes patentability and thus any means of protection. Strong IPR thus creates an incentive for confidential handling of innovative knowledge and the development in several scientific fields would be adversely affected by such a policy. Knowledge share or exchange is of substantial importance for technological innovation in several fields of public relevance (e.g. pharmaceuticals, medicine, surgery, agriculture etc.). It becomes therefore a substantial task for IPR policy to avoid such side-effects.

Distorts technological choices

The introduction of strong IPR may also result in a focusing of efforts on only economically valuable knowledge. This would eventually result in neglecting several technological fields, causing a "market-niche development"

Need for detailed case studies on IPR impact on growth

It is difficult to make predictions as to the nature of IPR-related implications. Defining the role of IPR or predicting any retarding or driving effect thereof on international trade relations or technological development is extremely difficult. IPR regimes are one of numerous circumstances varying over time and across countries correlated to economic structures. The impacts will vary where countries with different levels of development or diverging economic structures are subject to mutual IPR policy. Hence, there is a considerable need for specific national case studies on how various types of IPR systems affect growth.

Whether private incentives would be sufficient to generate the optimal flow of additions to the stock of scientific and technological knowledge must be considered. Practice shows that private contracts for technology transfer often succeed in sustainable knowledge exploitation coupled with sufficient public availability of innovative information. However, in the absence of special provisions, any result remains unpredictable and substantially subject to the equitable trade-related conduct of concerned parties. The abuse of strong market positions remains an inherent risk. Public intervention thus remains necessary.

The aforementioned aspects are of relevance in developing international and domestic IPR regimes.

IPR and their Relation with Plant Genetic Resources

At first sight, it might appear difficult to make a connection between IPR and plant genetic resources (PGR). However, the importance of biomaterials for industrial purposes is expanding significantly. The constantly expanding domains of the pharmaceutical, biotechnological and agricultural industries, establish a direct link between economic and environmental issues.

Possibility of gene typing made patenting possible

Genetic resources were originally recognized as the "common heritage" of mankind committing the global community to its preservation and sustainable exploitation and excluding living material from the concept of private property. This ideal reflected the impossibility of genetically defining plant and animal species, thus eliminating any possibility of specification, a prerequisite for protection under IPR. Developments in genetic science and technology now make possible the determination of single species, opening the doors for IP related to living matter. The ethical aspects of this issue are not discussed in this framework. However, the environmental impacts should not be neglected.

Advantages of patenting difficult to realize in gene-rich countries

Biological and genetic resources, including plant germplasm, clearly have an economic value. Genetic resources comprise a vast reservoir of hereditary characteristics that have actual or potential economic use. The employment of IPR in relation to living matter may create a direct financial incentive for countries rich in genetic resources to protect, preserve and control the exploitation thereof, where the beneficiaries would be mainly developing countries. Although positive, this ideal is deemed difficult to achieve. A precise attribution of rights as to specific genetic material could result in complicated disputes between countries, especially where the origin of plant species or genetic material thereof cannot be defined within politically settled geographical areas (an issue peculiar to most plant species). Further complications arise searching for an unambiguous right holder, since genetic variability was mainly bestowed by community efforts or the interaction between populations. The granting of IPR related to genetic material implies the right to exclude others from its use, which may eventually act as an active deterrent to the exchange of genetic material within the global genetic pool, leading to unpredictable environmental consequences.

While drawbacks in terms of loss of genetic diversity are evident

Industries relying on PGR undertake expensive research and development efforts to create technologies that allow the effective utilization of natural species' genes and to market the improved products that result. The same industries will thus make intensive use of any IPR regime in force to recover their investments. Applying IPR to plant genetic material could therefore result in constraining farmers to use specific, mostly genetically engineered seeds over longer periods. In this way, breeding alternatives for local and indigenous farmers and communities will be substantially limited to genetic material of artificial origin. This eventually results in increasing genetic uniformity in crops or other extremely commercialized plant species. The latter are alarmingly vulnerable to pests and diseases and genetic variability is important to ensure continued survival. Strong IPR systems eventually promote genetic erosion thus bearing fatal implications for the genetic pool. It is necessary to ensure sustainable exploitation and conservation of biological diversity when designing an IPR system.

IPR and the North-South Debate Over Plant Genetic Resources

- Due to the widespread diffusion of biotechnology within the different industrial domains, the industrialized world has massively increased its demands for IP protection of living organisms.
- In fact, the genes of living organisms, i.e. PGR, are the basic "raw material" of the new biotechnologies. The **'Gene Rush'** has thus become a new version of the old 'Gold Rush' in the scramble for future profits (Khor, n.d.).
- According to the World Resources Institute, more than half of the world's plant and animal species live in the rainforests of the Third World and nowhere else (Kimbrell, 1997). The primary holders of this resource are thus found in the southern hemisphere, a situation with a massive potential for disputes between the gene-poor but technology-rich global north and the gene-rich but technology-poor global south.

Conflicts between indigenous resources and biotech companies

- Many companies have been granted patents on domestically bred plants in developing countries and often also on indigenous knowledge about their use. This is occurring in many parts of the developing world, where the folk wisdom of indigenous peoples is used to locate and understand the use of medicinal plants and then exploit them commercially by systematically patenting the results. IPR are being claimed on plant species long used by Africans, pain killers used by Chinese, Andean crop species and traditional rice varieties nurtured by Indian farmers.

Developing countries call that kind of patenting "**biopiracy**" (GRAIN, n.d.). Under their domestic legislation industrial breeders can get the commercial gains of exploited PGR without necessarily sharing them with the indigenous communities of the countries from which the genetic material originates.

- Farmers and indigenous communities are protesting against IPR being granted on plants used for nutritional, medicinal or other functions, identified and developed by them. These communities also face the threat that in future they might become totally dependent on industrial supplies for vital inputs such as seeds.

Pros and cons of possible solutions to this conflict

IPR systems allow gene technology to be rewarded, but there is no equivalent legal system whereby the indigenous-related knowledge be rewarded. Several suggestions have been made by different proponents in an attempt to alleviate this controversy (Greaves, 1994). One approach is to protect the rights of indigenous farmer-breeders using industrial world IPR concepts and methods such as contracts, trademarks and patents. This faces great complications due to recurrent genetic likelihood in different plant species. In fact, because of the widespread seed sharing between neighboring peoples, there may be little or no genetic differences between folk varieties of some plant species. The legal complication increases from the fact that the contributions are often made by entire communities and therefore cannot be attributed to distinct groups or even individuals. If a particular gene, critical to the success of a new variety, has origins in different areas, it is not possible to determine priority in recognition and reward. This could foster competition between communities in a region, raising potential for economy-based ethnic conflicts. Thus, current IPR mechanisms for indigenous peoples seem to be inherently unsuitable. Different procedures should be developed to recognize and reward community contributions to genetic resources.

A second approach is to balance private plant breeder's rights with farmers' rights defined as a group. For example, a fund could be created to be used to finance national or regional level genetic conservation and utilization programmes, particularly, but not exclusively, in developing countries, possibly involving some sort of technology transfer in exchange of the "raw materials" supplied.

A further suggestion is the antithesis to the first two in that it opposes control of crop genetic resources for profit, and aims to keep them freely available for everyone's use. This suggestion meets vigorous opposition from industrialized governments as conflicting with notions of private property, undermining economic interests of western industries.

An additional approach removes folk varieties from the same playing field as plant breeder's varieties and attempts to force the global community to recognize indigenous people's values as equal to those of the dominant society. This is deemed feasible by defining protection of natural, cultural and intellectual property as part of the overall protection of indigenous people's human rights. This implies protecting local communities' IP by asserting collective ownership of their land base establishing sovereign rights to manage their own traditional knowledge and resources. The sharing of these knowledge and resources with the rest of the world should therefore be on the condition that their rights to define and control their knowledge are internationally recognized and protected.

A last approach is that of bilateral agreements as a pathway for sharing of benefits. This would offer a case by case solution thus optimizing the management of the specific features in each situation. Notwithstanding this advantage, in the absence of legally determined conditions, the abuse of strong market positions would remain an inherent risk.

TRIPS Agreement leaves resolution of these conflicts open

The TRIPS Agreement will have the most decisive influence over national laws' approach to these questions. Still, the Agreement has an ambiguous language in the relevant clause on living organisms (Article 27.3). As intended by its architects, this provides a broad field of action. The outcome of the implementation and review processes will thus be of crucial importance for the recognition and solution of such disputes.

IPR protects Indian agriculture Liberty Institute Newsletter March 1999, India Monday, March 01, 1999

The consequences of IPR (Intellectual Property Rights) in agriculture, and in areas where knowledge is said to have been traditionally handed down the generations is a well debated topic.

However, a proper IPR regime can help protect not only the foreign inventors, but also the domestic players.

A lot has been said over the consequences of IPR (Intellectual Property Rights) in agriculture, and in areas where knowledge is said to have been traditionally handed down the generations. However, a proper IPR regime can help protect not only the foreign inventors, but also the domestic players. Following is a summary of the present status of IPR in some of the most prominent instances.

Turmeric

In 1993, two American scientists of Indian origin filed a patent for use of turmeric to heal wounds. Initially the specification contained information on the prior knowledge of the various uses of turmeric and it also acknowledged that the pharmaco-kinetics involving the safety toxicity dosage and biological properties of turmeric are well known.

It also said the main object of the invention is the use of turmeric powder at the site of injury by topical application and/or oral intake. The other claims related to optional features of the powder – oral and topical.

The US Patent Office in 1995, granted the patent claim, revesing its earlier ruling that had held the invention not novel. CSIR (Council for Scientific and Industrial Research) challenged it, providing 32 references from ancient texts to show prior knowledge on the use of turmeric as a powder. The US Patent office re-examined the claim, maintained that the invention lacked novelty and revoked the patent.

Neem

There are more than 40 patents, mainly for neem pesticides, held by laboratories in the US and Europe. A US company, W R Grace, was granted a patent for neem as a pesticide. India couldn't challenge it because the patent was for a formulation which increased the shelf life of neem extract. It was a valid patent and so protests had no effect. Indian scientists had been experimenting with this but since they had not applied for a patent for this process, they couldn't challenge it.

Incidentally, even India has granted more than 80 patents on neem, 48 of them of Indian origin. None of these patents have prevented Indians from using neem nor has anyone paid royalty.

Cotton

In October 1992, Agracetus Inc, a subsidiary of W R Grace, got a patent for rights on all forms of genetically engineered cotton, no matter what techniques or genes are used to create them. It was challenged in court and the US patent office revoked it in 1994.

Baruch S Blumberg, a US virologist, obtained a patent for a formulation of phyllanthus amarus, a plant used for centuries in south India for the treatment of jaundice. On the face of it, this does not satisfy the criterion of novelty, but Blumberg obtained the patent for viral hepatitis, not mentioned in earlier texts. So his innovation satisfied the criterion of novelty.

Basmati

In September 1997, Ricetec obtained a patent for calling aromatic rice grown outside India as basmati and selling it under any brand name. Ricetec has already been selling brands like Kasmati and Texmati claimed to be basmati type rice. With the patent, Ricetec will be able to label its exports of rice also as basmati.

Company claimed basmati is a generic term and that it had invented certain novel basmati lines and grains wh

In the Action plan for Development of Post-tsunami "New Andamans" - report submitted to Andaman and Nicobar Administration by M.S. Swaminathan Research Foundation (MSSRF), due emphasis was given to establish Rural Knowledge Centre in these Islands with the vision : to harness the power of ICT in the knowledge, skill, economic and social empowerment of rural families on the principle of 'reaching the unreached and voicing the voiceless'.

For the development of Agriculture in this Union Territory, farmers have to be motivated, guided, assisted, supported and helped to have access to the latest research findings in the areas of Agriculture, Animal Husbandry and Fisheries by making available all the information that is required in their village so that they may not waste their time, energy and meager resources in transporting the inputs required and their produce to the different locations where the information of goods that they seek is available. With this motive 'Rural Knowledge Centres' is established throughout the Islands at identified locations which will function as "Rural Knowledge Bureau" for the respective area. They will have connectivity with the computer systems to be set up in the Blocks by the IT Department as a part of IT Plan of Andaman and Nicobar Administration. The information to be available in the Centres will include various Govt. schemes, contact persons for getting services, technical information on Agriculture, Animal Husbandry, Fisheries, Industries, Information on Disaster Management, Civil Supplies, Social Welfare Pension Schemes, Health, Education, Meteorological, Shipping Services etc. Wherein the Department of Agriculture will be the 'Nodal Department' to run these Centres and these centres will be operationalized on community basis. Initially the Centers will be operated by the Department al staff during 2007-08, under the plan scheme the Department has kept provision to engage computer literate candidates to operate the computers and other equipments set up in Rural Knowledge Centres by paying them Rs.5000/- as honorarium. Two candidates will be appointed in each of the Rural Knowledge Centres and they will work in two shifts - one in the morning hours from 8 am to 2 pm and another in the afternoon from 2 pm to 9 pm which will enable the village people to visit the Rural Knowledge Centres during their free time.

For appointment of candidates, only those compute literate candidates who have enrolled their names in the Employment Exchange, local candidates of the particular area and who have been given training by the Tata Institutes of Social Science (a NGO) will be given preference at the time of selection.

Rural Knowledge Centres is to tone up the efficiency of Administration as well as implementation of various plan programs of the Department of Agriculture and allied sectors of Andaman and Nicobar Administration with the ultimate objective of delivering the benefits intended to the farmers as quickly as possible. It deals with issues like observation and prevalence of crop diseases, monitoring of natural disasters and corresponding relief measures, monitors availability of basic inputs such as seeds, fertilizers and pesticides etc. and analyzing the qualities of the soil for farmers.

The Rural Knowledge Centers is established to fulfill the following **objectives.**

- Setting up of villages Information Center that enable rural families access a basket of modern information, internet connectivity through dial up telephones and communication technologies.

Training educated youth, especially women, in rural areas in operating information center.

- Training the rural youth in the organization and maintenance of a system that generates locally relevant information from generic information.
- Maintenance, updating and dissemination of information on entitlements to rural families using an appropriate blend of modern and existing channels of communication.
- Relief measures taken by the Government during disaster time.
- Conducting impact assessment based on organization of surveys, participatory rural appraisal, and other appropriate methods of data gathering.
- Building of a model in information dissemination and exchange in rural areas that uses advanced information and communication technologies.
- Value addition to the raw information, use of various local languages by the people of different Islands.
- Conducting research on formation of multi-sectoral partnerships (private-public/government NGOs) with rural communities to form a sustainable model of ICTs for rural areas.

Benefits to the Farmers

- Availability of Agriculture resources from the Administration.
- Farmers even in the remotest villages will be having access to the computers in the Department of Agriculture and thereby the database. Thus, there is a better interface between the farmer and Agricultural office in areas like knowledge about status of soil conditions, analysis of the farmer's land etc.
- Ground level information for the higher officials like land pattern, land holdings, crop pattern, input usage for better decision making.
- The details of various Government schemes are made available to the farmers from the computerized database in the Department of Agriculture.
- Relief measures taken by the Government during disaster time.

- Details regarding various relief measures to sustain Agriculture activity during times of natural calamities, crop diseases etc. can be accessed by the farmer.

Establishment of Rural Knowledge Centre

Action taken so far

- Basic infrastructures like furniture TVs, DVDs, Public Address System, Display boards, Notice boards and also the requisite stationary items have been procured and dispatched to the respective RKCs.
- The information in the mode of books, booklets, leaflets, brochure, charts, posters, etc. were also provided. The procurement of computer systems is in final stage expected to reach within a week time. Telephone connection is also expected within this week in all the
- Rural Knowledge Centre Once telephone is connected and computer system is installed, immediate action will initiate to get internet connectivity, wherever possible broadband or else dial up connectivity will be established.
- Hub center for the Rural Knowledge Centre will be established in the Directorate of Agriculture, Haddo, Port Blair.
- Computer literate personal for handling the electronic items in all the Rural Knowledge Centre have to be engaged shortly for the smooth function.

Village Knowledge Centre (VKC)

What is VKC?

Village Knowledge Cantres serve as information dissemination centre providing instant access to farmers to latest information/ knowledge available in the field of agriculture, starting from crop production to marketing. Every VKC is manned by a "VKC In-charge" who looks after the operations of the VKC.

It's Importance

Development in a country as large and diverse as India poses special challenges. As India embarks on a trajectory of high rates of economic growth, it is imperative to focus on the impact of such growth on society. Economic growth will translate to sustainable development only when it is inclusive, creating capacity for better lifestyle among those living at the bottom of the pyramid - majority of whom are residing in the 6 lacs plus villages. Our Bank has taken various initiatives to promote inclusive growth and achieving sustainable development as its contribution towards nation building. VKC is one of the many pioneering initiatives taken by our Bank that aims at impacting rural livelihoods to build resourceful and progressive villages.

Mission

The mission of the VKCs is to harness the power of information and communication technologies in the knowledge, skill, economic and social empowerment of rural families.

Objectives

- To inculcate reading habits among the children through our library in VKC.
- To make the children familiar with computers.
- To connect the villagers to the external world through newspapers, magazines, computers, transistor and television that are set up in our VKC.

This project has been taken up in lines with the Village Knowledge Cantres established by MS Swaminathan Research Foundation. As a part of VKCs, two village libraries have been established and running successfully in Raila and Dhandar villages of Pilani chapter through this project. Sankalp, an organization from US University is funding for two more libraries and for upgrading one library to a full fledged VKC. These are planned to up nearby Pilani, Hyderabad and at Warangal.

The System of Rice Intensification (SRI) is a method of increasing the yield of rice produced in farming. It was developed in 1983 by the French Jesuit Father Henri de Laulanie in Madagascar. However full testing of the system did not occur until some years later. The productivity of SRI is under debate between supporters and critics of the system.

History and main ideas of SRI

- Assembly of the practices that culminated in SRI began in the 1960s based on **Fr. de Lausanne's** observation of **'positive-deviant'** farmer practices, starting with planting single seedlings instead of multiple seedlings in a clump, and not keeping irrigated paddy fields flooded during the rice plants' vegetative growth stage. Planting with wider spacing in a square pattern, rather than randomly or in rows, followed, as did controlling weed growth by use of a soil-aerating push-weedler (rotating hoe).

- In 1983, the beneficial effect of transplanting very young seedlings, less than 15 days old, was discovered serendipitously. Subsequently, when fertilizer prices increased, compost made from any decomposed biomass turned out to give even better results than chemical fertilizer. SRI concepts and practices continue to evolve as they are being adapted to rain-fed (unirrigated) conditions and with transplanting being superseded by direct-seeding sometimes.

The synthesis of SRI has proceeded empirically, but the central principles for getting best results are :

- Rice field soils should be kept moist rather than continuously saturated, minimizing anaerobic conditions, as this improves root growth and supports the growth and diversity of aerobic soil organisms;

- Rice plants should be planted singly and spaced optimally widely to permit more growth of roots and canopy and to keep all leaves photo synthetically active; and

- Rice seedlings should be transplanted when young, less than 15 days old with just two leaves, quickly, shallow and carefully, to avoid trauma to roots and to minimize transplant shock.

These changes from conventional practice when managing plants, soil, water and nutrients induce more productive phenotypes from any rice genotype, although some varieties respond better than others to SRI methods. Increased yield is achieved with 80-90% reductions in seed requirements (lower plant population) and 25-50% less irrigation water. Chemical fertilizer and agrochemical crop protection can be used, but best results can be attained without use of purchased inputs.

Rotary weeder

Weed growth will be higher in these fields due to the absence of standing water. But since the rice plants are planted with wider spaces between rows, a rotary weeder can be drawn in between the rows of plants to remove the weeds and also aerate the soil. On the 10 day after planting weeds may not be prominent, but its better to weed so that the weeds do not get a chance to assimilate the nutrients. Modified weeders such as the conoweeder are available. In Kerala, India the technique has been adopted by some people owing to the pressure from the Department of Agriculture and Mithraniketan (an NGO Krishi Vignan Kendra located at Trivandrum). Failures are attributed to poor water management, non use of rotary weeder and use of older seedlings.

Spread of SRI

The spread of SRI from Madagascar to around the globe has been credited to Norman Uphoff, director of the International Institute for Food, Agriculture and Development at Cornell University, Ithaca, New York from 1990 to 2005. In 1993, Uphoff met officials from Association Tefy Saina, the non-governmental organisation set up in Madagascar in 1990 by de Laulanie to promote SRI. After seeing the success of SRI for three years when Malagasy farmers previously averaging 2 tons/ hectare averaged 8 tons/hectare with SRI, Uphoff became persuaded of the merits of the system, and in 1997 started to promote SRI in Asia. As of 2007, the beneficial effects of SRI methods had been documented in 28 countries, most recently in Bhutan, Iraq, Iran and Zambia. Governments in the largest rice-producing countries (China, India and Indonesia) are now supporting SRI extension. In India, SRI concepts and practices have been extrapolated successfully to other crops such as sugar cane, finger millet and wheat. To help farmers in efficient and easy weeding, KGVK is now marketing a Single Row Weeder, This innovative Weeder, developed by the Engineers of Usha Martin's Machinery Division, in close collaboration with KGVK Agriculture professionals and Grassroots Farmers. The new Weeder is robust and well engineered with features, which make the work of Farmers relatively much easier, requiring less effort.

Cases of Success

Several cases of success are reported in Kerala, India. The program has proven successful in Nellanad Gramapanchayat in Thiruvananthapuram. Agricultural Officers including Sreekala, Sreekumari, and Geethakumari have several cases of success.

Criticism

There are criticisms of SRI. While supporters of SRI report many benefits in addition to yield increase—resistance to pests and diseases, resistance to biotic stresses like drought and storm damage, more output of polished rice (in kg) when SRI paddy (unmilled rice) is processed (bushels), less chemical pollution of soil and water resources—critics have focused on yield suggesting that claims of increase are due to "poor record keeping and unscientific thinking". Some critics have suggested that SRI success is unique to soil conditions in Madagascar, a point that is disputed by supporters.

Critics have objected that there is a lack of details on the methodology used in trials and a lack of publications in the peer-reviewed literature. This latter deficiency is now being remedied by such publications, but systematic trials that will satisfy scientific critics remain to be done. A global field trial is planned for 2009-2011, to be conducted by researchers from Cornell University, Wageningen University, and the International Rice Research Institute

Precision Farming

Precision farming or precision agriculture is a farming management concept based on observing and responding to intra-field variations. It relies on new technologies like satellite imagery, information technology, and geospatial tools. It is also aided by farmers' ability to locate their precise position in a field using satellite positioning system like GPS.

Precision agriculture issues

Precision agriculture aims to optimize field-level management with regard to :

- *Crop science* : by matching farming practices more closely to crop needs (e.g. fertilizer inputs);
- *Environmental protection* : by reducing environmental risks and footprint of farming (e.g. limiting leaching of nitrogen);
- *Economics* : by boosting competitiveness through more efficient practices (e.g. improved management of fertilizer usage and other inputs).

Precision agriculture also provides farmers with a wealth of information to :

- Build up a record of their farm;
- Improve decision-making;
- Foster greater traceability
- Enhance marketing of farm products
- Improve lease arrangements and relationship with landlords
- Enhance the inherent quality of farm products (e.g. protein level in bread-flour wheat)

Stages and tools

Precision agriculture is a four-stage process using techniques to observe spatial variability :

Relocation of data

Geolocating a field enables the farmer to overlay information gathered from analysis of soils and residual nitrogen, and information on previous crops and soil resistivity. Geolocation is done in two ways:

- The field is delineated using an in-vehicle GPS receiver as the farmer drives a * The field is delineated on a baseman derived from aerial or satellite imagery. The base images must have the right level of resolution and geometric quality to ensure that geolocation is sufficiently accurate.

Characterizing variability

Intra- and inter-field variability may result from a number of factors. These include climatic conditions (hail, drought, rain, etc.), soils (texture, depth, and nitrogen levels), cropping practices (no-till farming), weeds and disease. Permanent indicators—chiefly soil indicators—provide farmers with information about the main environmental constants. Point indicators allow them to track a crop's status, i.e., to see whether diseases are developing, if the crop is suffering from water stress, nitrogen stress, or lodging, whether it has been damaged by ice and so on. This information may come from weather stations and other sensors (soil electrical resistivity, detection with the naked eye, satellite imagery, etc.). Soil resistivity measurements combined with soil analysis make it possible to precisely map agro-pedagogical conditions.

Decision-making – two strategies for dealing with variability

Using soil maps, farmers can pursue two strategies to adjust field inputs :

- *Predictive approach* : based on analysis of static indicators (soil, resistivity, field history, etc.) during the crop cycle.
- *Control approach* : information from static indicators is regularly updated during the crop cycle by :
- *Sampling* : weighing biomass, measuring leaf chlorophyll content, weighing fruit, etc.
- *Remote sensing* : measuring parameters like temperature (air/soil), humidity (air/soil/leaf), wind or stem diameter is possible thanks to Wireless Sensor Networks
- *Proxy-detection* : in-vehicle sensors measure leaf status; this requires the farmer to drive around the entire field.
- *Aerial or satellite remote sensing* : multispectral imagery is acquired and processed to derive maps of crop biophysical parameters.

Decisions may be based on decision-support models (crop simulation models and recommendation models), but in the final analysis it is up to the farmer to decide in terms of business value and impacts on the environment.

Implementing practices to remedy variability

New information and communication technologies (NICT) make field-level crop management more operational and easier to achieve for farmers. Application of crop management decisions calls for farming equipment that supports variable-rate technology (VRT), for example varying seed density along with variable-rate application (VRA) of nitrogen and phytosanitary products.

Precision agriculture uses special equipment on board the farmer's tractor :

- Positioning system (e.g. GPS receivers that use satellite signals to precisely determine a position on the globe);
- Geographic information systems (GIS), i.e., software that makes sense of all the available data;
- Variable-rate farming equipment (seeder, spreader).

Precision agriculture around the world

The concept of precision agriculture first emerged in the *United States* in the early 1980s. In 1985, researchers at the University of Minnesota varied lime inputs in crop fields. It was also at this time that the practice of grid sampling appeared (applying a fixed grid of one sample per hectare). Towards the end of the 1980s, this technique was used to derive the first input recommendation maps for fertilizers and pH corrections. The use of yield sensors developed from new technologies, combined with the advent of GPS receivers, has been gaining ground ever since. Today, such systems cover several million hectares. In the American Midwest (US) it is associated not with sustainable agriculture but with mainstream farmers who are trying to maximize profits by spending money only in areas that require fertilizer. This practice allows the farmer to vary the rate of fertilizer across the field according to the need identified by GPS guided Grid or Zone Sampling. Fertilizer that would have been spread in areas that don't need it can be placed in areas that do, thereby optimizing its use. Around the world, precision agriculture developed at a varying pace. Precursor nations were the United States, Canada and Australia. In Europe, the United Kingdom was the first to go down this path, followed closely by France. In France, precision agriculture first appeared in 1997-1998. The development of GPS and variable-rate spreading techniques helped to anchor precision farming management practices. Today, less than 10% of France's farmers are equipped with variable-rate systems. Uptake of GPS is more widespread. But this hasn't stopped them using precision agriculture services, which supplies field-level recommendation maps. [4]

Economic and environmental impacts

Precision agriculture management practices can significantly reduce the amount of nutrient and other crop inputs used while boosting yields. Farmers thus obtain a return on their investment by saving on phytosanitary and fertilizer costs. The second, larger-scale benefit of targeting inputs—in spatial, temporal and quantitative terms—concerns environmental impacts. Applying the right amount of inputs in the right place and at the right time benefits crops, soils and groundwater, and thus the entire crop cycle. Consequently, precision agriculture has become a cornerstone of sustainable agriculture, since it respects crops, soils and farmers. Sustainable agriculture seeks to assure a continued supply of food within the ecological, economic and social limits required to sustain production in the long term. Precision agriculture therefore seeks to use high-tech systems in pursuit of this respectable and worthy goal.

Genetically Modified (GM) Foods

What are genetically modified (GM) organisms and GM foods?

- These questions and answers have been prepared by WHO in response to questions and concerns by a number of WHO Member State Governments with regard to the nature and safety of genetically modified food.
- Genetically modified organisms (GMOs) can be defined as organisms in which the genetic material (DNA) has been altered in a way that does not occur naturally. The technology is often called "modern biotechnology" or "gene technology", sometimes also "recombinant DNA technology" or "genetic engineering". It allows selected individual genes to be transferred from one organism into another, also between non-related species.
- Such methods are used to create GM plants - which are then used to grow GM food crops.

Why are GM foods produced?

- GM foods are developed - and marketed - because there is some perceived advantage either to the producer or consumer of these foods. This is meant to translate into a product with a lower price, greater benefit (in terms of durability or nutritional value) or both. Initially GM seed developers wanted their products to be accepted by producers so have concentrated on innovations that farmers (and the food industry more generally) would appreciate.
- The initial objective for developing plants based on GM organisms was to improve crop protection. The GM crops currently on the market are mainly aimed at an increased level of crop protection through the introduction of resistance against plant diseases caused by insects or viruses or through increased tolerance towards herbicides.
- Insect resistance is achieved by incorporating into the food plant the gene for toxin production from the **bacterium Bacillus thuringiensis (BT).** This toxin is currently used as a conventional

insecticide in agriculture and is safe for human consumption. GM crops that permanently produce this toxin have been shown to require lower quantities of insecticides in specific situations, e.g. where pest pressure is high.

- Virus resistance is achieved through the introduction of a gene from certain viruses which cause disease in plants. Virus resistance makes plants less susceptible to diseases caused by such viruses, resulting in higher crop yields.
- Herbicide tolerance is achieved through the introduction of a gene from a bacterium conveying resistance to some herbicides. In situations where weed pressure is high, the use of such crops has resulted in a reduction in the quantity of the herbicides used.

Are GM foods assessed differently from traditional foods?

- Generally consumers consider that traditional foods (that have often been eaten for thousands of years) are safe. When new foods are developed by natural methods, some of the existing characteristics of foods can be altered, either in a positive or a negative way National food authorities may be called upon to examine traditional foods, but this is not always the case. Indeed, new plants developed through traditional breeding techniques may not be evaluated rigorously using risk assessment techniques.
- With GM foods most national authorities consider that specific assessments are necessary. Specific systems have been set up for the rigorous evaluation of **GM organisms and GM foods** relative to both human health and the environment. Similar evaluations are generally not performed for traditional foods. Hence there is a significant difference in the evaluation process prior to marketing for these two groups of food.
- One of the objectives of the WHO Food Safety Programme is to assist national authorities in the identification of foods that should be subject to risk assessment, including GM foods, and to recommend the correct assessments.

How are the potential risks to human health determined?

The safety assessment of GM foods generally investigates :

- Direct health effects (toxicity),
- Tendencies to provoke allergic reaction (allergen city);

- Specific components thought to have nutritional or toxic properties;
- The stability of the inserted gene;
- Nutritional effects associated with genetic modification; and
- Any unintended effects which could result from the gene insertion.

What are the main issues of concern for human health?

- While theoretical discussions have covered a broad range of aspects, the three main issues debated are tendencies to provoke allergic reaction (allergen city), gene transfer and out crossing.
- Allergen city. As a matter of principle, the transfer of genes from commonly allergenic foods is discouraged unless it can be demonstrated that the protein product of the transferred gene is not allergenic. While traditionally developed foods are not generally tested for allergen city, protocols for tests for GM foods have been evaluated by the Food and Agriculture Organization of the United Nations (FAO) and WHO. No allergic effects have been found relative to GM foods currently on the market.
- Gene transfer. Gene transfer from GM foods to cells of the body or to bacteria in the gastrointestinal tract would cause concern if the transferred genetic material adversely affects human health. This would be particularly relevant if antibiotic resistance genes, used in creating GMOs, were to be transferred. Although the probability of transfer is low, the use of technology without antibiotic resistance genes has been encouraged by a recent FAO/WHO expert panel.
- Out crossing. The movement of genes from GM plants into conventional crops or related species in the wild (referred to as "out crossing"), as well as the mixing of crops derived from conventional seeds with those grown using GM crops, may have an indirect effect on food safety and food security. This risk is real, as was shown when traces of a maize type which was only approved for feed use appeared in maize products for human consumption in the United States of America. Several countries have adopted strategies to reduce mixing, including a clear separation of the fields within which GM crops and conventional crops are grown.

- Feasibility and methods for post-marketing monitoring of GM food products, for the continued surveillance of the safety of GM food products, are under discussion.

How is a risk assessment for the environment performed?

Environmental risk assessments cover both the **GMO concerned** and the potential receiving environment. The assessment process includes evaluation of the characteristics of the GMO and its effect and stability in the environment, combined with ecological characteristics of the environment in which the introduction will take place. The assessment also includes unintended effects which could result from the insertion of the new gene.

What are the issues of concern for the environment?

Issues of concern include : the capability of the GMO to escape and potentially introduce the engineered genes into wild populations; the persistence of the gene after the GMO has been harvested; the susceptibility of non-target organisms (e.g. insects which are not pests) to the gene product; the stability of the gene; the reduction in the spectrum of other plants including loss of biodiversity; and increased use of chemicals in agriculture. The environmental safety aspects of GM crops vary considerably according to local conditions.

Current investigations focus on : the potentially detrimental effect on beneficial insects or a faster induction of resistant insects; the potential generation of new plant pathogens; the potential detrimental consequences for plant biodiversity and wildlife, and a decreased use of the important practice of crop rotation in certain local situations; and the movement of herbicide resistance genes to other plants.

Are GM foods safe?

- Different GM organisms include different genes inserted in different ways. This means that individual GM foods and their safety should be assessed on a case-by-case basis and that it is not possible to make general statements on the safety of all GM foods.

- GM foods currently available on the international market have passed risk assessments and are not likely to present risks for human health. In addition, no effects on human health have been shown as a result of the consumption of such foods by the general population in the countries where they have been approved. Continuous use of risk assessments based on the Codex principles and, where appropriate, including post market monitoring, should form the basis for evaluating the safety of GM foods.

How are GM foods regulated nationally?

The way governments have regulated GM foods varies. In some countries GM foods are not yet regulated. Countries which have legislation in place focus primarily on assessment of risks for consumer health. Countries which have provisions for GM foods usually also regulate GMOs in general, taking into account health and environmental risks, as well as control- and trade-related issues (such as potential testing and labeling regimes). In view of the dynamics of the debate on GM foods, legislation is likely to continue to evolve.

What kind of GM foods are on the market internationally?

All GM crops available on the international market today have been designed using one of three basic traits : resistance to insect damage; resistance to viral infections; and tolerance towards certain herbicides. All the genes used to modify crops are derived from microorganisms.

What happens when GM foods are traded internationally?

- No specific international regulatory systems are currently in place. However, several international organizations are involved in developing protocols for GMOs.
- The Codex Alimentations Commission (Codex) is the joint FAO/ WHO body responsible for compiling the standards, codes of practice, guidelines and recommendations that constitute the Codex Alimentations : the international food code. Codex is developing principles for the human health risk analysis of GM foods. The premise of these principles dictates a premarket assessment, performed on a case-by-case basis and including an evaluation of both direct effects (from the inserted gene) and unintended effects (that may arise as a consequence of insertion of the new gene). The principles are at an advanced stage of development and are expected to be adopted in July 2003. Codex principles do not have a binding effect on national legislation, but are referred to specifically in the Sanitary and Phytosanitary

Agreement of the World Trade Organization (SPS Agreement), and can be used as a reference in case of trade disputes.

- The Cartagena Protocol on Biosafety (CPB), an environmental treaty legally binding for its Parties, regulates transboundary movements of living modified organisms (LMOs). GM foods are within the scope of the Protocol only if they contain LMOs that are capable of transferring or replicating genetic material. The cornerstone of the CPB is a requirement that exporters seek consent from importers before the first shipment of LMOs intended for release into the environment. The Protocol will enter into force 90 days after the 50th country has ratified it, which may be in early 2003 in view of the accelerated depositions registered since June 2002.

Have GM products on the international market passed a risk assessment?

The GM products that are currently on the international market have all passed risk assessments conducted by national authorities. These different assessments in general follow the same basic principles, including an assessment of environmental and human health risk. These assessments are thorough; they have not indicated any risk to human health.

Why has there been concern about GM foods among some politicians, public interest groups and consumers, especially in Europe?

- Since the first introduction on the market in the mid-1990s of a major GM food (herbicide-resistant soybeans), there has been increasing concern about such food among politicians, activists and consumers, especially in Europe. Several factors are involved.

- In the late 1980s – early 1990s, the results of decades of molecular research reached the public domain. Until that time, consumers were generally not very aware of the potential of this research. In the case of food, consumers started to wonder about safety because they perceive that modern biotechnology is leading to the creation of new species.

- Consumers frequently ask, "what is in it for me?". Where medicines are concerned, many consumers more readily accept biotechnology as beneficial for their health (e.g. medicines with improved treatment potential). In the case of the first GM foods introduced onto the European market, the products were of no apparent direct benefit to consumers (not cheaper, no increased shelf-life, no better taste). The potential for GM seeds to result in bigger yields per cultivated area should lead to lower prices. However, public attention has focused on the risk side of the risk-benefit equation.

- Consumer confidence in the safety of food supplies in Europe has decreased significantly as a result of a number of food scares that took place in the second half of the 1990s that are unrelated to GM foods. This has also had an impact on discussions about the acceptability of GM foods. Consumers have questioned the validity of risk assessments, both with regard to consumer health and environmental risks, focusing in particular on long-term effects. Other topics for debate by consumer organizations have included allergen city and antimicrobial resistance. Consumer concerns have triggered a discussion on the desirability of labeling GM foods, allowing an informed choice. At the same time, it has proved difficult to detect traces of GMOs in foods : this means that very low concentrations often cannot be detected.

How has this concern affected the marketing of GM foods in the European Union?

- The public concerns about GM food and GMOs in general have had a significant impact on the marketing of GM products in the European Union (EU). In fact, they have resulted in the so-called moratorium on approval of GM products to be placed on the market. Marketing of GM food and GMOs in general are the subject of extensive legislation. Community legislation has been in place since the early 1990s. The procedure for approval of the release of GMOs into the environment is rather complex and basically requires agreement between the Member States and the European Commission. Between 1991 and 1998, the marketing of 18 GMOs was authorized in the EU by a Commission decision.
- As of October 1998, no further authorizations have been granted and there are currently 12 applications pending. Some Member States have invoked a safeguard clause to temporarily ban the placing on the market in their country of GM maize and oilseed rape products. There are currently nine ongoing cases. Eight of

these have been examined by the Scientific Committee on Plants, which in all cases deemed that the information submitted by Member States did not justify their bans.

- During the 1990s, the regulatory framework was further extended and refined in response to the legitimate concerns of citizens, consumer organizations and economic operators (described under Question 13). A revised directive will come into force in October 2002. It will update and strengthen the existing rules concerning the process of risk assessment, risk management and decision-making with regard to the release of GMOs into the environment. The new directive also foresees mandatory monitoring of long-term effects associated with the interaction between GMOs and the environment.
- Labeling in the EU is mandatory for products derived from modern biotechnology or products containing GM organisms. Legislation also addresses the problem of accidental contamination of conventional food by GM material. It introduces a 1% minimum threshold for DNA or protein resulting from genetic modification, below which labeling is not required.
- In 2001, the European Commission adopted two new legislative proposals on GMOs concerning traceability, reinforcing current labeling rules and streamlining the authorization procedure for GMOs in food and feed and for their deliberate release into the environment.
- The European Commission is of the opinion that these new proposals, building on existing legislation, aim to address the concerns of Member States and to build consumer confidence in the authorization of GM products. The Commission expects that adoption of these proposals will pave the way for resuming the authorization of new GM products in the EU.

What is the state of public debate on GM foods in other regions of the world?

- The release of GMOs into the environment and the marketing of GM foods have resulted in a public debate in many parts of the world. This debate is likely to continue, probably in the broader context of other uses of biotechnology (e.g. in human medicine) and their consequences for human societies. Even though the issues under debate are usually very similar (costs and benefits, safety issues), the outcome of the debate differs from country to country. On issues such as labeling and traceability of GM foods as a way to address consumer concerns, there is no consensus to date. This has become apparent during discussions within the Codex Alimentations Commission over the past few years. Despite the lack of consensus on these topics, significant progress has been made on the harmonization of views concerning risk assessment. The Codex Alimentations Commission is about to adopt principles on premarket risk assessment, and the provisions of the Cartagena Protocol on Biosafety also reveal a growing understanding at the international level.

- Most recently, the humanitarian crisis in southern Africa has drawn attention to the use of GM food as food aid in emergency situations. A number of governments in the region raised concerns relating to environmental and food safety fears. Although workable solutions have been found for distribution of milled grain in some countries, others have restricted the use of GM food aid and obtained commodities which do not contain GMOs.

Are people's reactions related to the different attitudes to food in various regions of the world?

Depending on the region of the world, people often have different attitudes to food. In addition to nutritional value, food often has societal and historical connotations, and in some instances may have religious importance. Technological modification of food and food production can evoke a negative response among consumers, especially in the absence of good communication on risk assessment efforts and cost/benefit evaluations.

Are there implications for the rights of farmers to own their crops?

Yes, intellectual property rights are likely to be an element in the debate on GM foods, with an impact on the rights of farmers. Intellectual property rights (IPRs), especially patenting obligations of the TRIPS Agreement (an agreement under the World Trade Organization concerning trade-related aspects of intellectual property rights) have been discussed in the light of their consequences on the further availability of a diversity of crops. In the context of the related subject of the use of gene technology in medicine, WHO has reviewed the conflict between IPRs and an equal access to genetic resources and the sharing of benefits? The review has considered potential problems of monopolization and doubts about new patent regulations in the field of genetic sequences in human medicine. Such considerations are likely to also affect the debate on GM foods.

Why are certain groups concerned about the growing influence of the chemical industry on agriculture?

Certain groups are concerned about what they consider to be an undesirable level of control of seed markets by a few chemical companies. Sustainable agriculture and biodiversity benefit most from the use of a rich variety of crops, both in terms of good crop protection practices as well as from the perspective of society at large and the values attached to food. These groups fear that as a result of the interest of the chemical industry in seed markets, the range of varieties used by farmers may be reduced mainly to GM crops. This would impact on the food basket of a society as well as in the long run on crop protection (for example, with the development of resistance against insect pests and tolerance of certain herbicides). The exclusive use of herbicide-tolerant GM crops would also make the farmer dependent on these chemicals. These groups fear a dominant position of the chemical industry in agricultural development, a trend which they do not consider to be sustainable.

What Further Developments can be Expected in the Area of GMOs?

Future GM organisms are likely to include plants with improved disease or drought resistance, crops with increased nutrient levels, fish species with enhanced growth characteristics and plants or animals producing pharmaceutically important proteins such as vaccines. At the international level, the response to new developments can be found in the expert consultations organized by FAO and WHO in 2000 and 2001, and the subsequent work of the Codex ad hoc Task Force on Foods Derived from Biotechnology. This work has resulted in an improved and harmonized framework for the risk assessment of GM foods in general. Specific questions, such as the evaluation of allergen city of GM foods or the safety of foods derived from GM microorganisms, have been covered and an expert consultation organized by FAO and WHO will focus on foods derived from GM animals in 2003.

What is WHO doing to improve the evaluation of GM foods?

WHO will take an active role in relation to GM foods, primarily for two reasons?

- On the grounds that public health could benefit enormously from the potential of biotechnology, for example, from an increase in the nutrient content of foods, decreased allergen city and more efficient food production; and
- Based on the need to examine the potential negative effects on human health of the consumption of food produced through genetic modification, also at the global level. It is clear that modern technologies must be thoroughly evaluated if they are to constitute a true improvement in the way food is produced. Such evaluations must be holistic and all-inclusive, and cannot stop at the previously separated, non-coherent systems of evaluation focusing solely on human health or environmental effects in isolation.

Work is therefore under way in WHO to present a broader view of the evaluation of GM foods in order to enable the consideration of other important factors. This more holistic evaluation of GM organisms and GM products will consider not only safety but also food security, social and ethical aspects, access and capacity building. International work in this new direction presupposes the involvement of other key international organizations in this area. As a first step, the WHO Executive Board will discuss the content of a WHO report covering this subject in January 2003. The report is being developed in collaboration with other key organizations, notably FAO and the United Nations Environment Programme (UNEP). It is hoped that this report could form the basis for a future initiative towards a more systematic, coordinated, multi-organizational and international evaluation of certain GM foods.

GM Products : Benefits and Controversies

Benefits

- Crops
  - Enhanced taste and quality
  - Reduced maturation time
  - Increased nutrients, yields, and stress tolerance

- Improved resistance to disease, pests, and herbicides
- New products and growing techniques

❒ Animals
- Increased resistance, productivity, hardiness, and feed efficiency
- Better yields of meat, eggs, and milk
- Improved animal health and diagnostic methods

❒ Environment
- "Friendly" bio herbicides and bio insecticides
- Conservation of soil, water, and energy
- Bio processing for forestry products
- Better natural waste management
- More efficient processing

❒ Society
- Increased food security for growing populations

Controversies

❒ Safety
- Potential human health impacts, including allergens, transfer of antibiotic resistance markers, unknown effects
- Potential environmental impacts, including : unintended transfer of transgenic through cross-pollination, unknown effects on other organisms (e.g., soil microbes), and loss of flora and fauna biodiversity

❒ Access and Intellectual Property
- Domination of world food production by a few companies
- Increasing dependence on industrialized nations by developing countries
- Bio piracy, or foreign exploitation of natural resources

❒ Ethics
- Violation of natural organisms' intrinsic values
- Tampering with nature by mixing genes among species
- Objections to consuming animal genes in plants and vice versa
- Stress for animal

- Labelling
  - Not mandatory in some countries (e.g., United States)
  - Mixing GM crops with non-GM products confounds labelling attempts
- Society
  - New advances may be skewed to interests of rich countries

United Nations Food and Agricultural Organization

- Report of the Panel of Eminent Experts on Ethics in Food and Agriculture (May 2001). A discussion of basic ethical concerns, including globalization and the right to food, farmers' rights.
- The United Nation's Food and Agriculture Organization's price index topped off at an all-time high this January, reflecting the skyrocketing global food prices and teetering stability of the market. These increases pose a severe risk to developing countries that depend upon agriculture for both subsistence farming and as the sole source of income for many communities. Developing countries' vulnerability to extreme weather attributed to climate change, crop diseases, pests, and the characteristic isolation of these communities puts them at a disadvantage. The Bill & Melinda Gates Foundation has teamed up with the UK's Department for International Development (DFID) to jointly fund $102 million in research, development, and on the ground assistance for agricultural communities in developing countries.
- "Along with health, I think agricultural productivity is one of the most basic things that can help the poor move to a more prosperous life."
- The two groups will identify projects jointly, and then the Gates Foundation's Agricultural Initiative will manage the selected projects.
- The collaboration comes at a key time in agricultural history, as food prices spike forcing millions of people to go hungry or malnourished and as the economic stability of the marketplace remains questionable. Yet this venture is not simply about increasing agricultural productivity, it is about the livelihood and

potential for sustainable agriculture that combines the best knowledge of current practice with ancient farming techniques. The hope is to leverage scientific research that identifies farming systems with fewer inputs and greater outputs, as well as increased access to tools, training, stable markets, and policies.

- A large portion of the funds, $40 million, is designated for Cornell's Durable Rust Resistant Wheat project. After developing a wheat strain that is resistant to stem rust, a global threat to wheat worldwide, the University has now dispersed this seed to seven countries with high food security risks. Another $3 million has been allotted for Diagnostics for All (DFA), to address dairy testing. This work is specifically designed to create an affordable diagnostic test that is accessible to small dairy farmers. Costing only a few cents, the test will check the milk quality, bovine pregnancy, and a common toxin found in grain. The results will help increase food security, food safety, and milk production.

- This collaboration marks a significant advancement in funds for small-scale sustainable agriculture in developing countries. It also formally recognizes climate change and biodiversity loss's potentially devastating impact on subsistence farming, hunger, and poverty. The Gates Foundation, contributing $70 million to this project, already has extensive agriculture programs in Africa and Southeast Asia that address soil quality, dairy productivity, micro irrigation, and business training for farmers. The DFID is committed to the Millennium Development Goals through the UK Government, which have targeted to halve world poverty by 2015. This work is part of their efforts to address the issue of poverty on the ground level through accessibility to safe and nutritious food along with stable livelihoods through agriculture. The specifics, such as the number of individuals aided and projected yield increases have not been released yet.

- Critics of the funding question whether the agriculture investments in hybridized crops and seeds from the developed world further push developing countries toward dependency on developed nations. Some seed saving advocates believe that investments in genetically modified crops are short-sighted and insufficient in addressing the ecological system in its entirety, costing more money in the long-term to "solve." By manipulating natural systems, critics fear that humans have created a never-ending technological rat race. For example, insects have relatively short life cycles. Spraying pesticides to kill them or developing GMO plants that kill them after ingestion solves the immediate pest problem. Yet, due to a short lifecycle, insects are able build up resistance quickly, creating the need for harsher chemicals, and so the cycle continues. In this case the details about GMO crops as part of the solution have not been fully disclosed outside of the wheat seeds.

- Proponents of GMO crops believe the seeds are not only technological wonders, but necessary components of properly addressing rising rates of hunger and poverty. This difference is often attributed to different worldviews. The upstream approach is to work with the natural system to understand limits and opportunities based on the natural system today and in the future. Others argue that an immediate need that requires an immediate solution on a large-scale developed by humans (GMO advocates).
- Instead of pitting these two opinions against one another, the approach taken by The Gates Foundation has leveraged benefits of both kinds of thinking. This has lead to the creation of an integrative approach that values the urgency to address the issue, and the necessity to lay the foundation for future sustainability and success. As the program evolves, it will be interesting to see what types of solutions the collaboration will choose to invest in to address the pressing issue of agricultural instability and world hunger.

The genetically modified foods controversy is a dispute over the relative advantages and disadvantages of genetically modified (GM) food crops and other uses of genetically-modified organisms in food production. The dispute involves biotechnology companies, governmental regulators, non-governmental organizations and scientists. The dispute is most intense in Japan and Europe, where public concern about GM food is higher than in other parts of the world such as the United States. In the United States GM crops are more widely grown and the introduction of these products has been less controversial.

The key areas of political controversy related to genetically engineered food are food safety, the effect on natural ecosystems, gene flow into non GE crops and corporate control of the food supply. While it is not possible to make general statements on the safety of all GM foods, to date, no adverse health effects caused by products approved for sale have been documented, although two products failed initial safety testing and were discontinued, due to allergic reactions.

Most feeding trials have observed no toxic effects and saw that GM foods were equivalent in nutrition to unmodified foods, although a few non-peer-reviewed reports speculate physiological changes to GM food. Although there is now broad scientific consensus that GE crops on the market are safe to eat ,some scientists and advocacy groups such as Greenpeace and World Wildlife Fund call for additional and more rigorous testing before marketing genetically engineered food.

Health risks and benefits

Present knowledge on GM food safety

Worldwide, there is a range of perspectives within non-governmental organizations on the safety of GM foods. For example, the US pro-GM group Ag Bio World has argued that GM foods have been proven safe, while other pressure groups and consumer rights groups, such as the Organic Consumers Association and Greenpeace. Claim the long-term health risks which GM could pose, or the environmental risks associated with GM, have not yet been adequately investigated. In Japan, Consumers Union of Japan are opposed to GMO foods. They also claim that truly independent research in these areas is systematically blocked by the GM corporations which own the GM seeds and reference materials.

The European Commission Directorate-General for Research and Innovation 2010 report on GMOs noted that "The main conclusion to be drawn from the efforts of more than 130 research projects, covering a period of more than 25 years of research, and involving more than 500 independent research groups, is that biotechnology, and in particular GMOs, are not per se more risky than e.g. conventional plant breeding technologies." A 2008 review published by the Royal Society of Medicine noted that GM foods have been eaten by millions of people worldwide for over 15 years, with no reports of ill effects. Similarly a 2004 report from the US National Academies of Sciences stated : "To date, no adverse health effects attributed to genetic engineering have been documented in the human population." A 2004 review of feeding trials in the *Italian Journal of Animal Science* found no differences among animals eating genetically modified plants. A 2005 review in *Archives of Animal Nutrition* concluded that first-generation genetically modified foods had been found to be similar in nutrition and safety to non-GM foods, but noted that second-generation foods with "significant changes in constituents" would be more difficult to test, and would require further animal studies..

However, a 2009 review in *Nutrition Reviews* found that although most studies concluded that GM foods do not differ in nutrition or cause any detectable toxic effects in animals, some studies did report adverse changes at a cellular level caused by some GM foods, concluding that "More scientific effort and investigation is needed to ensure that consumption of GM foods is not likely to provoke any form of health problem".A review published in 2009 by Dona and Arvanitoyannis concluded that "results of most studies with GM foods indicate that they may cause some common toxic effects such as hepatic, pancreatic, renal, or reproductive effects and may alter the hematological, biochemical, and immunologic parameters".

However responses to this review in 2009 and 2010 note that the Dona and Arvanitoyannis concentrated on articles with an anti-GM bias that have been refuted by scientists in peer-reviewed articles elsewhere - for example the 35S promoter, stability of transgenic, antibiotic marker genes and the claims for toxic effects of GM foods. In 2007, a review by Domingo of the toxicity by searching in the Pub limed database using 12 search terms, cited 68 references, found that the "number of references" on the safety of GM/transgenic crops was "surprisingly limited" and questioned whether the safety of genetically modified food has been demonstrated; the review also remarked that its conclusions were in agreement with three earlier reviews by Zdunczyk (2001), Bakshi (2003), and Primed and Lembcke (2003). However, an article in 2007 by Vain found 692 research studies focusing on GM crop and food safety and identified a strong increase in the publication of such articles in recent years. Vain commented that the multidisciplinary nature of GM research complicates the retrieval of GM studies and requires using many search terms (he used more than 300) and multiple databases.

Safety assessments

- The starting point for the safety assessment of genetically engineered food products is to assess if the food is "substantially equivalent" to its natural counterpart.

- The issue of GM food safety was first discussed at a meeting of the Food and Agriculture Organization (FAO), the World Health Organization (WHO) and biotech representatives in 1990. The "substantial equivalence" concept was proposed by the FAO in 1993 and endorsed by the FAO and WHO in early 1996 as a means to reassure consumers by obtaining official approval for genetically modified foods. The testing normally required for new food products can cost millions of dollars and take years of testing before a product gains approval for marketing which was also seen as inhibiting the development of biotechnology companies. The adoption of the concept of substantial equivalence permitted marketing of new foods without any safety or toxicology tests as long as they were not grossly different in chemical composition to foods already on the market.
- "Substantial equivalence embodies the concept that if a new food or food component is found to be substantially equivalent to an existing food or food component, it can be treated in the same manner with respect to safety (i.e., the food or food component can be concluded to be as safe as the conventional food or food component)"
- The rationale for this approach is that it would be impossible to test all the new crop varieties every year for food safety. Only a few food crops on the market have been shown to cause adverse health effects and all of these were the result of conventional genetic modification, not from genetic engineeringTo decide if a modified product is substantially equivalent, the product is tested by the manufacturer for unexpected changes in a limited set of components such as toxins, nutrients or allergens that are present in the unmodified food. If these tests show no significant difference between the modified and unmodified products, then no further food safety testing is required. The manufacturer's data is then assessed by an independent regulatory body, such as the U.S. Food and Drug Administration.

However, if the product has no natural equivalent, or shows significant differences from the unmodified food, then further safety testing is carried out A 2003 review in *Trends in Biotechnology* identified seven main parts of a standard safety test :

1. Study of the introduced DNA and the new proteins or metabolites that it produces;
2. Analysis of the chemical composition of the relevant plant parts, measuring nutrients, anti-nutrients as well as any natural toxins or known allergens;
3. Assess the risk of gene transfer from the food to microorganisms in the human gut;
4. Study the possibility that any new components in the food might be allergens;
5. Estimate how much of a normal diet the food will make up;
6. Estimate any toxicological or nutritional problems revealed by this data;
7. Additional animal toxicity tests if there is the possibility that the food might pose a risk.

This process was examined further in a review published by Kuiper *et al.* 2002 in the journal *Toxicology*, which stated that substantial equivalence does not itself measure risks, but instead identifies differences between existing products and new foods, which might pose dangers to health. If differences do exist, identifying these differences is a starting point for a full safety assessment, rather than an end point.[25] The authors concluded that "The concept of substantial equivalence is an adequate tool in order to identify safety issues related to genetically modified products that have a traditional counterpart". However, the review also noted difficulties in applying this standard in practice, including the fact that traditional foods contain many chemicals that have toxic or carcinogenic effects and that our existing diets therefore have not been proven to be safe. This lack of knowledge on unmodified food poses a problem, as GM foods may have differences in anti-nutrients and natural toxins that have never been identified in the original plant, raising the possibility that harmful changes could be missed.

- The application of substantial equivalence has also been more strongly criticized. For example, in a speech in 1999, **Andrew Chesson** of the University of Aberdeen, stated that substantial equivalence testing "could be flawed in some cases" and that some current safety tests could allow harmful substances to enter the human food chain. In a commentary in *Nature*, Millstone *et al.* argued that all GM foods should have extensive biological, toxicological and immunological tests and that the concept of substantial equivalence based solely on chemical analyzes of the components of a food should be abandoned.[27] They stated that this is necessary since it is currently impossible to predict the biological properties of a substance only from knowledge of its chemistry. This commentary was controversial and was criticized for misleading presentation of dataand presenting an over-simplified version of safety assessments. For example, Kuiper *et al.* responded to this criticism by noting that equivalence testing does involve more than chemical tests and may include toxicity testing.

- Medical writer Barbara Keeler and Marc Lapped argued in a 2001 article in the *Los Angeles Times*that the differences between genetically modified and conventional foods challenge the presumption of equivalence. Using Roundup ready soy that has been on the market since 1995 as an example, they noted the differences when compared to its unmodified counterpart. Significantly lower levels of protein than unmodified soy. Significantly lower levels of phenylalanine, an essential amino acid and as a dietary supplement, the reason doctors advise the consumption of soy products. Levels of trypsin inhibitor were 27% higher and after toasting, lectin was double that found in conventional soy; both are known allergens. GM soy also has 29% less choline, a B-complex vitamin. Round up ready soy had also stunted the growth of rats in Monsanto's study but had not affected cattle although it had increased the fat content of their milk. The authors do not maintain that modified soy is a hazard but that the FDA accepting such significant differences as being *substantially equivalent* illustrates the need for more rigorous testing, and preferably not by the biotech industries themselves.
- However, a 2008 paper by Cheng *et al.* showed that genetic engineering of soybeans cause's smaller unintended changes than are seen with traditional breeding. A 2002 paper by Ridley *et al.* showed that genetically engineered maize was equivalent to conventional maize for proximates, fiber, amino acids, fatty acids, vitamin E, nine minerals, phytic acid, trypsin inhibitor, and secondary metabolites. Baudo *et al.* in a 2006 paper compared transgenic wheat with conventionally bred wheat and concluded that "...transgenic plants could be considered substantially

equivalent to untransformed parental lines." A 2008 paper by di Carli *et al.* compared genetically engineered Lycopersicon esculentum (a tomato) and Nicotiana benthamiana (a close relative of tobacco) with their untransformed counterparts and concluded that genetic engineering did not significantly affect the plants proteanic profile.

- The value of current independent studies is problematic as, due to restrictive end-user agreements, researchers are forbidden by law from publishing independent research in peer-reviewed journals without the approval of the agritech companies. Cornell University's Elson Shields, the spokesperson for a group of scientists who oppose this practice, submitted a statement to the United States Environmental Protection Agency (EPA) protesting that "as a result of restrictive access, no truly independent research can be legally conducted on many critical questions regarding the technology". *Scientific American* noted that several studies that were initially approved by seed companies were later blocked from publication when they returned "unflattering" results. While recognising that seed companies' intellectual property rights need to be protected, *Scientific American* calls the practice dangerous and has called for the restrictions on research in the end-user agreements to be lifted immediately and for the EPA to require, as a condition of approval, that independent researchers have unfettered access to GM products for testing.
- The Welsh pressure group GM Free Cymru argues that governments should use independent studies rather than industry studies to assess crop safety. GM Free Cymru has also stated that independently funded researcher, Professor Bela Darvas of Debrecen University was refused Mon 863 Bt corn to use in his studies after previously publishing that a different variety of Monsanto corn was lethal to two Hungarian protected insect species and an insect classified as a rare.

Allergen city

Worldwide, reports of allergies to all kinds of foods, particularly nuts, fish and shellfish, seem to be increasing, but it is not known if this reflects a genuine change in the risk of allergy, or an increased awareness of food allergies by the public. Some environmental organizations, such as the European Green Party and Greenpeace, have suggested that GM food might trigger food allergies, although other environmentalists have implicated causes as diverse as the greenhouse effect increasing pollen levels, greater exposure to synthetic chemicals, cleaner lifestyles, or more mold in buildings. A 2005 review in the journal *Allergy* of the results from allergen testing of current GM foods stated that "no biotech proteins in foods have been documented to cause allergic reactions".

A well-known case of a GM plant that did not reach the market due to it producing an allergic reaction was a new form of soybean intended for animal feed. The allergen was transferred unintentionally from the Brazil nut into genetically engineered soybeans, in a bid to improve soybean nutritional quality for animal feed use. This new protein increased the levels in the GM soybean of the natural essential amino acid methionine, which is commonly added to poultry feed. Investigation of the GM soybeans revealed that they produced immune reactions in people with Brazil nut allergies, since the methionine rich protein chosen by Pioneer Hi-Bred happened to be a major source of Brazil nut allergy. Although this soybean strain was not developed as a human food, Pioneer Hi-Bred discontinued further development of the GM soybean, due to the difficulty in ensuring that none of these soybeans entered the human food chain.[44]

In November 2005 a pest-resistant field pea developed by the Australian CSIRO for use as a pasture crop was shown to cause an allergic reaction in mice. Work on this variety was immediately halted. The protein added to the pea did not cause the reaction in humans or mice in isolation, but when it was expressed in the pea, it exhibited a subtly different structure which may have caused the allergic reaction. The immunologist who tested the pea noted that crops need to be evaluated case-by-case, and a Greenpeace activist commented that to his knowledge, no countries required feeding studies for approval of genetically modified foods.

- Plant scientist Maarten J Chris peels has made these comments about this example :
- The recent Prescott et al. paper in JFAC contains a very interesting study on the immunogenicity of amylase [starch digestion enzyme] inhibitor in its native form (isolated from beans) and expressed as a transgene in peas. First of all, amylase inhibitor is a food protein, but also a "toxic" protein because it inhibits our digestive amylases. This is one of the reasons you have to cook your beans! (The other toxic bean protein is phytohemagglutinin

and it is much more toxic). This particular amylase inhibitor is found in the common bean (other species have other amylase inhibitors). Even though it is a food protein, it is unlikely ever to be used for genetic engineering of human foods because it inhibits our amylases. What the results show is that the protein, when synthesized in pea cotyledons has a different immunogenicity than when it is isolated from bean cotyledons (the native form). This is somewhat surprising but may be related to the presence of slightly different carbohydrate chains.

- These cases of products that failed safety testing can either be viewed as evidence that genetic modification can produce unexpected and dangerous changes in foods, or alternatively that the current tests are effective at identifying any safety problems before foods come on the market.
- Genetic modification can also be used to remove allergens from foods, which may, for example, allow the production of soy products that would pose a smaller risk of food allergies than standard soybeans. A hypo-allergenic strain of soybean was tested in 2003 and shown to lack the major allergen that is found in the beans. A similar approach has been tried in ryegrass, which produces pollen that is a major cause of hayfever : here a fertile GM grass was produced that lacked the main pollen allergen, demonstrating that the production of hypoallergenic grass is also possible.[49]

Environmental risks and benefits

- The large scale growth of GM plants may have both positive and negative effects on the environment. These may be both direct effects, on organisms that feed on or interact with the crops, or wider effects on food chains produced by increases or decreases in the numbers of other organisms. As an example of benefits, insect-resistant Bt-expressing crops will reduce the number of pest insects feeding on these plants, but as there are fewer pests, farmers do not have to apply as much insecticide, which in turn tends to increase the number of non-pest insects in these fields. A 2006 study of the global impact of GM crops, published by the UK consultancy PG Economics, concluded that globally, the technology reduced pesticide spraying by 286,000 tons in 2006, decreasing the environmental impact of herbicides and pesticides by 15%. By reducing the amount of laughing needed, GM technology led to reductions of greenhouse gases from soil equivalent to removing 6.56 million cars from the roads. However, a 2009 study published by the Organic Center stated that the use of genetically engineered corn, soybean, and cotton increased the use of herbicides by 383 million pounds (191,500 tons), and pesticide use by 318.4 million pounds (159,200 tons). As an example of a concern about environmental risk, a lab at Cornell University published an article which caused worry in the US that Bt-corn pollen might affect the monarch butterfly. However this concern was disproved by six comprehensive articles in the Proceedings of the National Academy of Sciences in 2001. Monarch populations increased, despite increased Bt corn probably due to reduced pesticide use. Other possible effects might come from the spread of genes from modified plants to unmodified relatives, which might produce species of weeds resistant to herbicides. In some areas of the US "super weeds" have evolved naturally, these weeds are resistant to herbicides and have forced farmers to return to traditional crop management practices.

- There has been controversy over the results of a farm-scale trial in the United Kingdom comparing the impact of GM crops and conventional crops on farmland biodiversity. Some claimed that the results showed that GM crops had a significant negative impact on wildlife. They pointed out that the studies showed that using herbicide resistant GM crops allowed better weed control and that under such conditions there were fewer weeds and fewer weed seeds. This result was then extrapolated to suggest that GM crops would have significant impact on the wildlife that might rely on farm weeds. The President of the Royal Society, the body that had carried out the trials, stated that "To generalize and declare 'all GM is bad' or 'all GM is good' for the environment as a result of these experiments is a gross over-simplification", arguing that although the trials showed that the combination of some GM crops with long-lasting herbicides were bad for biodiversity, using other GM crops without these herbicides increased biodiversity.

- In July 2005 British scientists showed that transfer of a herbicide-resistance gene from GM oilseed rape to a wild cousin, charlock, and wild turnips was possible.

- Many agricultural scientists and food policy specialists view GM crops as an important element in sustainable food security and environmental management. This point of view is summarized in the ABIC Manifesto :
- On our planet, 18% of the land mass is used for agricultural production. This fraction cannot be increased substantially. It is absolutely essential that the yield per unit of land increases beyond current levels given that : The human population is still growing, and will reach about nine billion by 2040;70,000 km^2 of agricultural land (equivalent to 60% of the German agricultural area) are lost annually to growth of cities and other non-agricultural uses; Consumer diets in developing countries are increasingly changing from plant-based proteins to animal protein, a trend that requires a greater amount of crop-based feeds.
- Other scientists, such as Dr. Charles Benbrook, argue that improvement of global food security is hardly being addressed by genetic research and that a lack of yield is often not caused by insufficient genetic resources. Regarding the issues of intellectual property and patent law, an international report from the year 2000 states :
- If the rights to these tools are strongly and universally enforced - and not extensively licensed or provided pro bono in the developing world - then the potential applications of GM technologies described previously are unlikely to benefit the less developed nations of the world for a long time (ie until after the restrictions conveyed by these rights have expired).

Issues with Bt Maize

- A well publicized claim associated with Bt crops (or transgenic maize) was the concern that pollen from Bt maize might kill the monarch butterfly. This report was puzzling because the pollen from most maize hybrids contains much lower levels of Bt than the rest of the plant and led to multiple follow-up studies. One possible issue revealed in these studies is the possibility that the initial study was flawed; based on the way the pollen was collected, in that they collected and fed non-toxic pollen that was mixed with anther walls that did contain Bt toxin. A collaborative research exercise was carried out over two years by several groups of scientists in the US and Canada, looking at the effects of Bt pollen in both the field and the laboratory. This resulted in a risk assessment that concluded that any risk posed by the corn to butterfly populations under real-world conditions was negligible. The USDA has stated that the weight of the evidence is that Bt crops do not pose a risk to the monarch butterfly. An independent 2002 review of the scientific literature concluded that "the commercial large-scale cultivation of current Bt–maize hybrids did not pose a significant risk to the monarch population" and noted that despite large-scale planting of GM crops, the butterfly's population is increasing.

- In 2007 Andreas Lang, Éva Lauber and Béla Darvas criticized these studies, arguing that there can be a great difference in the effects between the acute exposure tested for and chronic exposure. Moreover, they stated that the "worst case conditions" performed were not in fact worst case scenarios, as laboratory conditions with ample food supply and a favorable climate ensure healthy subjects. They instead believe that in the wild, low temperatures, rain and parasites and disease might exacerbate a Bt effect on butterfly larvae. Their own experiments suggested that some butterfly species were negatively affected by such chronic exposure. Jörg Romeis, who conducted the original studies, replied that if species of Butterfly are affected as Darvas claims that a "more comprehensive assessment will be needed and, depending on the degree and nature of concern, this may extend to field testing"

- A 2001 report in *Nature* presented evidence that Bt maize was cross-breeding with unmodified maize in Mexico, although the data in this paper was later described as originating from an artifact and *Nature* stated that "the evidence available is not sufficient to justify the publication of the original paper". A subsequent large-scale study, in 2005, failed to find any evidence of contamination in Oaxaca. However, other authors have stated that they also found evidence of cross-breeding between natural maize and transgenic maize.

- There is also a risk that for example, transgenic maize will crossbreed with wild grass variants, and that the Bt-gene will end up in a natural environment, retaining its toxicity. An event like this would have ecological implications. However, there is no evidence of crossbreeding between maize and wild grasses.

- In 2009 it was reported that 82,000 hectares (200,000 acres) of Bt corn in South Africa failed to produce seeds.Monsanto claimed average yield was reduced by 25% in those fields affected, it compensated the farmers concerned and the corn varieties were affected by a mistake made in the seed breeding process. Marian Mayet an environmental activitist and director of the Africa-centre for biosecurity in Johannesburg called for a government investigation and asserted that the biotechnology was at fault, *"You cannot make a 'mistake' with three different varieties of corn"*.In 2009 South African farmers planted 1,900,000 hectares (4,700,000 acres) of GM maize (73% of the total crop).

- As of 2007, a phenomenon called Colony Collapse Disorder (CCD) was noticed in bee hives all over North America, and elsewhere. Although it is not certain if this is a new phenomenon, initial ideas on the possible causes ranged from poor nutrition, infections, parasites and pesticide use. More unusual speculations included radio waves from cell phone base stations, climate change, and the use of transgenic crops containing Bt. The Mid-Atlantic Apiculture Research and Extension Consortium published a report on 2007-03-27 that found no evidence that pollen from Bt crops is adversely affecting bees. Several researchers in the US have since attributed CCD to the spread of a new virus called Israeli acute paralysis virus,[81] although other parasites have also been implicated.

- In a 2011 study, the Department of Obstetrics and Gynaecology, at the Université de Sherbrooke Hospital Centre in Quebec, Canada found traces of Bt toxin in the blood of 93% of pregnant women and 80% of the umbilical cords. Researchers believe the toxin is getting into the human body as a result of eating meat, milk and eggs from farm livestock fed GM corn. A spokesman stated that Bt toxins were "clearly detectable and appears to cross the placenta to the foetus. Given the potential toxicity...and the fragility of the foetus, more studies are needed". The industry has always argued that when eaten, Bt toxins would be destroyed in the gut. The new study has triggered calls for a ban on GM imports and for a total overhaul of the safety standards for GM crops and food. The Agriculture Biotechnology Council, which speaks for the GM industry, questioned the reliability and value of the research.

Legal Issues in the US

Alfalfa

- On 21 June 2010, the US Supreme Court in *Monsanto Co. v. Geertson Seed Farms* issued its first ruling about GM crops. This case relates to Roundup Ready alfalfa and dates back to 2006, when a coalition of groups led by the Center for Food Safety raised concerns about environmental impacts that the United States Department of Agriculture (USDA) allegedly failed to address before approving the planting of Roundup Ready alfalfa. In response, in 2007, Judge Charles Breyer of the California Northern District Court ruled that USDA was in error when it approved the planting of Roundup Ready alfalfa.

- In June 2009, a divided three-judge panel on the 9th U.S. Circuit Court of Appeals upheld Judge Charles Breyer's decision, and Monsanto appealed to the Supreme Court.

- The impact of the 2009 US Supreme Court ruling was somewhat unclear, with both sides appearing to claim victory. While Monsanto claimed technical victory in the case (it struck down the 2007 injunction on the planting of GM alfalfa), various other issues remained open and the planting of GM alfalfa remained halted.

- In January 2011, Agriculture Secretary Tom Vilsack announced that the USDA had approved the unrestricted planting of genetically modified alfalfa. Agriculture Secretary Tom Vilsack commented "After conducting a thorough and transparent examination of alfalfa .. APHIS [Animal and Plant Health Inspection Service] has determined that Roundup Ready alfalfa is as safe as traditionally bred alfalfa." Christine Bushway, CEO of the Organic Trade Association said "A lot of people are shell shocked. While we feel Secretary Vilsack worked on this issue, which is progress, this decision puts our organic farmers at risk." About 20 million acres (8 million hectares) of alfalfa were grown in the US, the fourth-biggest crop by acreage, of which about 1% were organic. Some biotechnology officials forecast that half of the US alfalfa acreage could eventually be planted with GM alfalfa.

Sugar beets

- There is also a somewhat similar case involving sugar beets before the same California Northern District Court. This is a case involving Monsanto's breed of pesticide-resistant sugar beets. While Judge Jeffrey S. White (issuing his ruling in the spring of 2010) allowed the planting of GM sugar beets to continue, he also warned that this may be blocked in the future while an environmental review is taking place.
- On 13 August 2010, **Judge Jeffrey S. White** ordered the halt to the planting of the genetically modified sugar beets in the US. He indicated that "the Agriculture Department had not adequately assessed the environmental consequences before approving them for commercial cultivation."[91]
- In 2010, before the ruling, 95% of the sugar beet grown in the US was GM.[92] About half the sugar supply in the US came from sugar beet.
- In February 2011, a federal appeals court for the Northern district of California in San Francisco overturned a previous ruling by Judge Jeffrey S. White to destroy juvenile GM sugar beets, ruling in favor of Monsanto, the Department of Agriculture's Animal and Plant Health Inspection Service (APHIS) and four seed companies. The court concluded that "The Plaintiffs have failed to show a likelihood of irreparable injury. Biology, geography, field experience, and permit restrictions make irreparable injury unlikely."
- In February 2011, The USDA allowed commercial planting of GM sugar beet in the US under closely controlled conditions. Michael Gregorian from APHIS said "After conducting an environmental assessment, accepting and reviewing public comments and conducting a plant pest risk assessment, APHIS has determined that the Roundup Ready sugar beet root crop, when grown under APHIS imposed conditions, can be partially deregulated without posing a plant pest risk or having a significant effect on the environment." GM sugar beet opponents such as Earth justice said the USDA action circumvents court orders, and vowed they would fight the USDA in court.

Control of the market

Patent holder companies use their control of their own GMO to corner the market and gain profit. However, other companies often compete for the little market share available to GM foods worldwide. Detractors such as Greenpeace say that patent rights give corporations a dangerous amount of control over their product while corporations claim that they need product control in order to prevent seed piracy, fulfill financial obligations to shareholders and invest in further GM development. Governments have also sought to protect their commercial interests through punitive measures against countries resisting GM foods on ethical or scientific grounds : after moves in France to ban a Monsanto GM corn variety, the US embassy recommended that 'we calibrate a target retaliation list that causes some pain across the EU'.

Controversial Cases

Pusztai potato controversy

- A large media event occurred in 1998 when scientist Árpád Pusztai, who was considered to be a world expert on plant lectins, reported in a television interview that he had found that rats fed potatoes genetically modified by the English biotech company *Cambridge Agricultural Genetics* to contain lectin, a natural insecticide in snowdrop plants, had stunted growth and immune system damage. Initially the Rowett Institute's director, Philip James, congratulated Pusztai after the interview, but his attitude changed after confusion arose over whether the lectin was from a jackbean or a snowdrop (the jackbean lectin is poisonous to the intestines). A press release from the Rowett Institute stated that the lectin had come from a jackbean, but no experiments were made using lectins from the jackbean; in the confusion, it was alleged that Pusztai hadn't actually carried out the experiments. James suspended Pusztai and used misconduct procedures to seize the data. His annual contract was not renewed and both Pusztai and his wife were banned from speaking publicly. In October 1998 the Rowett Institute published an audit criticizing Pusztai's results, which, along with Pusztai's raw data, was sent to six anonymous reviewers who criticized Pusztai's work. Pusztai responded that the raw data was "never intended for publication under intense scrutiny".Pusztai sent the audit report and his rebuttal to scientists who requested it, and in

February 1999, twenty-one European and American scientists, mostly friends or acquaintancesof Pusztai, orchestrated by the environmental group Friends of the Earth, released a memo supporting Pusztai. Stanley Ewen, who worked with Pusztai, conducted a followup study supporting Pusztai's work and presented the work to a lectin meeting in Sweden.

- The potatoes were chosen because they were deemed substantially equivalent to non-GM desiree red potatoes. The Rowett study was intended to confirm the safety of the exemption. The study used two transgenic lines of potato, both with the same gene insertion, grown in the same identical conditions together with the non-GM parent plant. It was found that one of the GM lines contained 20% less protein than the other while the starch and sugar contents varied by up to 20%. It was also found that neither of the two GM lines were substantially equivalent to their non-GM parent.
- "We had two transgenic lines of potato produced from the same gene insertion and the same growing conditions...With our two lines of potato, which should have been substantially equivalent to each other...the two lines were not substantially equivalent to each other. But we also found that these two lines were not substantially equivalent to their parent. This could not be predicted. It demonstrates that the unpredictability is inherent in the GM process on a case by case basis - and also at the level of every single GM plant created."
- In October 1999 Pusztai's research was published in the journal The Lancet. Because of the controversial nature of his research the data in this paper, co-authored by Stanley Ewan, was seen by a total of six reviewers when presented for peer review; five of these reviewers judged the work acceptable, although one of the five "deemed the study flawed but favored publication to avoid suspicions of a conspiracy against Pusztai and to give colleagues a chance to see the data for themselves".[103] The paper did not mention stunted growth or immunity issues, but reported that rats fed on potatoes genetically modified with the snowdrop lectin had "thickening in the mucosal lining of their colon and their jejunum" when compared with rats fed on non modified potatoes. Three Dutch scientists criticized the study on the grounds that the unmodified potatoes were not a fair control

diet, and that any rats fed only on potatoes will suffer from a protein deficiency; Pusztai responded to these criticisms by stating that the protein and energy were comparable, and that "a sample size of six is perfectly normal in studies like this".

- The genetically modified potatoes were later destroyed by *Cambridge Agricultural Genetics* as were all details of their modification.

Bioequivalence Study of a Corn Cultivar

- A controversy arose around biotech company Monsanto's data on a 90-Day Rat Feeding Study on the MON863 strain of GM corn. In May 2005, critics of GM foods pointed to differences in kidney size and blood composition found in this study, suggesting that the observed differences raises questions about the regulatory concept of substantial equivalence. Anti-GM campaigner Jeffrey M. Smith, writing in Biophile Magazine, quoting comments from Pusztai and Seralini,has stated that nutritional studies typically use young, fast-growing animals with starting weights not varying by more than 2% from the average whereas Monsanto's research design used a mix of young and old animals with starting weights ranging from 198.4 to 259.8 grams. Seralini and two other authors published a study of these data, funded by Greenpeace, in 2007 making similar points.

- The raising of this issue prompted the European Food Safety Authority (EFSA) to reexamine the safety data on this strain of corn. The EFSA concluded that the observed small numerical decreases in rat kidney weights were not biologically meaningful, and the weights were well within the normal range of kidney weights for control animals. There were no corresponding microscopic findings in the relevant organ systems, and they stated that all blood chemistry and organ weight values fell within the "normal range of historical control values" for rats. In addition the EFSA review stated that the statistical methods used by Sealing et al. in the analysis of the data were incorrect. The European Commission has approved the ÌÍÍ863 corn for animal and human consumption. Food Standards Australia New Zealand reviewed the 2007 Seralini *et al.* study and concluded that "...all of the statistical differences between rats fed MON 863 corn and control rats are attributable to normal biological variation."

- Greenpeace stated in a 2007 press release that Séralini et al. had completed a similar analysis of the NK603 strain of corn and came to similar conclusions as they did in their previous study. Séralini *et al.* included this in a re-analysis of three existing rat feeding studies published in 2009.
- The European Food Safety Authority reviewed the 2009 Seralini *et al.* paper and concluded that the author's claims were not supported by the data in their paper. They noted that many of their fundamental statistical criticisms of the 2007 paper also applied to the 2009 paper. There was no new information that would change the EFSA's conclusions that the three GM maize types were safe for human, animal health and the environment The French High Council of Biotechnologies Scientific Committee (HCB) also reviewed the 2009 study and concluded that it "..presents no admissible scientific element likely to ascribe any hematological, hepatic or renal toxicity to the three re-analysed GMOs." The HCB also questioned the author's independence. Food Standards Australia New Zealand concluded that the results from the 2009 Séralini *et al.* study were due to chance alone.

Contamination issues

- In the 1990s genetically modified Flax tolerant to herbicide residues in soil was developed by the Crop Development Centre (CDC) at the University of Saskatchewan in Canada. Named Flax variety FP967, but commonly called *CDC Triffid,* research was controversially halted following protests from Canadian farmers who stood to lose up to 70% of their traditional export markets if it was introduced. GM Flax was deregistered, its sale was criminalized and in 2001 all modified seeds were destroyed. No modified crops had been planted and no seed had been sold but GM industry proponent Alan McHugh controversially passed out sample packets of seeds at presentations. In early September 2009, Flax imported into Germany was found to be contaminated with *CDC Triffid* causing the price of Canadian Flax to fall 32 percent. By mid November 35 countries reported contamination of imported Canadian Flax which has now been banned by the European Union. Canadian farmers are expected to be responsible for the cost of the cleanup and testing of future crops.
- In 2000, Aventis Star Link corn, which had been approved only as animal feed due to concerns about possible allergic reactions

in humans, was found contaminating corn products in U.S. supermarkets. An episode involving Taco Bell taco shells was particularly well publicized which resulted in sales of StarLink seed being discontinued. The registration for the Star link varieties was voluntarily withdrawn by Aventis in October 2000. Aid sent by the UN and the US to Central African nations was also found to be contaminated with StarLink corn and the aid was rejected. The US corn supply has been monitored for Star link Bt proteins since 2001 and no positive samples have been found since 2004.

- Gene Watch UK and Greenpeace International set up the GM Contamination Register in 2005.

Public perception

- Research by the Pew Initiative on Food and Biotechnology has shown that in 2005 Americans' knowledge of genetically modified foods and animals continues to remain low, and their opinions reflect that they are particularly uncomfortable with animal cloning. In one instance of consumer confusion, DNA Plant Technology's Fish tomato transgenic organism was conflated with Calgene's Flavr Savr **transgenic food product.** The Pew survey also showed that despite continuing concerns about GM foods, American consumers do not support banning new uses of the technology, but rather seek an active role from regulators to ensure that new products are safe.
- Only 2% of Britons were said to be "happy to eat GM foods", and more than half of Britons were against GM foods being available to the public, according to a 2003 study. However a 2009 review article of European consumer polls concluded that opposition to GMOs in Europe has been gradually decreasing. Approximately half of European consumers accepted gene technology, particularly when benefits for consumers and for the environment could be linked to GMO products. 80 % of respondents did not cite the application of GMOs in agriculture as a significant environmental problem. Many consumers seem unafraid of health risks from GMO products and most European consumers did not actively avoid GMO products while shopping.
- In Australia, GM foods that have novel DNA, novel protein, altered characteristics or has to be cooked or prepared in a different way compared to the conventional food have, since

December 2001, had to be identified on food labels. However, multiple surveys have shown that while 45% of the public will accept GM foods, some 93% demand all genetically modified foods be labelled as such. A 2007 survey by the Food Standards Australia and New Zealand found that 27% of Australians looked at the label to see if it contained GM material when purchasing a grocery product for the first time. Labelling legislation has been introduced and rejected several times since 1996 on the grounds of "restraint of trade" due to the cost of labelling. The controversy erupted again in 2009 when Graincorp, the nation's largest grain handler, announced it would mix GM Canola with its unmodified grain. Traditional growers, who largely rely on GM-free markets, had been told they would need to pay to have their produce certified GM free. Graincorp reversed its decision the same year. Critics such as Greenpeace and the Gene Ethics Network have renewed calls for more labelling.

- Opponents of genetically modified food often refer to it as "Frankenfood", after Mary Shelley's character Frankenstein and the monster he creates, in her novel of the same name. The term was coined in 1992 by Paul Lewis, an English professor at Boston College who used the word in a letter he wrote to the *New York Times* in response to the decision of the US Food and Drug Administration to allow companies to market genetically modified food. The term "Frankenfood" has become a battle cry of the European side in the US-EU agricultural trade war.

- Critics have protested in regards to the appointment of pro GM lobbyists to senior positions in the FDA. Michael R. Taylor has been appointed as a senior adviser to the FDA on food safety and Dennis Wolff is expected to take up the position of Under-Secretary of the newly created Agriculture for Food Safety. Taylor is a former Monsanto lobbyist credited as being responsible for the implementation of "substantial equivalence" in place of food safety studies and for his advocacy that resulted in the Delaney clause that prohibited the inclusion of "any chemical additive found to induce cancer in man. or animals" in processed foods being amended in 1996 to allow the inclusion of pesticides in GMOs. Wolff is the Pennsylvania Secretary of Agriculture who successfully lobbied to ban organic farmers from labeling their products as being GM free and was a proponent of the "ACRE" initiative which gave the Pennsylvania state attorney general's office the authority to sue municipalities that banned GMOs. Several anti-GMO organizations have organized petitions demanding Taylor's resignation and opposing Wolff's appointment and also conducted letter writing campaigns protesting the conflict of interest.

Opponents

- Soil Association
- Center for Food Safety
- Greenpeace
- Sierra Club

Advocates

- Council for Biotechnology Information
- AgBioWorld

Governmental

- German Federal Ministry of Education and Research
- US Biotechnology Regulators
- UK Food Standards Agency
- European Food Safety Authority
- Government of Canada BioPortal

Medical and scientific

- NIH National Library of Medicine
- Royal Society
- *Bacillus thuringiensis* (or Bt) is a Gram-positive, soil-dwelling bacterium, commonly used as a biological pesticide; alternatively, the Cry toxin may be extracted and used as a pesticide. *B. thuringiensis* also occurs naturally in the gut of caterpillars of various types of moths and butterflies, as well as on the dark surface of plants.
- During sporulation, many Bt strains produce crystal proteins (**proteinaceous inclusions**), called ä-endotoxins, that have insecticidal action. This has led to their use as insecticides, and more recently to genetically modified crops using Bt genes. There

are, however, many crystal-producing Bt strains that do not have insecticidal properties.[2]

Discovery and Study

- *B. thuringiensis* was first discovered in 1901 by Japanese biologist Shigetane Ishiwatari. In 1911, *B. thuringiensis* was rediscovered in Germany by Ernst Berliner, who isolated it as the cause of a disease called *Schlaffsucht* in flour moth caterpillars. In 1976, Robert A. Zakharyan reported the presence of a plasmid in a strain of *B. thuringiensis* and suggested the plasmid's involvement in endospore and crystal formation.[3][4] *B. thuringiensis* is closely related to *B.cereus*, a soil bacterium, and *B.anthracis*, the cause of anthrax : the three organisms differ mainly in their plasmids. Like other members of the genus, all three are aerobes capable of producing endospores. Upon sporulation, *B. thuringiensis* forms crystals of proteinaceous insecticidal ä-endotoxins (called crystal proteins or Cry proteins), which are encoded by *cry* genes. In most strains of *B. thuringiensis*, the *cry* genes are located on the plasmid.
- Cry toxins have specific activities against insect species of the orders Lepidoptera (moths and butterflies), Diptera (flies and mosquitoes), Coleoptera (beetles), Hymenoptera (wasps, bees, ants and sawflies) and nematodes. Thus, *B. thuringiensis* serves as an important reservoir of Cry toxins for production of biological insecticides and insect-resistant genetically modified crops. When insects ingest toxin crystals, the alkaline pH of their digestive tract activates the toxin. Cry toxin gets inserted into the insect gut cell membrane, forming a pore. The pore results in cell lysis and eventual death of the insect.

Use in pest control

- Spores and crystalline insecticidal proteins produced by *B. thuringiensis* have been used to control insect pests since the 1920s. They are now used as specific insecticides under trade names such as Dipel and Thuricide. Because of their specificity, these pesticides are regarded as environmentally friendly, with little or no effect on humans, wildlife, pollinators, and most other beneficial insects. The Belgian company Plant Genetic Systems was the first company (in 1985) to develop genetically engineered (tobacco) plants with insect tolerance by expressing *cry* genes from *B. thuringiensis*.

- *B. thuringiensis*-based insecticides are often applied as liquid sprays on crop plants, where the insecticide must be ingested to be effective. The solubilized toxins are thought to form pores in the midgut epithelium of susceptible larvae. Recent research has suggested the midgut bacteria of susceptible larvae are required for *B. thuringiensis* insecticidal activity.
- *Bacillus thuringiensis* serovar *israelensis*, a strain of *B. thuringiensis* is widely used as a larvicide against mosquito larvae, where it is also considered an environmentally friendly method of mosquito control.

Genetic Engineering for Pest Control

Bt toxins present in peanut leaves (bottom image) protect it from extensive damage caused by European corn borer larvae (top image).

Usage

- In 1995, potato plants producing Bt toxin were approved safe by the Environmental Protection Agency, making it the first pesticide-producing crop to be approved in the USA. By 1996, Bt maize, Bt potato and Bt cotton were being grown by farmers in the USA.
- Bt crops (in corn and cotton) were planted on 281,500 km^2 in 2006 (165,600 km^2 of Bt corn and 115900 km^2 of Bt cotton). This was equivalent to 11.1% and 33.6%, respectively, of global plantings of corn and cotton in 2006.[18] Claims of major benefits to farmers, including poor farmers in developing countries, have been made by advocates of the technology, and have been challenged by opponents. The task of isolating impacts of the technology is complicated by the prevalence of biased observers, and by the rarity of controlled comparisons (such as identical seeds, differing only in the presence or absence of the Bt trait, being grown in identical situations). The main Bt crop being grown by small farmers in developing countries is cotton, and a recent exhaustive review of findings on Bt cotton by respected and unbiased agricultural economists concluded, "the overall balance sheet, though promising, is mixed. Economic returns are highly variable over years, farm type, and geographical location".
- Environmental impacts appear to be positive during the first ten years of Bt crop use (1996–2005). One study concluded insecticide

use on cotton and corn during this period fell by 35.6 million kg of insecticide active ingredient, which is roughly equal to the amount of pesticide applied to arable crops in the EU in one year. Using the environmental impact quotient (EIQ) measure of the impact of pesticide use on the environment, the adoption of Bt technology over this ten-year period resulted in 24.3% and 4.6% reduction, respectively, in the environmental impact associated with insecticide use on the cotton and corn area using the technology.

Advantages

There are several advantages in expressing Bt toxins in transgenic Bt crops :

- The level of toxin expression can be very high, thus delivering sufficient dosage to the pest.
- The toxin expression is contained within the plant system, hence only those insects that feed on the crop perish.
- The toxin expression can be modulated by using tissue-specific promoters, and replaces the use of synthetic pesticides in the environment. The latter observation has been well documented worldwide.

Health and safety

- Overall, Bt-modified crops appear to be environmentally safe. The proteins produced by Bt have been used in sprays for agricultural weed control in France since 1938 and the USA since 1958 with seemingly no ill effects on the environment.
- Bt toxins are considered to be environmentally friendly by many farmers and may be a potential alternative to broad-spectrum insecticides. The toxicity of each Bt type is limited to one or two insect orders; it is nontoxic to vertebrates and many beneficial arthropods, because Bt works by binding to the appropriate receptor on the surface of midgut epithelial cells. Any organism that lacks the appropriate receptors in its gut cannot be affected by Bt.
- There is clear evidence from laboratory settings that Bt toxins can affect no target organisms. Usually, but not always, affected organisms are closely related to intended targets. Typically,

exposure occurs through the consumption of plant parts, such as pollen or plant debris, or through Bt ingestion by their predatory food choices. The methodology used by these researchers has been called into question. Due to significant data gaps, the real-world consequences of Bt transgenic remains unclear.

- Not all scientific reports on Bt safety have been positive. A 2007 study funded by the European arm of Greenpeace, suggested the possibility of a slight but statistically meaningful risk of liver damage in rats. While small statistically significant changes may have been observed, statistical differences are both probable and predictable in animal studies of this kind, (known as type I errors), that is, the probability of finding a false-positive due to chance alone. In this case, the number of positive results was within the statistically predicted range for type I errors.
- The observed changes have been found to be of no biological significance by the European Food Safety Authority.

Limitations of Bt crops

Kenyans examining insect-resistant transgenic Bt corn

- Constant exposure to a toxin creates evolutionary pressure for pests resistant to that toxin. Already, a diamondback moth population is known to have acquired resistance to Bt in spray form (i.e., not engineered) when used in organic agriculture. The same researcher has now reported the first documented case of pest resistance to biotech cotton.
- One method of reducing resistance is the creation of non-Bt crop refuges to allow some nonresistant insects to survive and maintain a susceptible population. To reduce the chance an insect would become resistant to a Bt crop, the commercialization of transgenic cotton and maize in 1996 was accompanied with a management strategy to prevent insects from becoming resistant to Bt crops, and insect resistance management plans are mandatory for Bt crops planted in the USA and other countries. The aim is to encourage a large population of pests so any genes for resistance are greatly diluted. This technique is based on the assumption that resistance genes will be recessive.

- This means that with sufficiently high levels of transgenic expression, nearly all of the heterozygote's (S/s),i.e. the largest segment of the pest population carrying a resistance allele, will be killed before they reach maturity, thus preventing transmission of the resistance gene to their progeny. The planting of refuges (i. e., fields of no transgenic plants) adjacent to fields of transgenic plants increases the likelihood that homozygous resistant (s/s) individuals and any surviving heterozygotes will mate with susceptible (S/S) individuals from the refuge, instead of with other individuals carrying the resistance allele. As a result, the resistance gene frequency in the population would remain low.
- Nevertheless, limitations can affect the success of the high-dose/refuge strategy. For example, expression of the Bt gene can vary. For instance, if the temperature is not ideal, this stress can lower the toxin production and make the plant more susceptible. More importantly, reduced late-season expression of toxin has been documented, possibly resulting from DNA methylation of the promoter. So, while the high-dose/refuge strategy has been successful at prolonging the durability of Bt crops, this success has also had much to do with key factors independent of management strategy, including low initial resistance allele frequencies, fitness costs associated with resistance, and the abundance of non-Bt host plants that have supplemented the refuges planted as part of the resistance management strategy.

Insect resistance

In November 2009, Monsanto scientists found the pink bollworm had become resistant to Bt cotton in parts of Gujarat, India. In four regions, Amreli, Bhavnagar, Jamnagar and Rajkot, the crop is no longer effective at killing the pests. This was the first instance of Bt resistance confirmed by Monsanto anywhere in the world .Monsanto confirmed field resistance of the worm to the *Cry1Ac* first generation Bollard cotton, which expresses a single Bt gene.

Secondary pests

Several studies have documented surges in "sucking pests" (which are not affected by Bt toxins) within a few years of adoption of Bt cotton. In China, the main problem has been with myriads, which have in some cases "completely eroded all benefits from Bt cotton cultivation".

Similar problems have been reported in India, with both mealy bugs and aphids.

Possible problems

Further information : Genetically modified food controversies

Lepidoptera toxicity

The most publicized problem associated with Bt crops is the claim that pollen from Bt maize could kill the monarch butterfly. This report was puzzling because the pollen from most maize hybrids contains much lower levels of Bt than the rest of the plant and led to multiple follow-up studies.

The initial study apparently was flawed by faulty pollen-collection procedure; researchers fed nontoxic pollen mixed with anther walls containing Bt toxin. The weight of the evidence is that Bt crops do not pose a risk to the monarch butterfly. Monarch butterflies have no innate relationship to maize crops in the wild, and are not believed to consume maize pollen (or pollen of related plants) in either life stage.

Wild maize genetic contamination

- A study in *Nature* reported that Bt-containing maize genes were contaminating maize in its center of origin. *Nature* later "concluded that the evidence available is not sufficient to justify the publication of the original paper." However, there still remains a controversy over the highly unorthodox retraction on the part of *Nature*. In 1998, Chapala, one of the original paper's authors, spoke out against Berkeley accepting a multimillion dollar research grant from the Swiss pharmaceutical company, Novartis.
- A subsequent large-scale study, in 2005, failed to find any evidence of contamination in Oaxaca. However, further research confirmed initial findings concerning contamination of natural maize by transgenic maize.
- However, further studies, such as that published in *Molecular Ecology* in 2008, have shown some small-scale (about 1%) genetic contamination (by the 35S promoter) in sampled fields in Mexico. One meta-study has found evidence for and against Bt contamination of maize, concluding that the preponderance of evidence points to Bt maize contamination in Mexico.

Possible link to colony collapse disorder

As of 2007, a new phenomenon called colony collapse disorder (CCD) is affecting bee hives all over North America. Initial speculation on possible causes ranged from new parasites to pesticide use to the use of Bt resistant transgenic crops.The Mid-Atlantic Apiculture Research and Extension Consortium published a report in March 2007 that found no evidence that pollen from Bt crops is adversely affecting bees. The actual cause of CCD remains unknown, and scientists believe that it may have multiple causes.

Food Security

Hungry for change : Transforming Indian agriculture to ensure food security

- India is said to be a land of contradictions and there is perhaps no greater one than the gap between its agricultural potential and food security.
- India has the largest bank of cultivable land, after only the United States, and enjoys growing conditions suited to most crops, vegetables and fruit. It also has adequate water resources, even if they are not effectively harnessed. Yet food security eludes its grasp; about 380 million Indians suffer from hunger and malnourishment today and, with rising food prices, up to 150 million more are expected to face the same fate by 2020.
- This need not be the case, as experience shows, not just of other countries but of India itself. In the 1960s, India turned Punjab into the country's bread basket by bringing water, seed, and fertilizer and extension services to farmers via the Food Corporation of India (FCI). A decade later, Thailand transformed its agriculture sector, commercially, by providing high quality agricultural inputs and two decades later China did so, collectively, by liberalizing pricing and marketing of agricultural products to increase farmers' access to markets.
- The government took the lead in Punjab and will need to do so now as well to bring not just food security but also prosperity to its people. The creation of the National Food Security Mission is a commendable step in this direction. The key to success will be recognizing the limitations of the current approach to agriculture and food production and making the radical change needed, preferably through public-private partnerships.

India has high but unrealized agricultural potential

- India has all the elements critical to a healthy agricultural system : 160 million hectares of cultivable land, second only to the United States' holdings and more than China's, and abundant water resources that can be tapped more efficiently to increase the proportion of irrigated land. Most states have the potential to double farm yields through better application of known cropping best-practices and available technology.

- However, this potential has never been realized. One outcome is that food prices keep rising due to product shortages. Indian consumers have suffered annual increases between 2000 and 2010 in the price of staple crops such as rice and pulses. Years of drought or flood, such as 2009, have seen worse price spikes : a three-fold rise in the price of toor dal and a doubling of the price of urinal, both common items on Indian dining tables. This had made simple meals unaffordable, even for those above the poverty line.

- The paradox is stark and simply explained. Over the last 25 years, India has not invested enough in raising agricultural productivity. As the largest investor in Indian agriculture, the government sells seeds, fertilizer, water and extension services to farmers and also buys their products. But the government's investments have focused on subsidizing not developing agriculture. For instance, the government has provided heavy discounts through FCI to make rice more affordable rather than investing in breakthrough rice hybrids, which would raise productivity and hence lower prices. In the financial year 2008-09, the government invested Rs. 10,000 crore in agriculture (for infrastructure, markets, roads, and so on) but spent 15 times that amount, Rs. 1,65,000 crore, on its fertilizer and food subsidies.

- The lack of investment in raising productivity has been the biggest obstacle to food security in India. Other developing countries such as Indonesia and China recognized the importance of raising productivity and have long since transformed their agriculture sectors to the enduring benefit of their people.

The next green revolution : Experience shows the way

- The challenge is not a new one; nor is the solution unknown. As mentioned earlier, a holistic solution for Punjab in the 1960s, bringing together water, seeds, extension services and the market through FCI doubled the per hectare productivity of wheat in five years. Similar success was reached with rice in Thanjavur. But India's green revolution took a long pause right after and is yet to see a second wave. Agriculture-intensive states such as UP, Bihar, West Bengal and Orissa, which have the potential to double their farm yields simply by applying known and existing technology effectively, are yet to receive the benefit of such a concerted effort.

- But it is never too late. India can still build on its own experience and that of other countries to transform agriculture and hence its economy.
- Countries usually have three objectives in transforming agriculture : boost GDP growth, increase smallholder incomes, and ensure food and nutrition security. They have tried to achieve these objectives in many ways, mainly policy reform and investments in broad programmes such as infrastructure creation, fertilizer subsidies and capacity building (institutional reform; specialized government agencies).
- McKinsey's extensive global experience in transforming agriculture, suggests that for such large scale change to occur, it is imperative to priorities and sequence programmes, and implement them as holistic mini-transformations of specific geographic units or value chains, harnessing the change energy of individual entrepreneurs.
- As the single largest investor in a sector that has seen little private investment, the government will need to trigger the transformation and keep up its momentum.

The five actions below will be crucial to success :

- First, pursue a holistic solution, if required in a few states. States with a supportive and eager political leadership, potential for yield improvement and positive investment climate could qualify as the transformation-ready ones. Sequence efforts rather than do several little things across the country.
- Second, strengthen critical value chains in transformation-ready states : Conduct a rapid assessment of the right crop mix for the region based on local soil and water conditions as well as market needs. Then choose if the state must fix all the value chains

simultaneously or if a few should be prioritized. McKinsey's global experiences suggest that a mix usually works better as the focus on staple crops is essential but high-value crops generate significant employment opportunities improve rural productivity.

- Further, the effort conceived should be comprehensive in nature i.e. it should be envisioned as the redesign of initiatives across various frontiers and deliver a 'package' of solutions for the farmer e.g. inputs, water, market linkages as well as for the private sector investors.
- Third, attract investors : Design policies to attract serious and committed private sector participants (few hundred) to the value chain. This is critical for the long term health of the sector. In particular, providers of agricultural inputs, micro irrigation, financing and risk management services, fresh produce aggregation and marketing, food processing, poultry or dairy production and processing, and those that can help build extension services capabilities are vital to drive a transformation.
- Develop tailored propositions to attract players across social agriculture i.e. staple low value crops as well as commercial agriculture i.e. high-value crops and food chains. The key here is designing effective aggregation models (we have usually found as many as eight work depending on crop and region); defining the policy for investment (in core and infrastructure) and the ensuring availability of land for either demonstration farms or aggregation.
- Finally, drive, monitor and measure progress : Design a special delivery unit e.g., a project management office to focus on implementation by following a three - point agenda. First, cultivate effective change leaders from various tiers of government i.e. from the panchayat to the state government, to guide farmers, private sector participants and other stakeholders. (a) Identify a change agent—a private or social sector leader who is given the responsibility and incentives for success; (b) allocate adequate public resources; and (c) ensure timely implementation to make the projects into commercially viable opportunities.
- Such a programmed would at last realize India's potential as a leading agricultural producer in the world, bringing food security and prosperity to its people and improving rural incomes systematically.

Food security

- Food security refers to the availability of food and one's access to it. A household is considered food-secure when its occupants do not live in hunger or fear of starvation. According to the World Resources Institute, global per capita food production has been increasing substantially for the past several decades. In 2006, MSNBC reported that globally, the number of people who are overweight has surpassed the number who is undernourished - the world had more than one billion people who were overweight, and an estimated 800 million who were undernourished.[2] According to a 2004 article from the BBC, China, the world's most populous country, is suffering from an obesity epidemic.[3] In India, the second-most populous country in the world, 30 million people have been added to the ranks of the hungry since the mid-1990s and 46% of children are underweight.
- Worldwide around 852 million people are chronically hungry due to extreme poverty, while up to 2 billion people lack food security intermittently due to varying degrees of poverty (source: FAO, 2003). Six million children die of hunger every year - 17,000 every day. As of late 2007, export restrictions and panic buying, US Dollar Depreciation, increased farming for use in biofuels, world oil prices at more than $100 a barrel, global population growth, climate change, loss of agricultural land to residential and industrial development, and growing consumer demand in China and India are claimed to have pushed up the price of grain. However, the role of some of these factors is under debate. Some argue the role of biofuel has been overplayed as grain prices have come down to the levels of 2006. Nonetheless, food riots have recently taken place in many countries across the world.
- The ongoing global credit crisis has affected farm credits, despite a boom in commodity prices. Food security is a complex topic, standing at the intersection of many disciplines.
- A new peer-reviewed journal of *Food Security : The Science, Sociology and Economics of Food Production and Access to Food* began publishing in 2009. In developing countries, often 70% or more of the population lives in rural areas. In that context, agricultural development among smallholder farmers and landless people provides a livelihood for people allowing them the opportunity to stay in their communities. In many areas of the world, land

ownership is not available, thus, people who want or need to farm to make a living have little incentive to improve the land.

- In the US, there are approximately 2,000,000 farmers, less than 1% of the population. A direct relationship exists between food consumption levels and poverty. Families with the financial resources to escape extreme poverty rarely suffer from chronic hunger, while poor families not only suffer the most from chronic hunger, but are also the segment of the population most at risk during food shortages and famines.
- Two commonly used definitions of food security come from the UN's Food and Agriculture Organization (FAO) and the United States Department of Agriculture (USDA) :
- Food security exists when all people, at all times, have physical, social and economic access to sufficient, safe and nutritious food to meet their dietary needs and food preferences for an active and healthy life.
- Food security for a household means access by all members at all times to enough food for an active, healthy life. Food security includes at a minimum (1) the ready availability of nutritionally adequate and safe foods, and (2) an assured ability to acquire acceptable foods in socially acceptable ways (that is, without resorting to emergency food supplies, scavenging, stealing, or other coping strategies). (USDA).

The stages of food insecurity range from food secure situations to full-scale famine. "Famine and hunger are both rooted in food insecurity. Food insecurity can be categorized as either chronic or transitory. Chronic food insecurity translates into a high degree of vulnerability to famine and hunger; ensuring food security presupposes elimination of that vulnerability. [Chronic] hunger is not famine. It is similar to undernourishment and is related to poverty, existing mainly in poor countries."

The Role of Extension in the Transfer and Adoption of Agricultural Technologies

To alleviate rural poverty by ensuring food security and sustainable improvement in People's well-being, research-based technologies must reach and be widely adopted by farmers.

1. Change agents may know the solution to problems confronting farmers, yet be unable to communicate these solutions if they lack effective communication skills, and do not apply sound extension education principles.
2. These principles emphasize two-way communication between farmers and change agents, focus on farmers' practices and priorities, consider gender and land tenure issues, and
3. Enable agents to better understand how farmers make farm and marketing decisions
4. Developing countries are particularly susceptible to the problems of low agricultural
5. Productivity and environmental degradation.
6. These problems can be alleviated if extension education systems have a sound research and technology base, and understand and apply
7. Extension education principles to enable farmers to adopt new and improved technology, so that they can reap the economic benefit flowing from these technologies.
8. Adverse consequences of soil erosion such as decreased land resale value, loss of soil fertility, sedimentation of streams and lakes, and contamination of water supplies with pesticides and fertilizers.
9. In contrast, extension education programs have been found to enhance soil and water conservation, improve farming systems, aid in transfer of appropriate technology, and develop rapport with and persuade farmers to adopt improved farm practices
10. It is the purpose of this paper to present extension education principles, and the characteristics of new technologies which facilitate transfer and adoption of technology among farmers.
11. Examples from the experiences of the author in extension work are provided to show what happens when these principles are observed or violated.

Agricultural Extension for Sustainable Development

A major challenge of the 21st century will be that of implementing sustainable development and meeting the food needs of an increasing world population. Technical experts from a range of disciplines have analysed the links between population and food and one of the conclusions reached is that, in view of the limited scope and high economic and environmental costs of expanding crop land, 80 percent of food production increases will have to come from increased yields on lands already in production (IFPRI, 1994).

The overriding problem is how to improve yields, especially of basic food crops, without further deterioration of the natural resource base. Adverse environmental effects of intensified agriculture are all too visible throughout the world (soil erosion, water pollution, loss of biodiversity). The technologies and management practices that slow down degradation and protect the resource base are often not available in many regions, or are not adopted when they are available.

Based on FAO projections for food demand and supply, it is clear that national and international investment is required in :

- Research to generate productivity-enhancing technologies without adverse environmental consequences,
- Extension so that there is widespread dissemination and efficient use of these technologies.

Yet with the recession in aid funding in recent years, investment in agricultural research and extension has declined drastically. Without such investment the challenges of sustainable agricultural development and increased food production will not be met.

Extension for sustainable development

The challenge extension faces is that of promoting sustainable development while at the same time ensuring the sustainability of extension agencies themselves in the face of budget cuts. Donor funds, when available, are increasingly being directed to help countries decentralize and streamline their public extension services.

It is important to view extension for sustainable development in an institutional strengthening context, including the enhancement of extension organizations in both the public and private sectors. These organizations include extension agencies of ministries of agriculture, non-governmental organizations (NGOs) and farmers' organizations. The tasks facing them in achieving sustainable agricultural and rural development require that their organizational capability be strengthened. For example, farmers' organizations need to be strengthened not only to receive and provide services but as a desirable objective itself. To be sustainable they require the capability to assume autonomous management of *their* development goals.

The strengthening of public agricultural extension organizations is critical given that many of these organizations have suffered debt- and adjustment-induced budget cuts in the last decade. At the same time that these organizations faced serious cut-backs in funding, the national, regional and internal agricultural sectors were rapidly changing with increased emphasis on environmental concerns. The solution, however, is not to abolish public extension services but to strengthen their capability to collaborate with other organizations, both public and private, that can help develop sustainable livelihoods for rural families.

One result of public-sector budgetary constraints has been the trend towards an increased extension role for NGOs. However, many workers have low levels of formal agricultural education; less than 20 percent of NGO staff in Honduras working on agricultural projects has university training (Kaimowitz, 1993). NGOs often have little capacity for testing agricultural technologies and few ties with research organizations that do have such capacity. A low level of formal agricultural education may help in developing rapport with farmers who have little formal schooling, but it can limit the capability of NGOs to help farmers put research results into practice. There is clearly a need to link farmers' groups, NGOs, research organizations and public extension services. Such an approach recognizes that there are multiple sources of technology development and dissemination which require integration.

Integrating Sustainable Development Themes into Agricultural Extension

In December, 1993, FAO held an expert consultation in Rome on the "Integration of Environmental and Sustainable Development Themes into Agricultural Education and Extension Programmes." Background papers were prepared based on 10 country case studies describing the current situation in selected agricultural education institutions and 10 case studies looking at national (public) extension systems. The agricultural extension case studies were conducted in Brazil (Paraña State), Burkina Faso, Egypt, India (Andhra Pradesh), Indonesia, Ireland, Malawi, Mexico Syria and the United States (Nebraska). Some key findings were :

1. Current environmental roles and activities

Differences were evident between case study countries in how environmental themes are translated into extension roles and activities. Some countries (e.g., Syria and Malawi) place emphasis on production and technology transfer, with environmental issues added to general extension activities. Others (e.g., Ireland and Brazil) have undergone extensive change to incorporate a mandate which highlights environmental protection. Clearly, there is a need to integrate production and environmental protection activities, demonstrating how sustainable development can be economically rational and ecologically sound.

2. Environmental content and messages

The scope of the environmental content and messages of extension agencies is large, reflecting the interests of the particular country and agency, the physical environment and the extent of environmental awareness. Overall, there appeared to be a lack of input from environmental agencies (e.g., NGOs) at the national and local levels. Also, much of the content seemed geared to "negative messages" - telling farmers what they are doing wrong instead of messages that emphasize how sustainable development benefits both farmers and the general public. The needs of women farmers appeared to be neglected despite their crucial roles in natural resource management.

3. Extent of integration

All agencies claimed that environmental concerns had been "integrated" into existing extension programmes instead of the creation of separate programmes; however, the number of issues addressed was small and in most cases environmental matters had been "added" to on-going activities. Overall, environmental concerns made up a small proportion of time, subject matter and budget of extension agencies (ranging from 5% to 25%); the number of environmental specialists was typically less than 10% of total staff. There was little evidence that potential beneficiaries (e.g., farmers) were participating in identifying environmental concerns and planning programmes.

Conclusion

A number of conclusions can be drawn from the experiences of the ten extension case studies. The case studies suggest that the majority of environmental issues being addressed by extension programmes are conventional topics closely linked to increasing agricultural productivity. For example, the safe and efficient use of "farm inputs" (e.g., pesticides) was the most common "environmental" topic. The implication is that the content of extension messages and practices has hardly changed and the concept of sustainable development is used more for rhetoric than reality. In fact, extension staff as well as their farm clients often seem to view environmental concerns as limiting production and adding to farm costs and restrictions. There has been a broadening of the extension agenda to include environmental topics, but primarily those subjects directly related to productivity or compliance with environmental legislation.

One question that arises is how extension agencies can be organizationally adapted to promote environmental and sustainable development themes. The experiences of a number of agencies show that these themes can be effectively addressed without major organizational changes, increases in funding or additional staff. Many environmental topics can be incorporated into existing programmes and staff can be trained in environmental matters as part of regular training. There is also considerable scope for collaboration with NGOs which can expand the reach and strengthen extension environmental activities.

Promoting sustainable agricultural development requires that extension agencies shift emphasis from the maximization of short-term production outputs, which may be attractive in economic terms but which cannot be sustained, to the promotion of technologies and practices that are productivity-enhancing without negative environmental consequences. Too often environmental concerns have been merely "tacked on" to standard programmes of increased production and technology transfer. The ultimate purpose should be promotion of environmental concerns as inseparable from the production goals of farmers.

However, it is unlikely that the promotion of sustainable agricultural development can be achieved within the confines of a single agency. This means that extension agencies will have to strengthen linkages and coordination with research organizations, environmental agencies, educational institutions, NGOs, farmers groups and others who recognize that agriculture is sustainable only when it is ecologically sound as well as economically viable.

❑❑❑

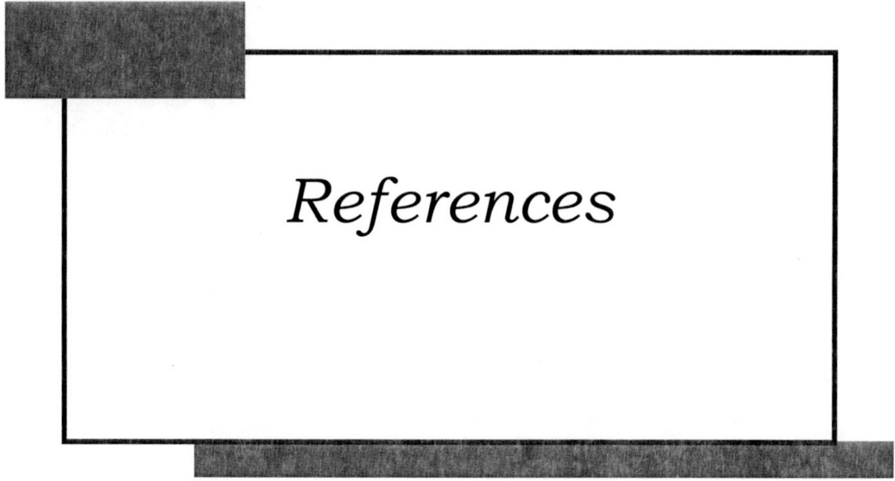

References

Adhikarya, R. (1989). Strategic Extension Campaigns : A Case Study of FAO's Experiences. Rome: Food and Agriculture Organization.

Agbamu, J.U. (2000). 'Agricultural research-extension linkage systems: An international perspective'.

Agricultural Research and Extension Network Paper No. 106a. London: ODI.

Albrecht, H. (1954). Organization und Arbeitsweise derlandwirtschaftliche Beratung in Denmark and Holland, Landwirtschaft - Angewandte Wissenschaft, No. 15, Hilt up, Landwirtschaftsverlag.

Ambalavanan, K. (1999). Personal communication.

Anderson, J.R., Hazell, P.B.R. and Evans, L.T. (1987). Variability of cereal yields—Sources of change and implications for agricultural research and policy. Food Policy, 12: 199-212.

Asana, R. (1976). Physiological approach to breeding of drought resistant crops. ICAR Tech. Bull., New Delhi.

Axinn, G.H. (1988). Guide on Alternative Extension Approaches. Rome : Food and Agriculture Organization.

Bahn, H.M. (1999). 'The challenge of serving unwilling learners: implications for extension education methods' in J. Kania and M. Drygas, (eds) *Proceedings of the 14th European Seminar on Extension Education: The role of extension education in the global world.* Krakow, Publishing House ofthe Agricultural University of Cracow.

Baumeister, Roy F. and Leary Mark R. (1995). The Need to Belong: Desire for Interpersonal Attachments as a Fundamental Human Motivation. *Psychology Bulletin*, 117(3): 497-529.

Bell, A. (1998). 'Changing from adviser to consultant: the consultant's experience' in I. Wallace, (ed.) *Rural Knowledge Systems for the 21st Century; rural extension in Western, Central and Eastern Europe.* Reading: AERDD pp. 199–203.

Bennett, C.F. (1996). 'Rationale for public funding of agricultural extension programs'. *Journal of Agricultural and Food Information*, 3: 3–25.

Berne E. (1961). Transactional Analysis in Psychotherapy Grove Press. New York.

Beynon, J., Akroyd, S., Duncan, A. and Jones, S. (1998). Financing the Future: Options for research and extension in Sub-Saharan Africa. Oxford: Oxford Policy Management.

Black, A. and Champion. J (1976). Methods and Issues in Social research. John Wiley and Sons, New York.

Blackenburg, P. Von. (1988). Agricultural Extension Systems in Some African and Asian Countries. Economic and Social.

Burkman, E. (1987). Factors Affecting Utilization. In R.M. Gagne (ed). Instructional technology: Foundations. Hillsdale, NJ: Lawrence Erlbaum.

Candler, W., and Kumar, N. (1998). India: The dairy revolution. The impact of dairy development in India and the World Bank's contribution. Washington D.C.: World Bank.

Central Bank (2006). Annual Central Bank Report, 2006.

Collion, M-H., and Rondot, P. (1998). 'Partnerships between agricultural services institutions and producers' organizations: myth or reality?'*Agricultural Research and Extension Network Paper*No. 80. London: ODI.

Coursers, C.J. (1981). Guide for Agricultural Extension Agents. Rome : Food and Agriculture Organization.

Crews F. (1997). Freud's Legacy in Dispute. NYRB Collections, New York.

De Koning, J., Hagelaar, G. and Keet, E. (1997). 'Privatization in the Netherlands: a reflection on the Dutch way of handling privatization'. *Strategic Change* 6: 227–236.

December, J. (1993). Characteristics of an oral culture in discourses on the net. Paper presented at the twelfth annual Pennsybrania State Conference on Rhetoric and Composition, University Park, Pennsylvania, July 8, 1993 [WWW document]. URL http://www.december.com/john/papers/pscrc93.txt

Development. Paper No. 46. Rome : Food and Agriculture Organization.

Duijsings, P. (1998). 'Information and communication management in a privatized extension organisation: the case of DLV (The Netherlands Extension Service)' in 'Assessing the impact of information and communication on institutional performance: proceedings of a CTA workshop', Wageningen.

Dusay, J. (1972). Egograms and the Constancy Hypothesis. Transactional Analysis Journal, 2: 3.

Edwards, M. and Hulme, D. (eds) (1996). Beyond thematic Bullet: NGO performance and accountability in the Post-Cold War world. West Harford: Kumarian Press.

Evenson, R.E. (1986). "The Economics of Extension". In : *Investing in Rural Extension Strategy and Goals*, G.E. Jones (ed.). NewYork : Elsevier.

FAO (1990). Report of the global consultation on agricultural extension. Rome: FAO.

FAO, (1987). FAO Production Year Book, 1986.

Farquhar, J.F, and Surry, D.W. (1994). Adoption analysis: An additional tool for instructional developers. Education and Training Technology International, 31(1): 19-25.

Farrington, J. (1994). 'Public sector agricultural extension: Is there life after structural adjustment'. *Natural Resource Perspectives* No. 2. London: ODI*Perspectives on agricultural extension* 19Farrington, J. (1997) 'The role of non-governmental organizations in extension' in B.E. Swanson.

Fastener Land Katn D. (1953). Research Methods in Behavioural Sciences.Holt, Rinehart and Winston Inc., New York.

Food and Agricultural Organization (FAO), Rome (1959). The Agricultural Division of FAO : A Summary of the Organization, Development, and Accomplishments from 2 December 1946 to 31 December 1958.

Garcia, R. (1981). Drought and Man : The 1972 Case History. Vol. 1, Nature Pleads Not Guilty, Pergamon, New York.

Geoghegan, W.H. (1994). Whatever happened to instructional technology? Paper presented at the 22nd Annual Conference of the International Business Schools Computing Association, Baltimore, Maryland July 17-20.

Gifford, R.M. (1987). Exploiting the fertilizing effect of high levels of atmospheric carbon dioxide. Proc., Int. Workshop on Climate Variability and Food Security in Developing Countries, February 6-9, 1987, New Dehli, India.

Goode W.J.and Hatt P.K. (1981). Methods in Social Research. McGraw Hill Book Company, Singapore.

Goulding R., Goulding M. (1997). Changing Lives Through Redecision Therapy. New York; Grove-Atlantic.

Hansraj (1990). Theory and practice in social research. Asia Publishing House. Culcutta.

Harris, J. (1994). Teaching teachers to use telecomputing tools. Electronic reprint from October 1994 "Mining the Internet" column, The Computing Teacher. International Society for Technology in Education.

Hazell, P.B.R. (1984). Sources of increased instability in Indian and U.S. Cereal production. *Am. J. Agric. Econ.*, 66: 302-311.

Heinssen, R. K. P G. Levendusky, R H. Hunter (1995). Client as Colleague: Therapeutic Contracting With the Seriously Mentally Ill. *American Psychologist*, 50(7): 522-532.

House, E. (1974). The politics of educational innovation. Berkley, CA: McCutchan Publishing Company.

IBRD (1987). World Bank experience with rural development, 1965-1986. Washington, DC: The World Bank.

International Food Policy Research Institute (IFPRI), (1994). Population and "Food in the Early 21st Century : Meeting Food Needs of an Increasing World Population" Washington, D.C.

Jaffee, D. (1998). Institutionalized resistance to asynchronous learning networks. [WWW document]. URL http://www.aln.org/alnweb/journal/vol2_issue2/jaffee.htm

Kaimowitz, D. (1993). "The Role of Non-Governmental Organizations in Agricultural Research and Technology Transfer in Latin America". World Development, 21(7).

Karpman, S. (1968). "Fairy Tales and Script Drama Analysis." *TA Bulletin*, 7: 26.

Kates, R.W., Ausubel, J.H. and Berberian, M. (Eds.), (1985). Climate Impact Assessment. SCOPE 27, Wiley, London.

Kellogg, W.W. and Schware, R. (1981). Climate Change and Society—Consequences of increasing atmospheric carbon dioxide. Westview Press, Boulder, Colo.

Kerlinger F.N. (1973). Foundations of Behavioral Research. Holt, Rinehart and Winston Inc.. New York.

Korthari C.R. (1996). Research Methodology: Methods and techniques. wishwa Prakashan, New Delhi.

Kumarawadu, P. (2003). Assesing and Benchmark our digital readiness for e-Government.

Lapitan J.A, Shire David, Bell Mark and Emitiyagoda S. (2006). Accelerating impact of agricultural research through information and communication technology, paper presented for Annual Symposium of Department of Agriculture, Sri Lanka.

Levendusky, P.G., Berglas, S., Dooley, C.P. & Landau, R.J. (1983). Therapeutic contract program: A preliminary report on a behavioral alternative to the token economy. *Behavior Research and Therapy,* 21: 137-142.

Levendusky, P.G., Willis, B.S. & Ghinassi, F.A. (1994). The therapeutic contracting program: A comprehensive continuum of care model. *Psychiatric Quarterly,* 65: 189-208.

Levine, S. Stimulation in Infancy. *American Psychologist,* May 1960.

Loffredo *et al.* (2004). The Ego State Questionnaire Revised. *Transactional Analysis Journal,* 34: 1.

Lynch J. (1988). The broken heart; the medical consequences of loneliness. New York; Basic Books.

Management Study Guide.Com

Markley, O.W. and Hurley, T.J., III, (1983). A brief technology assessment of the carbon dioxide effect, *Technol. Forecasting Soc.,* 23: 185-202.

Maru, Ajit, P. (2003). Information technology and Efficient Agricultural Extension in the Asia Pacific Region status, Issue and strategies, study meeting on IT for Agriculture Extension 8-14 Jan, 2003, India 27.

Matlin M. and Stang D. *The Pollyanna Principle.* Shenkman. Cambridge MA, 1978.

McAdams, D.P., Reynolds, J., Lewis, M.L., Patten, A. Bowman, P.T. (2001). When bad things turn good and good things turn bad: Sequences of redemption and contamination in life narrative, and their relation to psychosocial adaptation in midlife adults and in students. *Personality and Social Psychology Bulletin,* 27: 472-483.

McKenzie K.R. *et al.* (1999). Guidelines for the Psychotherapies in Comprehensive Psychiatric Care. *Canadian Journal of Psychiatry,* 44.

Moore, G.A. (1991). Crossing the chasm: Marketing and selling technology products to mainstream customers. New York: Harper Business.

Moving Towards e-Sri Lanka, Proceeding of the 22nd National IT conference, Colombo, Sri Lanka.

Mulay Sumati and Sabarathanam V.E. (1980). Research Methods in Extension Education. Manasayan. New Delhi.

North, T. (1995). Masters Thesis: Carlin University. Internet and Usenet Global Computer Networks: An Investigation of their culture and its effect on new users in higher education [WWW document]. URL http://foo.curtin.edu.au/thesis/default.html

Novey, T. (2002). Measuring the effectiveness of transactional analysis; An International study. *Transactional Analysis Journal,* 32: 1.

Oblinger, D.G. and Maruyama, M.K. (1996). Distributed learning. Cause Professional Paper Series, #14. Boulder, Colorado: CAUSE, the Association for Managing and Using Information Resources in Higher Education.

Ornish, D. (1988). Love and Survival. HarperCollins, New York.

Parry M.L., Carter, T.R. and Konijn, N.T. (Eds.), (1987). The Impact of Climatic Variations on Agriculture, Vols. 1 & 2, Reidel, Dordrecht, Holland.

Persons, J.B., Joan Davidson and Michael A. Tompkins (2001). Essential Components of Cognitive-Behavior Therapy for Depression. Washington DC American Psychological Association.

Pimental, D. and Pimental, M. (1979). Food, Energy and Society. Edward Arnold Ltd.

Rathnayaka, R.M.M.S. (2006). Evaluation of the effectiveness of cyber extension - user farmers' perspective. Thesis, Department of Agricultural Extension, Faculty of Agriculture, University of Peradeniya.

Riggs, M. (2003). Information management for food security and sustainable agricultural development in the Asian –Pacific Region - Proceeding of study meeting on ICT for Agricultural Extension – India. 7-15 Jan, 2003.

Rogers, E.M. (1986). Communication: The new media in society. New York: The Free Press.

Rogers, E.M. (1995). Diffusion of innovations (4th ed.). New York: The Free Press.

Rosenberg, N.J. (1987). Drought and climate change : For better or worse, In: Planning for Drought- toward a Reduction of societal vulnerability. Wilhite, D.A. and Easterling, W.E. (Eds.), Westview Press, Boulder, Colo., 297-316.

Rowan, J and M. Cooper. (1999). The plural self; Multiplicity in everyday Life. Sage Publications, London.

Rowntree, P.R and Bolten, J.A. (1983). Effects of soil moisture anomalies over Europe in summer. In : Variations of the Global Water Budget. Street-Perrot, A., Beran, M. and Ratcliffe, R.A.S. (Eds.), Reidel, Dordrecht, Holland, pp. 447-462.

Sanderson, F.H. (1984). World food prospects to the year 2000. Food Policy, 9: 363-373.

Seligman M.and M. Csikszentmihalyi (2000). Positive Psychology *American Psychologist*, 55(1): 5-14.

Shapiro, F. Eye Movement Desensitization and Reprocessing: Basic Principles, Protocols and Procedures (2nd edition). New York: Guilford Press 2001.

Siegel D.J. (1999). The Developing Mind: How Relationships and the Brain Interact to Shape Who We Are. Guilford, New York.

Siegel S. (1956). Nonparametric statistics for the behavioral sciences. McGraw-Hill Book Company Inc., New York.

Sinha, S.K. (1986). Energy balance in agriculture : The developing world. In : Global Aspects of Food Production. Swaminathan, M.S. and Sinha, S.K. (Eds.), Tycooly International, London, 57-84.

Sinha, S.K. (1987). The 1982-83 drought in India : Magnitude and impact. In : The Societal Impacts Associated with the 1982-83 Worldwide Anomalies, Glantz, M., Katz, R. and Krenz, M. (Eds.), Stan Grossfield/Boston Globe, 37-42.

Slesser, M. (1986). Energy balance in agriculture : The developed countries. In : Global Aspects of Food Production. Swaminathan, M.S. and Sinha, S.K., (Eds.), Tycooly International, London.

Steiner, C. (1971). Scripts People Live. Grove Press. New York.

Stockdill, S.H. and Morehouse, D.L. (1992). Critical factors in successful adoption of technology: A checklist of TDC findings. Educational Technology, 32(1): 57-58.

Suomi, S J.; Collins, M L.; Harlow, H F. and G. Ruppenthal. (1976). Effects of maternal and peer separations on young monkeys. *Journal of Child Psychology & Psychiatry & Allied Disciplines,* 17(2): 101-112.

Swaminathan, M.S. (1986). Building national and global nutrition security systems. In : Global Aspects of Food Production, Swaminathan, M.S. and Sinha, S.K. (Eds.), Tycooly International, London, 417-449.

Swanson, B.E., Farmer, B.J., & Bahal, R. (1990). The current status of agricultural extension worldwide. In FAO, *Report of the global consultation on agricultural extension,* p. 43-76. Rome: FAO.

Tessmer, M. (1990). Environmental analysis: A neglected stage of instructional design. *Educational Technology Research and Development,* 38(1): 55-64.

Tiger, L. (1979). Optimism; The Biology of Hope. New York. Simon and Schuster.

Umali, D.L., & Schwartz, L. (1994). Public and private agricultural extension: Beyond traditional frontiers. Washington, DC: The World Bank.

UNDP (1991). Agricultural extension programme advisory note. New York: United Nations Development Programme.

WECD, (1987). Food 2000 : Global Policies for sustainable agriculture. Rep. to the World Commission on Environment and Development. Zed Books Ltd., London.

Wijekoon, R.R.A. (1999a). Strategies for the use of interactive multimedia to traind Extension Workers in Developing Countries, Ph.D. Thesis, University of Western Sydney, Nepean, Australia.

Wijekoon, RRA. and S. Newton, (1999b). The use of Interactive Multimedia (CDROM) in Training Extension Workers on Flip Chart Technology in Peiris S.E. (ed). *Tropical Agricultural Research*, 11: 154-173.

Wijekoon, R.R.A. (2003). Concept paper presented to Sri Lanka council for Agricultural Research Policy.

Wijekoon, R.R.A. and S. Emitiyagoda, (2006). Rice Knowledge Bank for Cyber extension in Sri Lanka, paper presented to workshop on Linking Extension and Research Needs through Information Technology, Bangkok, 6-8 Nov. 2006.

Wijekoon, R.R.A. and S. Newton, (1998). Multimedia support for Extension Trainers in Developing Countries in proceeding of 7th International Conference on Computer for Agriculture, Orland, Florida.

Wilson, B., Ryder, M., McCahan, J. and Sherry, L. (1996). Cultural assimilation of the Internet: A case study. M. Simonson (Ed.), Proceedings of selected research and development presentations. Washington, D.C.: Association for Educational Communications and Technology [WWW document]. URL http://www.cudenver.edu/~bwilson/cultass.html

Witter, S.H. (1986). Research and technology needs for the twenty-first century. In : Global Aspects of Food Production Swaminathan, M.S. and Sinha, S.K. (Eds.), Typology International, London, 85-116.

Yoshida, S. (1978). Tropical climate and its influence on rice. IRRI Res. Paper Ser. 20, Int. Rice Res. Inst., Manila.

Young F.V. (1996). Scientific Social Surveys and Research. Prentic Hall of India Pvt. Ltd. New Delhi.

Young, J.E. (1999). Cognitive therapy for personality disorders: A schema-focused approach. Sarasota, FL: Professional Resource Exchange.

❏❏❏

About the Authors

Dr. Dileep Kumar, born in Jhunjhunu district of Rajasthan did his B.Sc. Agriculture (Hons.) and M.Sc. (Ag.) Extension from S.K.R.A.U. Bikaner (Rajasthan), India, in the year 2001 and 2004 respectively. His area of research during Masters Degree in the University was on *adoption of sprinkler system of irrigation in the Jhunjhunu District of Rajasthan*. He pursued his Ph.D. degree from N.D.R.I. Karnal, (Haryana), on *Critical Analysis of Dairy Co-operatives and Private Dairies in Rajasthan*. He is closely associated with many national and international professional societies of Extension Education and published many research and popular articles. He remained a meritorious student during his academics. He is also *pursuing* PGJMC (Master in Journalism and Mass Communication) from IGNOU, New Delhi. He is founder chairman of Ram Mangalam Agri Rural Research foundation. A Rajasthan based NGO working for the betterment of lives of people.

Dr.Khajan Singh did his B.Sc. Dairying (DH), M.Sc. (Dairy Extension) and Ph.D. (Dairy Extension Education) from National Dairy Research Institute, Karnal, (Haryana). He joined Central Soil Salinity Research Institute, Karnal as ARS (Agricultural Extension) Scientist in February 1986. Presently working as Principal Scientist in the Division of Dairy Extension and has been engaged in teaching, research and extension for more than 9 (Nine) years. His area of research specialization includes dissemination of dairy innovations/information/technologies under rural conditions, impact assessment and identification of training needs of dairy farmers etc. Till today he has completed *eight* research projects as Principal Investigator and *ten* as a Co-Principal Investigator. He also worked as a group leader of inter-disciplinary team to under-take the transfer of dairy innovations under field conditions and technologies were popularized among farmers, students, subject-matter-specialists and high dignitaries through various methods, tools and techniques such as training, visits, radio talks and exhibitions etc. He has published more than 50 research papers, popular articles and technical bulletins in the field of dairy and agricultural extension. Dr. Khajan Singh has authored three books, two teaching manuals and also edited two books (Hindi). He is active member of four Professional Societies of Agricultural Extension at National level. Developed two knowledge tests for farmers one related to afforestation of salt affected soils and second regarding micro-irrigation technology for poor quality water areas. Guided several students (Both M.Sc. and Ph.D.) in the field of extension education concerning several issues in different agro climatic conditions of India. Delivered several lectures on AIR and actively participated in institution building activities.